Human Movement-a field of study

THE HUMAN MOVEMENT SERIES
General Editor: H.T.A. Whiting

Readings in Sports Physchology.

Edited by: H.T.A. WHITING

Readings in the Aesthetics of Sport.

Edited by: H.T.A. WHITING

Personality and Performance in Physical Education and Sport.

By: H.T.A. WHITING, K. HARDMAN, L.B. HENDRY and MARILYN G. JONES

Concepts in Modern Educational Dance.

By H.B. REDFERN

Techniques for the Analysis of Human Movement.

By O.W. GRIEVE, DORIS MILLER, D. MITCHELSON, T. PAUL and A.J. SMITH

Educational Gymnastics Themes.

By: JEAN WILLIAMS

Proprioceptive Control of Human Movement.

By: J. DICKINSON

Expression in Movement.

By DAVID BEST

Human Movement – a field of study

Edited by J.D. BROOKE (University of Salford)

& H.T.A. WHITING (University of Leeds)

Henry Kimpton Publishers
London 1973

For I do take the consideration in general and at large of human nature to be fit to be emancipated and made a knowledge by itself: not so much in regard of those delightful and elegant discourses which have been made of the dignity of man, of his miseries, of his state and life, and the like adjuncts of his common and undivided nature; but chiefly in regard of the knowledge concerning the sympathies and concordances between the mind and body, which being mixed cannot be properly assigned to the sciences of either.

Sir Francis Bacon (1605)
Advancement of learning
Book II, III Philosophy

Standard Book Number 85313 773 0

Computerised origination by Autoset, Brentwood.
Printed by Unwin Bros. Ltd., Woking.

CONTENTS

STUDY GROUP MEMBERS AND CONTRIBUTORS

* G.T. Adamson
Dept. of Physical Education
University of Leeds

J.D. Brooke
Human Performance Laboratory
Physical Education Section
University of Salford

**R. Carlisle
Physical Education Dept.
Aberdeen College of Education

Jean Carroll
Anstey College of Physical
Education
Sutton Coldfield

Hilary Corlett
Chelsea College of Physical
Education
Eastbourne

* G.F. Curl
Nonington College of Physical
Education
Kent

J.C. Evans
Carnegie School of Physical
Education
City of Leeds & Carnegie College

**B. Goodman
Carnegie School of Physical
Education
City of Leeds & Carnegie College

**A. Guy
Loughborough College of Education
Loughborough

* L.B. Hendry
Dept. of Education
University of Aberdeen

June Layson
Dept. of Physical Education
University of Leeds

* D.W. Masterson
Physical Education Section
University of Salford

Elizabeth Mauldon
City of Leeds & Carnegie College

Study Group Members and Contributors (cont.)

A.G. Roche
Carnegie School of Physical
Education
City of Leeds & Carnegie College

H.T.A. Whiting
Dept. of Physical Education
University of Leeds

Ida Webb
Anstey College of Physical
Education
Sutton Coldfield

---

* Contributor but not a member of 'Think Tank'
**Member of 'Think Tank' but not a contributor

In addition, acknowledgement is given to help received at various stages from:-

G.J.K. Alderson
Dept. of Physical
Education
University of Leeds

F.H. Sanderson
Dept. of Physical
Education
University of Leeds

J.W.R. Hutt
Carnegie School of
Physical Education
City of Leeds & Carnegie
College

PREFACE

The introduction of a new series of books is generally in response to a demand for particular kinds of information which are not currently being met by the existing literature. This may be because such texts are out of date, or because a comparatively new subject area has been or is being developed. The series on Human Movement of which this is a source text, reflects the progress of a new field of study from the embryo stage to that of early rapid development. While it would not be possible to pinpoint with any degree of accuracy its emergence as a separate discipline, it is clear that the past five years in particular has seen an acceleration both in interest and published material from diverse sources. For perhaps obvious as well as historical reasons, such a field of study has hitherto largely been subsumed under the umbrella term 'physical education'. Growing pains have however become increasingly intense as an interest in Human Movement Studies and the application of knowledge about human movement extends far beyond the bounds of the practical field of physical education. The current trend to establish Human Movement Studies as a discipline in its own right, is both a recognition of the existence of a field of knowledge on which such studies can be based and a realisation of its breadth of interest and application.

Although the material presented in this and subsequent texts is not orientated towards a particular section of the community, it will by its very nature have particular interest for students of physical education. In addition, it is anticipated that much of the material will be of interest to aestheticians, biomechanics, choreographers, educationists, ergonomists, paediatricians, psychologists and other students of human movement.

Without attempting to be definitive, this source text presents structured areas of Human Movement Studies which have been developed in various institutions in Great Britain. It is hoped that analyses of this kind will serve a useful purpose for the growing number of lecturers who are finding the need to formulate courses in Human Movement Studies and at the same time will enable students of human movement quickly to obtain an overview of a particular area, a synopsis of the main concepts and useful leads into the literature. This kind of orientation to the areas designated has been uppermost in the minds of the contributors. While a common format has been attempted, it is not wished at this stage to confine such diverse topics as are presented within a particular strait-jacket. Differences will therefore be apparent in the way such topic areas are structured and developed. It will be of interest to see how modifications are brought about at some future date. Variations in the quantity of material presented in the different topic areas in no way represents value judgements as to their importance but does to some extent reflect both the amount of information currently available and the stage of development reached.

Structure of the book

The book is divided into four stages.

Stage 1   is an attempt to justify Human Movement as a field of study.

Stage 2   consists simply of the titles of the topic areas to be covered prefixed by a code letter for ease of cross-referencing. The assignment of a particular code is purely arbitrary and absences from the alphabetical sequence represent areas which were contemplated but not developed*.

Stage 3   structures each of the main topic areas designated in Stage 2 in sub-heading form. This represents the main concepts which are being dealt with under the particular topic heading.

Stage 4   represents an expansion of the sub-headings under Stage 3 together with comprehensive literature references. Where possible, these have been classified as source texts, journal overviews and specific references.

While the four stages as outlined would seem to represent a logical development, the reader may find his own particular requirements satisfied by entering the book at any of the stages presented.

Subsequent developments within the Human Movement Series will be diverse in nature and will not simply represent a collection of books covering each of the topic areas. While it is possible that a single book may in fact make a useful overview, it seems more likely that a number of books from different viewpoints reflecting on the subject matter of a particular topic area will represent the most fruitful line of development. It is anticipated that some texts will be concerned with human movement per se while others will have an applied orientation making them more suitable for students in the various professions subserved by Human Movement Studies.

H.T.A. Whiting
Series Editor

---

*e.g. philosophical concepts.

INTRODUCTION

This text identifies many of the important areas of Human Movement Studies. By virtue of its scope, it is not detailed. The intention is to provide an overview, to act as a gazetteer stating important concepts and major references so that detail of particular interest may be pursued in depth through relevant specialist works.

On the first of January 1970, twelve physical educationists met at City of Leeds & Carnegie College, to convene a 'Think Tank'. Two days were spent in discussion and it was agreed to reconvene at Anstey College in March 1970. At that time, a further two days' discussion ensued, to be followed by a third meeting at Chelsea College in September where there were two more days of thinking and talking. Finally, in February of the following year a one-day meeting was held at the University of Salford. Between each of the sessions, papers were prepared by the participants and circulated around the group, together with the transcripts of the previous session's discussions. It is beneficial to record some of the forces in the text's embryonic development over this period.

From the outset, attempts to consider issues from many points of view and attempts at flexibility characterised the approach of the participants to each other. Little gain was seen in bothering to defend or entrench positions that had been proposed but subsequently were seen to be unsound or incomplete. Of course, marked differences did appear and resulted eventually in both resignations and limited contributions. This had to be accepted as an outcome of the process.

To quote from the transcript of the first meeting, the discussions overall centred mainly around:

the value of the terms 'Physical Education' and 'Human Movement' in delineating the field of knowledge we were to consider.

Although some contributors were satisfied with their ability to differentiate and relate the two terms, others were less happy. The problem for these people was the juxtaposition of the professional use of information about human movement with that information structured as a field of study in its own right.

Fig. 1, derived from an earlier paper (Brooke 1969) summarises this position and illustrates diagrammatically some of the aspects involved.

1

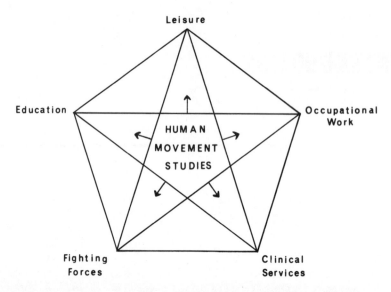

Following the meeting at Anstey College, papers were prepared by study group members which reflected individual attempts to structure parts of a Human Movement Studies domain. It was clear that individual responsibility would have to be taken for particular areas and that only limited coverage could be achieved.

The meetings and papers at Chelsea College and at the University of Salford refined these initial ideas about the preliminary structure of the field, brought in additional expertise from individuals outside the original group and resulted in the present text. As the parts of the domain were uncovered, it was exciting to see the problem of relevance for such studies—which had resulted previously when they were forced into the framework of the professional practice of physical education (Brooke 1969) — being gradually overcome. One problem never resolved was the question, 'should we include all movement?'. It is considered by Curl in Stage 1 and partly was circumvented by the adoption of a level of analysis that was primarily behavioural and with reference to British culture. However, there were essential substrates that were perceived to be part of the human movement field of knowledge and these can be seen in the papers analysising organic and systemic function, the study of non-human movement and cross-cultural comparisons.

In the practice of drawing the sections together as a whole text, a problem arises which illustrates one of the central characteristics of human movement study: all should be first and none last. That is, men moving are multi-factorial organisms with such factors interacting with each other within the man, between men and in the relationships between the men and their inanimate environment. Whatever position one adopts about the hierarchical ordering of these factors and their interrelationships in determining the movement characteristics observed, it is clear that a network of interactions is being vitalised constantly and by such a process changing constantly. This poses problems for the scholar, who can no longer be restricted by the defined boundaries such as molecular biology, psychology, physiology, aesthetics, etc., within which to study separate aspects of human existence, for they are not separate in this life situation. They participate together in the interacting systems of men and their environments, which systems neglect imposed divisions such as art-science, mind-body or social-individual dichotomies. It poses problems also for the writer, for a single temporal dimension is implicit in the act of communication and the attempt is to communicate a multi-dimensional, rapid succession of events. Nothing can be done but to present one concept and then the next, one section and then the next and trust that the reader's growing awareness elicits for him the synthesis into this multi-dimensional whole. For the student the problems probably are the most difficult, for the breadth of knowledge and range of concepts required are very great if the movement of humans in common life situations is to be understood. Some members of the Think Tank questioned whether such studies are possible. Yet the demands are only as great as the problems facing students in many modern interdisciplinary integrations that have been made because of the relevance there appears to be to life situations. Of course, not all sections are going to be of pressing importance to all students: however, an awareness of the nature of *all* sections and others we may have omitted might be held necessary for the scholar.

The statement that this text makes of the field of knowledge that is Human Movement Studies, is a pragmatic attempt to cast some light, despite the problems. There are inadequacies that occur with this type of solution. The domain is incomplete, partly from lack of expertise, partly from lack of vision in the early stages. It was felt necessary, however, to move to a statement now so that a start could be made and deficiencies could be seen. This text is that attempt. We hope it is a framework both to build upon and to build beside.

J.D. Brooke
H.T.A. Whiting (Editors)

Reference
   BROOKE J.D. (1969). Human Biology, physical education and ergonomics—the future for the active human animal. *Bull. Phys. Ed.,*7,56-63.

1

AN ATTEMPT TO JUSTIFY HUMAN MOVEMENT AS A FIELD OF STUDY

AN ATTEMPT TO JUSTIFY HUMAN MOVEMENT AS A FIELD OF STUDY

BY GORDON F. CURL

It is perhaps true to say that the 'logical geography' of human movement as a field of knowledge has yet to be mapped. Be that as it may, a bird's eye view of the territory reveals some well-staked areas and some on-going surveying by researchers in universities and colleges who are concerned to establish a unified territorial framework.

If there is any truth in the saying that it is a philosophical event that generates young, exciting, if blundering sciences, then perhaps one should not at this early stage spurn a little philosophical reflection in an effort to establish a field of knowledge in human movement. The discussion that follows, therefore, has the modest aim of bringing into focus one or two philosophical considerations against a background of current epistemological thinking.

The Concept of Human Movement

It has been suggested that 'to have a concept' is to be in possession of a principle of unity according to which a number of things may all be regarded as being the same, or as being of one kind (Dearden 1968). Now if we are to possess such a principle in respect of human movement it follows that we must at the outset try to determine the conditions which we consider both necessary and sufficient for our use of the term. Failure to find these to our satisfaction need not however rule out the possibility of conceiving of a unified field of knowledge, for it is possible that some notion of 'family resemblance' or 'organic unity' may serve to give cohesion to the many manifestations of human movement.

At a first superficial glance, a concept of human movement appears to be easily come by; it is seemingly uncomplicated and the principles that govern its use seem simple and straightforward. Human movement is most obviously a particular case of movement in a more general sense, and this we define as a 'change of position' in reference to such phenomena as solids, liquids, gases. Human movement then, is but the change of position of the human body or its parts. This on the face of it would seem to be a simple and sturdy enough definition whose conditions are both necessary and sufficient. But are these conditions sufficient? Do they in fact enable

7

us, for example, to distinguish between the movement of a 'live' as distinct from a 'dead' human body in motion? If we take as our defining characteristics criteria according to the canons of physics, have we successfully identified *human* movement in particular, or have we only returned to the definition of movement in general—with the human body merely as an object? Does our definition provide us with some feature or features by which we can recognise distinctly human qualities—qualities we should wish to consider as necessary?

If, for example, we confine ourselves to a definition couched in the language of physics or mechanical laws, how would we make the distinctions which Ayer so cogently makes in his classic illustration of the different meanings which can be ascribed to the act of raising an arm in order to drink a glass of wine. This simple act, he suggests, might be construed as:

> . . . an act of self-indulgence, an expression of politeness, a proof of alcholism, a manifestation of loyalty, a gesture of despair, an attempt at suicide, the performance of social rite, a religious communication, an attempt to summon up one's courage, an attempt to seduce or corrupt another person, the sealing of a bargain, a display of professional expertise, . . . an act of expiation, the response to a challenge . . . etc. (Ayer 1964)

Clearly any attempt to reduce this battery of different interpretations of a simple arm movement to an objective mechanical formula would be to ignore some crucial factors in the determination of specifically human meaning. Human movement as a concept cannot then be restricted to a purely physical account—to regard it as such would be to adopt a narrowly self-contained approach, looking for defining characteristics in a rigid and stereotyped way, in fact with the paradigm of just one model before us. The concept of human movement requires that we recognise human *meanings* and therefore the total 'context' or 'form of life' of which that human movement is a part. Psychological, physiological, mechanical as well as social factors must necessarily be built into our concept—to say nothing of the aesthetic, religious and moral factors involved.

Such a conclusion—that among the necessary conditions for a concept of human movement must reside a respect for human meanings—suggests that even a behaviouristic approach will be inadequate. Peters (1966) for example, is eloquent on the limitations of such an approach. The determination of behaviourists, he says:

> to confine themselves to observable data and to eschew introspective reports must severely limit what they can possibly investigate. If the inner life of man is banned from investigation, actions which necessarily involve intentions, emotions which necessarily involve appraisals of a situation, together with imagination, memory, perception, dreaming, and pain must all be ruled out as scientifically proper objects of investigation; for none of these phenomena can be described or identified without reference to consciousness. There is

precious little left of human behaviour to investigate.

Any account of human movement, therefore, as an aspect of human behaviour must necessarily acknowledge conditions which involve the recognition of imagination, memory, motivation, emotion, perception and intention.

One further illustration might serve before we leave our exploration of necessary conditions and conclude that any purely physical account fails completely to recognise human movement as a vehicle of human meaning; our illustration highlights the role played by illusion and imaginative perception in the apprehension of aesthetic objects. What, one asks, can Langer (1957) mean when she says quite seriously :-

> In watching a dance, you do not see what is physically before you—people running around or twisting their bodies; what you see is a display of interacting forces, by which the dance seems to be lifted, driven, drawn, closed, or attenuated, whether it be solo or choric, whirling like the end of a dervish dance, or slow, centred, and single in its motion. One human body may put the whole play of mysterious powers before you. But these powers, these forces that seem to operate in the dance, are not the physical forces of the dancer's muscles, which actually cause the movements taking place. These forces we seem to perceive most directly and convincingly are created for our perception; and they exist only for it.

Clearly, no account of such 'forces' in the language of physics or bio-mechanics will do anything to illuminate the nature of the 'mysterious powers' which are perceived 'most directly and convincingly'; we are here in a dimension of aesthetic perception and imagination—modes of apprehension which interpret human movement in their own terms. Perhaps this is what Meredith (1963) implies when he states that:

> A physicist can calculate the horse power of a ballet dancer but he cannot interpret the dance. This pattern has a cultural geometry which has personal dimensions . . . Motion as studied by a physicist and movement as studied by choreographer . . . would appear to be two different things . . . .

But we have said enough to indicate that any comprehensive account of human movement will admit of interpretations which depend upon human perception and take their meaning from forms of human culture of which they are an essential part.

Forms and Functions of Human Movement

It becomes increasingly evident that human movement is not a uniform phenomenon; like language it is deeply stratified and requires that its many layers be analysed and clearly identified. From the many and complicated aspects of organic movement to the complex range of locomotor movements, gross operational movements, fine skilled movements, emotional and expressive movements, playful movements, indicative, descriptive, prescriptive movements, ritualistic and religious movements, social movements, athletic, balletic and dramatic movements—we recognise many forms whose functions range from the most basic organic functions to the most sophisticated symbolic functions. Each form has its own particular structure and its own particular functions—albeit with much interaction and interdependence.

To ask what *function* a particular movement serves is to start an enquiry into the context of the movement and to try to understand the 'form of life' from which it springs and of which it is a part. Such a functional approach tends to by-pass all-embracing theories and broad generalisations and concentrates upon the phenomenon itself—in its total setting, be it lung-functioning during exercise, or highly skilled performance under competitive conditions. Taking an example from yet another sphere, say, the role of human movement in social interaction, we find that it becomes necessary to understand the particular social conditions, conventions—in fact the whole cultural setting in order to identify the specific meaning of a gesture. The significance of bodily contact, proximity, posture, facial as well as gestural movement become vital factors in non-verbal communication (Argyle 1969; Layson 1969). And yet again, to ask what function a phenomenon such as social dance plays is to be introduced into the concepts of the social philosopher—concepts which illuminate function, namely: *pattern maintenance, adjustment, integration, role differentiation, tension management* (Rust 1969).

Perhaps the concept of 'function' will play an increasing analytical role in the study of human movement, and perhaps this is not inappropriate. It is a concept which springs from biological science and as such is no stranger to students of human movement. Certainly theories which concern themselves with 'forms of life' have seized upon the organic model with some profit. Social theory for example finds analogous features to organic functions in social systems and structures, and 'functionalism' as a means of social analysis has provided some illuminating analyses in its *functional imperatives, functional alternatives, manfiest and latent functions* (Cotgrove 1967). Similarly in the sphere of art—another 'form of life', we find the organic model of considerable significance. Langer (1957) states:

> ... the more you study artistic composition, the more lucidly you see its likeness to the composition of life itself, from the elementary biological patterns to the great structures of human feeling and personality that are the import of our crowning works of art.

But to utilise a model to good effect is not to fall victim of any literal transposition; each form of life has its own structure and functions, and whilst the organic model may throw light on these, their principles are not identical, they are analogous.

Forms and Fields of Knowledge

We have insisted that human movement is not a uniform phenomenon and that its many forms must needs be clearly identified; we would go further, and suggest that its many distinct forms must be *analysable and describable in the distinct language which characterises quite different forms of knowledge.* This requirement demands that we be familiar with the distinct concepts in which each form of knowledge is expressed. To describe the dance in Langer's illustration (above) in the language of physics would of course be to miss the aesthetic significance; likewise to describe the strict mechanical technicalities of the sophisticated skill of pole-vaulting in the language of aesthetics would be quite inappropriate. And yet when we survey the literature concerned with human movement we are aware that such category mistakes have occurred quite systematically. The result, although perhaps not fully recognised, is what Ryle (1949) calls 'myth', which he describes as 'the presentation of facts belonging to one category in the idioms appropriate to another'. But he adds significantly, that to explode a myth is 'accordingly not to deny the facts but to re-allocate them'. It is certain that before the logical geography of human movement as a field of knowledge can be effectively mapped, many myths may well have to be exploded. We shall return to this question later.

Apart from models and metaphors to which we have referred, what conceptual categories are there by which we might characterise the many different forms of movement—categories which have their own criteria for truth and validity? The answer to such a question would seem to lie in the *nature of knowledge itself,* for we can only *know* the different forms of movement according to the established public conceptual schemes which govern the basic forms of knowledge themselves. We must enlarge.

Thanks to the work of recent epistemologists we are now very much clearer concerning the necessary features of 'forms of knowledge'. These 'forms' provide us with a framework with which we can determine the logical structure of existing knowledge of human movement. Cassirer (1955) in his classification of forms of knowledge has nominated language, myth, religion, art, science; these he would suggest are the symbolic forms which determine the circle of humanity. More recently, forms of knowledge have been classified in terms of logical, empirical, moral and aesthetic categories. Phenix (1964) refers to six realms of meaning, namely: Symbolics, Empirics, Esthetics, Synoetics, Ethics and Synoptics, whilst Reid (1961) distinguishes the arts, religion, morals, science and knowledge of persons. Most recently, Hirst (1966) has named seven distinct forms of knowledge each with its own uniqueness of concepts and criteria for validity; these he names as science, mathematics, history, morals, aesthetics, philosophy and religion.

Now is it possible from these many suggested categories to find an inclusive scheme for unifying our knowledge of the different forms of human movement? Most of the examples of human movement we have so far cited can be seen on examination to have their descriptions and explanations in such recognised disciplines as anatomy, physiology, neurology, psychology, physics, bio-mechanics, bio-chemistry, ergonomics, sociology, and aesthetics. But does the possibility of a unifying framework exist which will hold together these various disciplines in some integrated fashion? Here again we might appeal to Hirst (1966) and his valuable description of 'fields of knowledge'. A field of knowledge we are informed can be profitably organised and unified 'by some central object, phenomenon, abstract entity, or other interest'. As an example of such a central theme Hirst suggest 'aspects of power, natural as well as social and political'. He emphasises that such fields can be of considerable value in promoting the growth of knowledge in certain areas; such organisations are of value in themselves and 'are not necessarily developed simply for teaching purposes'. Geography is cited as an example; this is a field of knowledge in that its concepts derive from a number of basic forms of knowledge i.e. historical, scientific, and so on. Of importance is the fact that 'all questions with which the geographers deal seem to be intelligible and answerable only within the canons of the several forms'.

Now if we apply this principle to the field of human movement study, we might well conclude that, having no distinct concepts of its own, all questions concerned with human movement are intelligible and answerable *only* within the several forms of knowledge. This would seem to place studies in human movement in very much the same category as say geography. Undoubtedly, human movement is a lively 'centre of interest' around which a number of forms of knowledge 'collect' although we would have reservations concerning its status when compared with such territorial themes as "The Neighbourhood", "Man and his Environment", "Rural Studies" or "African and Asian Studies"; these undoubtedly are fields of study which provide some unity to a range of subjects. But human movement is not in the same sense 'regional'; it is a unitary phenomenon—a phenomenon which in contrast to the artificial unity of the regional themes has a natural 'organic' unity. *It might in fact be claimed that human movement is a field of knowledge in which the various 'forms' of knowledge converge and cohere in an organic fashion, in a mode analogous to the organism itself.* Our earlier recognition of the appeal of the organic model for other spheres of knowledge, must surely have its prototype here in human movement studies (Phenix 1964).

It follows that any claim for human movement as a field of knowledge carries with it a commitment to a complex range of disciplines, and for those who operate in such a field a necessary mastery of many conceptual schemes and criteria which govern each distinct form of knowledge. Hirst (1966) insists that 'to master a field of knowledge is a complex and difficult matter'.

Physical Education and Human Movement Studies

To have recognised the complexity of human movement as a field of knowledge and the responsibilities of those who will work in such a field, leads one to acknowledge that up to the present time study in human movement has been nurtured mainly within the confines of physical education. University departments and colleges of education have, under the aegis of physical education, pursued and developed programmes of research in many aspects of human movement. But to have been nurtured within physical education does not in any way commit us to believing that the study of physical education and the study of human movement are in the same logical category (Hinks et.al 1971). These two studies do in fact carry the distinction that Hirst has made between 'fields' of knowledge and 'practical theories'. Physical education as part of education, we acknowledge, shares the criteria of education and its theory therefore has all the characteristics of educational theory—which in turn has undeniably practical functions. Physical education theory, like educational theory, serves to determine what ought to be done in educational processes in which motor activity predominates; it consists of an assemblage of different forms of knowledge which are used in the formulation of principles for practice. Human movement as a study on the other hand, is not tied to any practical functions, it is free to develop its own standards of relevance which are intrinsic to it.

It is natural enough that those who pursue physical education as a study look to practical outcomes, whereas those who pursue human movement as a field of knowledge will do so for its own sake irrespective of any immediate field of application. In this latter case we recognise a freedom from extrinsic purposes—a freedom which can be crucial to the manner in which the studies themselves are undertaken. Such a distinction between 'free' and 'enslaved' studies is well illustrated by Griffiths (1965) who protests in respect of restrictions placed upon studies for utilitarian purposes that:

> . . . we are treating students like slaves who have to be equipped with hoes and hammers. If psychology is studied in so far as it might be of use to those who later on will have to advise advertisers about the most efficient methods of deceiving the general public, it has ceased to be a universal object. What makes the distinction is not the field of study but the way it is pursued. It cannot be pursued as a valuable universal object if arbitrary limits are set on what determines the importance and relevance of any questions within it; it loses its fecundity of reciprocity which makes it valuable.

But Griffiths also adds significantly that if subjects are pursued as ends in themselves that is not to say that they are all useless, 'it is only to say that their use does not determine the way in which they are studied'.

Human movement as a field of study, distinct from physical education, would then, seem to be an attempt to foster the necessary freedom for the pursuit of a 'universal object' with its own intrinsic standards of relevance. It is the way in which it will be pursued that will distinguish it from physical education with its essentially extrinsic aims and objectives.

One is of course only too aware that the term physical education has for many years been stretched beyond its logical limits—stretched in fact by the development of something within it i.e. the growth of a field of knowledge concerned with human movement. Applied studies such as kinesiology, body mechanics, the physiology of muscular exercise, neuro-muscular co-ordination in physical skills, motor learning and transfer, emotional and personality factors in performance, expressive, social and aesthetic functions of physical activity—studies which receive but scant treatment within the traditional disciplines from which they derive—have tended to gather together under the aegis of the over-extended title of physical education. But manifestly many of these studies have, as such, little to do with educational activity; their advance as areas of research would be severely limited if they were permanently shackled to the prospective application to work in education. Only through a freedom of approach can these studies develop their own appropriate concepts and criteria for validity. On the other hand, physical education in attempting to embrace a field of study as well as fulfil its practical functions now recognises its dilemma and will doubtless welcome the emancipation of human movement as an independent field of study from its confines. The study of physical education with its own problems of research we now recognise as a valuable area of knowledge—in fact a set of practical theories which are used in the formulation of principles for practice.

Sources of knowledge of human movement—knowledge or not?

In any projected mapping operation of a new field of knowledge our first priority would seem to be to determine what does in fact constitute knowledge of human movement and to what category such knowledge belongs. How do we determine what constitutes knowledge of human movement and with what forms of knowledge can we identify it? Many writers are clear and insistent upon the answer to our first question: in order to qualify as knowledge there must be *concepts which can be shared in a public world.* Hirst & Peters (1970) would say that 'Only where there is public agreement about the classification and categorisation of experience and thought can we hope for objectivity within them'. They would go further and insist that in addition there must be 'objective tests for what is claimed is experienced, known or understood'. Knowledge according to another epistemologist 'is the outcome of disciplined enquiry' and is to be contrasted with mere opinion which lacks critical examination and justification. Knowledge is 'tested by criteria of justification developed by organised companies of specialists' (Phenix 1964). These pre-requisites for determining what constitutes knowledge are unequivocal; they provide some stringent standards by which we might assess the wide ranging and diffuse literature in human movement as well as severe standards by which we might assess the non-discursive forms of knowledge to be found in practical activities from skilled operations to aesthetic forms of movement.

If we accept these standards how then do we proceed to determine to which basic forms of knowledge our knowledge of human movement belongs? Again we must appeal to the epistemologist:

> The only way in which we can successfully distinguish different forms of thought is in fact by reference to the particular set of terms and relations which each of the distinct forms of thought employs.
>
> (Hirst & Peters 1970)

If this is the case, then in order to identify the status of the various kinds of movement knowledge we are driven to examine carefully the terms in which they are expressed. If we are successful we should be able to place them into their scientific, historical, aesthetic etc., contexts (allowing of course for the servicing of one form of knowledge by another). Now such an examination does of course pre-suppose that we are thoroughly familiar with the distinct concepts of each form of knowledge, and even more important, that we can detect the use of 'mythical' terms which involve—to repeat Ryle's phrase—'the presentation of facts belonging to one category in the idioms appropriate to another'. Perhaps we should enlarge on this point a little.

To anyone familiar with the growing literature on human movement it becomes abundantly clear that some of it does in fact play freely across the line between one form of knowledge and another, resulting in what we have called myth. Facts belonging to one category of knowledge are systematically presented in the concepts of another, and writers are often unaware of the metaphorical status of the terms they use. As an example of such a tendency we would take the movement theories, which have found wide acceptance particularly in educational circles, of Laban (Thornton 1971). Here we find such concepts as *space, time* and *weight* made to do unqualified duty in so called scientific as well as aesthetic contexts. The result is of course extravagant pseudo-claims for science and confused perceptual categories for aesthetics. Laban's 'movement theories' propound a doctrine of 'universal movement forms' and as such resist the necessary and fruitful differentiation of movement into a number of distinct forms identifiable, on inspection, with the basic forms of knowledge themselves. On enquiry it is not surprising to find that Laban's concepts derive from a primitive Pythagorean philosophy of universal movement. According to this ancient cosmic theory the movements of the planets, molecules and man all reflect one vast parallelism—all movement is but the manifestation of an all-pervading undifferentiated universal power (Curl 1966-69). Such a universal theory with its mythical concepts is one such an example of theory which confuses physical fact with human meanings. If however we transpose Laban's specious account of the physical world into aesthetic concepts we do find—as with all myth—some valuable kernel of truth. Langer (1953) aptly describes Laban's dilemma when she says:

> The writings of the most thoughtful dancers are often hard to read because they play so freely across the line between physical fact and artistic significance. The complete identification of fact, symbol, and import, which underlies all literal belief in myth, also besets the discursive thinking of artists, to such an extent that their philosophical reflections are apt to be as confused as they are rich. To a careful reader with ordinary common sense they sound nonsensical; to a person philosophically trained they seem, by turns, affected or mystical, until he discovers that they are *mythical*. Rudolf von Laban offers a perfect instance: he has very clear ideas of what is created in dance, but the relation of the created 'tension' to the physics of the actual world involves him in a mystic metaphysics that is at best fanciful, and at worst rapturously sentimental.

If we have laboured our point it is because we feel forcibly that in our efforts to determine to which form of knowledge our movement theories belong that we should not only refer to the particular set of terms and relations which each use, but be ever on the alert to detect the mythical mode of thought which pervades our literature (Cassirer 1955).

In conclusion, as we survey and take stock of our present sources of knowledge in human movement, we should recognise that the history of the development of knowledge itself provides us with a model. It is, as Hirst (1965) reminds us:

> . . . the story of progressive development into a number of distinct forms each providing unique understanding because of the uniqueness of its concepts, its conceptual structure and its criteria for validity.

Human movement knowledge will in its development necessarily admit of progressive differentiation, but at the same time may well find support and cohesion in, we hope, a new unified *field* of knowledge.

References

ARGYLE M. (1967). "The Psychology of Interpersonal Behaviour". Harmondsworth: Penguin.

AYER A.J. (1964). "Man as a Subject for Science". London: Athlone Press.

CASSIRER E. (1955). "The Philosophy of Symbolic Forms. Vol. 2." Yale: University Press.

COTGROVE S. (1967). "The Science of Society". London: George Allen & Unwin.

CURL G.F. (1966-9). Philosophic foundations. *Laban Art of Movement Guild. Magazine* Nos. **37, 38, 39, 40, 41, 43.**

DEARDEN R.F. (1968). "The Philosophy of Primary Education". London: Routledge and Kegan Paul.

GRIFFITHS A.P. (1965). A Deduction of Universities. In R.D. Archambault (Ed.) "Philosophical Analysis and Education". London: Routledge and Kegan Paul.

HINKS E.M., ARCHBUTT S.E. & CURL G.F. (1971). H. Movement Studies—a new standing committee. *Bull. Univ. London Inst. Educ.* **No. 23.**

HIRST P.H. (1965). Liberal education and the nature of knowledge. In R.D. Archambault (Ed.) "Philosophical Analysis and Education". London: Routledge and Kegan Paul.

HIRST P.H. (1966). Educational Theory. In J.W. Tibble (Ed.) "The Study of Education". London: Routledge and Kegan Paul.

HIRST P.H. & PETERS R.S. (1970). "The Logic of Education". London: Routledge and Kegan Paul.

LANGER S.K. (1953). "Feeling and Form". London: Routledge and Kegan Paul.

LANGER S.K. (1957). "Problems of Art". London: Routledge and Kegan Paul.

LAYSON J. (1969). An introduction to some aspects of sociology of physical education. In G.F. Curl (Ed.) A.P.W.C.P.E. Conference Report: "Physical Education Social and Aesthetic Aspects."

MEREDITH G.P. (1963). "Mind and Movement." In Chelsea College of Physical Education Year Book 1969/70.

PETERS R.S. (1966). "Ethics and Education". London: George Allen & Unwin.

PHENIX P.H. (1964). "Realms of Meaning". New York: Mc.Graw Hill.

REID L.A. (1961). "Ways of Knowledge and Experience". London: George Allen & Unwin.

RUST F. (1969). "Dance in Society". London: Routledge and Kegan Paul.

RYLE G. (1949). "The Concept of Mind". Harmondsworth: Peregrine.

THORNTON S. (1971). "A Movement Perspective of Rudolf Laban". London: Macdonald and Evans.

2

HUMAN MOVEMENT - TOPIC AREAS

2 HUMAN MOVEMENT - TOPIC AREAS

Code	Topic	Contributor
A	Acquisition of skill	J.C. Evans
B	Movement impairment	H.T.A. Whiting
C	Central and peripheral nervous function	J.D. Brooke
D	Communication through movement	Elizabeth Mauldon
E	Aesthetics and human movement	June Layson
F	Human movement and the visual arts	D.W. Masterson
G	Human movement—a societal study	L.B. Hendry
H	Comparative studies	Jean Carroll
I	Personality and movement behaviour	H.T.A. Whiting
J	Physical work metabolism	J.D. Brooke
K	Physical growth, and degeneration	J.D. Brooke
L	Biomechanical analysis of human movement	G.T. Adamson
M	Movement analysis	Hilary Corlett
N	Movement ideas in perspective	Ida Webb
O	Play	A.G. Roche
P	Development of movement understanding in children	Hilary Corlett

3

TOPIC AREAS - OUTLINED

Stage

3 TOPIC AREAS - OUTLINED

A ACQUISITION OF SKILL
by J.C. EVANS

Theoretical frameworks for the study of perceptual-motor (P-M) skill

Systems analysis of P-M skills: functional models of P-M systems.
A taxonomy of skill.
Phases of skill acquisition.
Learning theory.
Cybernetic theory.
Information theory.
Individual differences.
The body concept.

Factors affecting the acquisition and performance of skill

Display characteristics.
Task difficulty and type: Discrete, serial and continuous.
Practice schedules: Massed and distributed practice:
                                   Whole and part learning:
                                   Guidance.

Mental practice and rehearsal:
Knowledge of results:
Transfer of training: Stimulus and response generalisation.
Motivation: Arousal, activation and the RAS.
                                   Competition and level of aspiration;
                                   Personality factors;
                                   Imitation and social facilitation;
                                   Vigilance and habituation;
                                   Mental fatigue.

25

Stress

Memory and retention: Long-term and operational memory;
                                  Short term memory—pre-and post-perceptual stores.
                                  Transformation and recoding of input.
                                  Psychological refractoriness and reaction time.

Perceptual aspects of skill

Visual, auditory and balance senses.
Kinaesthesis and proprioception.
Selective attention and discrimination.
Innate perceptual abilities and acquired perceptual analysers.
Personality and perceptual mode: psychological set.

Decision making aspects of skill

The single channel hypothesis.

S-R compatability: Choice reaction time.
Stimulus and response probability: expectation and anticipation;
                                              intrinsic and extrinsic;
                                              error correction and control.

Motor aspects of skill

Innate abilities underlying motor capacity.
Acquisition and modification of plans for action.

Timing and co-ordination   Speed and accuracy trade-off,
                                  Cerebral and cerebellar control systems.
                                    Movement time.

B MOVEMENT IMPAIRMENT
by H.T.A. WHITING

Observation of pathological movement behaviour in psychiatric diagnoses—neurasthenia, schizophrenia, autism etc.

Classification of movement disabilities at a gross level e.g. ataxia, chorea, spasticity etc.

Motor impairment—concept of clumsiness

Motor impairment—a misnomer?
Related to model of perceptual/motor performance

Motor impairment and information processing

Aetiology of motor impairment
    i.   brain damage
    ii.   neural dysfunction

Prenatal factors in the aetiology of motor impairment

    i.   prenatal anoxia.
    ii.   virus infection.
    iii.   rhesus factor.
    iv.   toxaemias of pregnancy.
    v.   climate and foetal damage.
    vi.   cigarette smoking.

Paranatal factors in the aetiology of motor impairment

    i.   prematurity.
    ii.   placenta praevia.
    iii.   precipitate labour.
    iv.   anomalies in presentation.
    v.   forceps deliveries.
    vi.   caesarian section.
    vii.   twin preganancies.

Possible effects of pre-and para-natal damage on motor ability

    i.   spasticity.
    ii.   athetosis.
    iii.   chorea.
    iv.   ataxia.
    v.   tremor.
    vi.   rigidity.

Postnatal factors in the aetiology of motor impairment

    i.   sensory, perceptual and movement deprivation.
    ii.   emotional disturbance.
    iii.   social class.

Motor impairment and intellectual development

  i.   motor ability and early cognitive development.
  ii.   motor and mental ability.
  iii.   motor ability and mental subnormality.
  iv.   'ability' concept.

Motor impairment and the socialisation process

  i.   selective attention/perception.
  ii.   tension in the mother.
  iii.   status significant motor skills.
  iv.   motor impairment and maladjustment.
  v.   brain damage, personality and socialisation.

C CENTRAL & PERIPHERAL NERVOUS FUNCTION
by J.D. BROOKE

All the topic areas fall back to this section. The areas A,B,I,K are involved primarily over all the sub-sections below. In addition very specific interactions have been noted as they occur.

Taxonomy

Peripheral somatic and visceral, specific and non-specific, afferent and efferent.
Central nervous hierarchy.
Brodman's areas of the cerebral hemispheres.
Relationship of anatomical and functional classifications.

Basic Neural Function

Excitability
Conductivity
Transmission

Peripheral Afferent Function

Somatic—   cutaneous, kinaesthetic
           and central lemniscal path;
             —temperature, pain, crude cutaneous

and central spinothalamic path.
Specific—   smell, taste, sight, sound.

Motor Systems

Innervation of Muscle —skeletal;
                       —cardiac;
                       —visceral;
Reflex Complexes   —spinal cord;
                       —hindbrain;
                       —midbrain;
                       —forebrain.

Maintenance of Homeostasis

Respiratory gases.
Blood pressure.
Electrolytes.
Feeding.
Blood sugar.
Hormones.
Acid-base state.
Temperature.
Brain metabolism.

Arousal, Emotion & Conditioning

Inadequate anatomical definition behind functional research.

Diffuse arousal—ascending and descending reticular activating systems of the hindbrain and midbrain.

More specific arousal—ascending and decending reticular activating systems of the forebrain.

Emotional addition—limbic system, hypothalamus and adrenal cortex, the orienting response.

Conditioning—   involvement of limbic and reticular systems, forebrain, brain electrical activity and behavioural response.

Learning & Memory

Inadequate anatomical definition and functional research behind observable
behaviour; complexity of organs and of interactions between structure
and input to cerebral cortex.

Electroencephalography.
Protein synthesis and memory.
Phylogenic and ontogenic
development of central grey vs cerebral grey matter.
Frontal lobe and memory.
Prefrontal areas and information retention.
Lobotomy and personality change.
Interrelated cerebral hemispheric systems.
Hemispheric interaction by
forebrain commissures and via midbrain and hindbrain paths.

D COMMUNICATION THROUGH MOVEMENT
by ELIZABETH MAULDON

The tactile channel

Tactual sensitivity the most primitive sensory process and the first to function.
Cultural and sex differences in spatial proximity, social distance, body zones and
degree of touch.
Self structure; body image; kinesthesis; body awareness; perceptual judgements
and cognitive development.
Tactual deprivation; tactual experiences and learning.
As reinforcement in skill learning; manual guidance.
Development of concept formation.

The auditory channel

Origins of words as 'vocal signs'; speech as a symbol-making activity.
Words as action; association of speech and action; sound and 'effort'; sound and
motion.
Inflexion, speed and tone in transmission of meaning.
Language as 'verbal behaviour'; 'rules of talk'; ritual in verbal interchange;
association between speech and non-verbal signals or cues.
Music as non-verbal, ordered sound produced by action; association of musical

sound with emotional mood and action; the significance of rhythms.
Identification of sound with movement; 'music as audible movement and movement as inaudible sound.'

The visual channel

Gesture as the sole means of communication; evolution of signals, signs and symbols.
Movement expression; effort and spatial theories; posture and gesture.
Personality and movement expression.
Relationship of gesture to verbal cues; congruency and incongruency; 'double-bind'; social interaction as 'information game'; facial expression; shadow movements; expressions 'given' and 'given-off'.
Gesture and leadership; kinesic leaders; charismatic movement leaders.
Visual cues in diagnosis of psychosomatic disorders.
The function of movement within the visual arts as communicating media.

E AESTHETICS AND HUMAN MOVEMENT
by JUNE LAYSON

Introduction
a) Aesthetics—definitions. Aesthetics as a branch of philosophy.
b) Perception—an element of aesthetics, psychology and aesthetics.
c) Cross-cultural studies.
d) Physical education.
e) Other contributory areas—general philosophy, literature, dance in the theatre, sport, etc.
f) Historical aspects.
g) Proposed structure of field.

The Activity
a) Movement activities and aesthetics.
b) Aesthetic accounts of movement in general—concept, continuum/spectrum, model and phenomenological approaches, terminology.
c) Aesthetic accounts of separate movement activities—sport as an art form, aesthetic content of sport, aquatics, gymnastics, boxing, dance as an art form.

The Performer
a) The performer's aesthetic experience of movement.
b) Skill/technique.

c) Kinaesthetic perception.
d) Individual differences.

The Spectator

a) The spectator's aesthetic experience of movement.
b) Aesthetic attitude.
c) Spectator typologies.
d) Films, television.

F HUMAN MOVEMENT AND THE VISUAL ARTS
by D.W. MASTERSON

General

The representation of human movement in art. Movement and energy as sources of aesthetic pleasure. Literature devoted to human movement as the *subject* of art and to the role of movement in the *production* of art works.
Aesthetic aspects of human motion.
The beauty of movement in sport and the dance.
Sport in art and sport as art.
Physical education and aesthetics.

Art and the Idea of Movement

The perception of movement in art objects.
'Pure movement'. Wertheimer's theory of phiphenomena. Retinal stimulation and after-imagery.
Relationships between movement and the framework in which it occurs.
The sensation of movement and kinesthesis.
Movement imagined from association.
Gestalt theory. Physiological mechanisms and psychological perception.
Movement and time. Individual sequences and 'timeless wholes'.
Movement and created space.

The Representation of Movement in Art

Modification and distortion in the depiction of motion.
Attitude and gesture. Arrested motion. Imbalance.
Repetition, rhythm and simultaneity.
Symbolic movement.

Dynamic form and line.
Multiple viewpoints.
Kinetic art.

Human Movement in Painting and Sculpture

Distortion in figurative art. Examples from the history of art.
Distortion in facial expression and its associations.
Movement in Classic art permitted by technical developments.
Evolution of 'the pregnant moment' in Fifth Century art.

Movement in Idealistic and Naturalistic Art

The characteristics of Ideal and Sensate art.
Classicism and the 'state of being'. Naturalism and the 'state of becoming'.
Movement in Naturalistic art. The emotional appeal of the Baroque and Romanticism.
Neo-classicism. Cubism.

Human Movement and Sport in Modern Art

The multiple viewpoints of Cubism. Cubism in theatrical décor and Cubist pictures of sport.
Futurism and the dynamism of sport. Photography of motion.
Sport in abstract art.
Sport and Neo-realism.
English art and modern sport. Exhibitions of sport in art.
Sport in American art from Eakins to Rauschenberg.
Sport in 'Pop' art. Artists' personal observations about sport as a subject of art.
Socialist Realism. Russian artists of the Twentieth Century and sport.
Sport and art in the German Democratic Republic.

HUMAN MOVEMENT : A SOCIETAL STUDY
by L.B. HENDRY

The Individual in the Family

Movements in utero.
Interaction of innate movements and environmental supports.
Early familial interactions and their influence on movement patterns and social skills.

Movement, perceptual input and total development.
Effects of deprivation on the child.
Imprinting, especially in relation to 'inappropriate' movement patterns.
Parental influences on movement behaviour.
Movement and status in childhood.
Motor indices of social development.
Bizarre experiences and self-image.

Sub-Cultural Influences

A social-class typology which may have implications for movement studies.
Speech, gesture and movement patterns within social classes.
Movement abilities within the context of peer groups in childhood and adolescence.
Adolescent interests in movement; sex and class differences.
Movement difficulties and maladjustment.
Use of 'play methods' and movement experiences in schools.

A Societal Classification

a) *Goal directed*—how everyday functional movements are influenced by various social conventions.

Man, the environment and efficiency of movement.

The effects of urban dwelling on movement behaviour.

b) *Excitatory*—patterns of sexual behaviour; ritualized and institutionalized sexual movement of an exaggerated, incomplete character—strip-tease etc.

Eye movements; sexual taboos; physical proximity and body accessibility.

The excitatory potential of games and dance.

c) *Bizarre movements*—movements as the overt manifestations of the influence of drugs, of mental states; mental health.

d) *Aesthetic movements*—the aesthetic elements of competitive sporting movements; tactile and kinaesthetic arts; dance.

e) *Communicative movement*—Interpersonal attitudes and emotions communicated by movements, techniques for measuring such gestures and movements.

Postures and group status, personality traits and movement patterns.

'Touch hunger' and sport, communicative movements in ballet and opera.

Crowd movements in social interactions.

f) *Group-Status*—Movement and group variables—size, affiliation, leadership potential dominante-submission. The participation of royalty and their movement involvement.

g) *Ritual*—Crowd movement at sporting occasions. Priest and congregation—religious movements and gestures.

h) *Protest*—'Pop' movement and the generation gap. Movement associated with protest marches, involvement theatre and 'pop' festivals.

i) *Stillness*—The containment of movement—by certain drugs and techniques, in various institutionalized areas e.g. parks, prisons, transport, for artistic reasons.

Cross-Cultural Classification

An examination of the previously utilised classification

a) Goal directed   f) Group-Status
b) Excitatory   g) Ritual
c) Bizarre   h) Protest
d) Aesthetic   i) Stillness
e) Communicative

Its application to other cultures, seeking movement comparisons, contrasts and universals; emphasising particularly the dangers of inactivity to modern western man, the association between movement patterns in sport, dance, and everyday activities to held cultural values.

H COMPARATIVE STUDIES
by JEAN CARROLL

Ethological studies in which movement behaviour is viewed in evolutionary perspective.

Vertebrates, other than mammals

Innate patterns of movement.
The concepts of 'imprinting' and 'critical period'.
Links with human studies.
Sensory stimulation, sensory deprivation, deficiency of stimulation, maternal deprivation, the sucking and smiling responses.
Redirected and displacement activities;
ritualized behaviour.

Mammals, other than primates

Play, expression and communication.
Food finding and storing.
Social organisation and territory.

Fighting, threat and appeasement behaviour.
Amicable and sexual behaviour.

Primates, other than humans

Early field studies.
Comparisons between species.
Varied and flexible behaviour in bodily contact.
Facial—visual signals.
Communication processes.
Repertoires of behaviour.

The use of the ethological method in human studies

Extrapolations between species.

Cross-cultural studies

Studies of everyday movement behaviour

Early studies and classifications

(a)**bodily aspects,** including—   physical appearance,
                                                           body parts,
                                                           tactile aspects,
                                                           perception and transmission of cues.

(b)**effort aspects,** including—   stress and timing in gesture,
                                                           changes in facial expression.
(c)**spatial aspects,** including—   orientation,
                                                           direction of gaze,
                                                           posture,
                                                           proximity.

(d)**relationship aspects,** including—   groupings,
                                                           leadership patterns.

Institutionalized forms of movement

(a)**traditional approaches**
—historical
—structuralist
—functionalist

(b) **contemporary work**
—continuation of traditional.
—narrow and wide range comparison.
—studies with movement classifications as the
basis for sociological or psychological hypotheses.

(c) **Physical Education as an institution within the
institution of the school**
Psychopathological studies in which movement behaviour is a dimension utilized in
the demarcation of the abnormal.

bodily aspects—
ego identity,
psychomotor reactions,
hand, head and leg gestures.

effort aspects—
qualitative stress in gestures,
of patients,
repetition.

spatial aspects—
proximity,
orientation.

relationship aspects—
co-operation,
communication,
the 'double-bind',
institutionalization.
Criticisms of the normal/abnormal criteria.

PERSONALITY AND MOVEMENT BEHAVIOUR
by H.T.A. WHITING

The nature/nurture hypothesis

Affects of inter-uterine environment (chemical, physical) on development.
Stress in pregnancy and personality change.
Genetic predispostion in mother and child.
Birth injury, brain damage and personality change.

Degree of spontaneity of movement behaviour.
Movement of foetus in utero.

Dimensional Analyses

Eysenck
> Extraversion/introversion.
> Neuroticism.
> Psychoticism.
> Physiological interpretation of above in
>   terms of excitation/inhibition balance.
> Concept of arousal. Yerkes-Dodson Law—inverted U curve.
> Conditionability. Fear conditioning and skilled performance.
> Inhibition and physical performance.

Cattell
> 16 personality factors
>   2nd, 3rd & 4th order factor derivatives including
>   creativity, independence, leadership.
> Use of personality profiles as a basis for
>   comparison of criterion populations.
> Movement behaviour sterotypes.

Witkin
et al
> Personality through perception.
> Perceptual systems.
> Selective attention/perception.
> Perceptual types.
> Field dependence/independence.
> Field dependence and creativity.
> Differentiation/ integration.

Physique and temperament

Historical development—Kretschmer, Sheldon, Parnell,
      Eysenck etc.

Expressive movement and personality

Studies in expressive movement.
Posture and gesture.
Non-verbal communication.
Effort characteristics.

Clinical aspects of personality study

Classification.
Movement characteristics in psychopathology.
Personality, body type and mental illness.
Bizarre movement patterns.

Movement requirements of different personality types

Self-concept

Terminology—'I', 'me', self, self-image, self-esteem, body-
         concept, body image, body-awareness sensotype.
Affective, cognitive and psychomotor aspects of body concept.
Body-concept and personality theory.
Historical connotations of body-concept.
Development of body-concept.
Athletic participation and the body-concept.
Effects of deprivation on body-concept.
Body-concept and body-type.
Body-concept and physical fitness.
Body-concept and laterality.
Movement behaviour and the body concept.
Body-concept and compensatory education.
Body-concept and kinesthesis.
Body-concept and information processing.
Body-concept and arousal.
Contribution of movement experience to development of body-concept.
Assessment of the body-concept.
Distortion of body-image under stress, disease, drugs and deprivation.
Body concept and movement impairment.

J PHYSICAL WORK METABOLISM
by J.D. BROOKE

Lung Function

Standard Pappenheimer nomenclature.
Resting and exercise analyses.
Work and power of lung.
Homeostasis, disequilibrium, ventilatory fatigue.
Gas exchange.

Effects of sensory stimulation, pain, arousal.

Digestion, Assimilation and Storage

Conversion of food into material for work metabolism.
Carbohydrates.
Proteins.
Fats.
Electrolyte balance.
Vitamins.
Minerals.
Diets for vigorous activity.
Arousal and digestive changes.

Blood
Oxygen/carbon dioxide transport.
Acid—base state.
Nutrient transport.
Humoral regulation.
Body heat dissipation.
Electrolyte balance.
Plasma proteins.
Amino acids.

Heart and Circulatory Dynamics

Work of the heart.
Electrophysiology.
Circulatory pressures and flow.
Centrifugal pressors, tachycardiacs.

Skeletal Muscle

Provision—oxygen, glycogen, electrolytes, amino acids.
Uptake   —aerobic and anaerobic reactions.
                          —effects of $[\%_H^+]$, P_{CO_2}, temp., arousal.
                          —respiratory enzymes.
Electrophysiology.
End product of muscle metabolism.

Behavioural Analysis

The Work Domain   —factors of physical work performance

—specificity.
Adaptation, training, progressive overload.

K PHYSICAL GROWTH AND DEGENERATION
by J.D. BROOKE

Incomplete Understanding

Growth-study key to life.
Continuum, fertilised ovum—death.

Levels of enquiry— molecular
cellular
behavioural
ecological
phylogenic and ontogenic.

Levels of information— data
principles
conceptual structure.

Micro Level

Cell complex— the unit,
the hierarchically ordered systems
of molecular control.

Quantification— growth and maintenance of a type.
Differentiation— growth of types, function and form.

Height, Weight, Physique

Embryo, child, adult, male, female.

Growth of Systems

Muscular (cardiac and skeletal).
Respiratory.
Skeletal.
Digestive.
Genito-urinary.
Nervous and endocrine.

Regulation of Growth

Genetic versus environmental.
Hormonal.
Dietary.

Aging

Degeneration, theories on aging.
Systemic changes—   muscular (cardiac and skeletal)
                                           respiratory
                                           skeletal
                                           digestive
                                           nervous.
Adaptation to environmental demand.

L BIOMECHANICAL ANALYSIS OF HUMAN MOVEMENT   by G.T.ADAMSON

An 'internal' and 'external' analysis under biostatic and biodynamic conditions.

Internal   **External**

Muscle mechanics.
Bone and joint kinematics.   Environmental reaction.
Intra-truncal pressures.
Haemodynamics.   Gravity
Body size and composition.
Work capacity and efficiency.   Inertia.

Mass, inertia, weight.
Centre of gravity and buoyancy.
Pressure, density, levers, torque.
Newton's laws.
Equilibrium.
Resolution of forces.
Friction, drag, flow, viscosity.
Elasticity.
Collision processes.
Force/distance/time relationships.

Uniform and accelerated motion.
Projectiles.
Impulse, momentum.
Power.
Potential and kinetic energy.
Work, efficiency.
Dynamics of rotation.
Moment of inertia.

MOVEMENT ANALYSIS
by HILARY CORLETT

Elemental analysis

Body — Body units.
Classification of bodily activities.
Postural/gestural analysis.
Co-ordination of body units in motion.
Kinesthesis.

Space — Kinesphere and environmental space.
Orientation; cross of axes.
Pathways, shapes and identification
of trace forms in choreutics.
Geometrical analysis of body movement
in space.
Individual spatial preferences.

Effort and Rhythm — Measurable principles.
Metric and free rhythms.
Effort rhythms: motion factors and
effort qualities.
Significance of rhythmic patterns in
movement communication: shadow movements.
Effort/shape analysis.

Relationship — In the body with objects, with persons.
Classifiable units and channels of
communication.
Movement responses in the interaction process.

Effort in work study

Analysis of job-sequence.
Observable features of motion.
Fusion of kinetographical and effort analysis.

Behavioural analysis

Movement/personality.
Movement as a non-verbal language.
Elements and patterns of social interaction
     —dyadic and group.

Aesthetic components

Dance, existential analysis, illusory qualities.
Rhythmic and spatial structures of dance.
Aesthetic qualities in man-made movement forms.

MOVEMENT IDEAS IN PERSPECTIVE
by IDA WEBB

Man as the centre of creation

i.	Aristotle	— motion, heavy, light, near, far, gravity, levity.
ii.	Plato	— principle of life, that which moves itself, that which is moved.

Organic evolution

i.	Darwin	— Biological determinants of movement patterns, upright locomotion, development of spine, limbs, feet, hands.

Force and the Laws of Motion

i.	Galileo	— freely falling bodies.
ii.	Newton	— inertia, acceleration, action and reaction, linear, angular, uniform, non-uniform motion.

Movement as a means of communication

i.	Delsarte	—	emotion, gesture, laws of trinity, correspondence, unity, laws of altitude, force, motion, sequence, direction, form, velocity, re-action, extension.
ii.	Lamb	—	gesture, posture.
iii.	Argyle	—	social interaction.
iv.	Gehlen	—	symbolic structure.
v.	Esalen	—	tactile sensitivity.
vi.	Birdwhistell	—	kinesics.

Objectivity of movement

i.	Noverre	—	movement expression.
ii.	Klages	—	self movement, dance, rhythm, surrender to image, cosmos.

Functional Activity

i.	Buytendijk	—	body at personal disposal, power, purposeful actions, opposition of subject and self, unity of function.
ii.	Weizsacker	—	identity of forms, Gestalt circle.

Scientific management

i.	Taylor	—	management as a science, time study, instructions card, functional foremanship, piece rates.
ii.	Gilbreth	—	reduction of number of movements.
iii.	Lawrence	—	effort, job analysis, economy of energy.

Movement as an art form

i.	Langer	—	abstraction.
ii.	Metheny	—	kinesis.
iii.	H'Doubler	—	organisation of experiences.

Elements of movement

i.	Laban	—	weight, space, time, flow.

Movement in an educational environment

i.   Athletics   —   Ancient Greece to modern olympic games,
                            cult, competition, style, skill.
ii.   Dance   —   traditional—   ethnic, style, dress.
                            educational—   Duncan, Laban.
                            theatrical—   ballet, classical, modern.
                                           contemporary—e.g. Graham,
                                           Cunningham, Nikolai, Nederland.
iii. Games   —   team, individual.
                            Public School cult.
                            National governing bodies.
                            Arnold, Osterberg.
iv.   Environmental—   recreation, leisure,
      pursuits and   outward bound.
      extracurricular
      activities.
v.   Gymnastics   —   drill, rhythmical, postural, medical,
                            aesthetic, educational, olympic.
                            Gutsmuths, Jahn, Spiess
                            Nachtegal, Ling, Osterberg
                            Morison.
vi.   Swimming   —   competitive, synchronised, long
                            distance, strokes.

Movement in the industrial situation

i.   Ergonomics   —   time and motion study,
                            job efficiency,
                            behavioural analysis, task orientation,
                            system approach, posture.

0 PLAY
by A.G. ROCHE

General

Definitions and concepts

Eutrapelia.

Positive educational agent.
Culture embodies the character of play.
Development in persons.
Non-instructional characteristics.

Some theories of play relevant to human movement

Surplus energy.
Recapitualation.
Instinct/practice.
Harmonious development.
Pleasure principle.
Repetition—compulsion.
Assimilation and accommodation.
Stimuli, responses and rewards.

Social play

Play and socialisation.
Adjustment.
Cultural influences.
Games and sports.

Motor performance

P THE DEVELOPMENT OF MOVEMENT UNDERSTANDING IN CHILDREN by HILARY CORLETT

The Infant (0—7 yrs.)

Pre-school, nursery and infant schoolchild.
Innate patterns of behaviour; emotional responses;
movement qualities of the pre-school child.
Physical growth and maturation—development of motor
abilities—general to specific.
Socio-emotional experiences with environment through play.
Cognitive development—relation to movement behaviour.

The Junior (7—11 yrs.)

General characteristics—skill hungry age.

Intellectual development—creative abilities.
Development of movement potential in inventive movement.
Physical abilities and specific psychomotor skills.
Socio-emotional influences—peer acceptance; friendship.

The Adolescent (11+)

Physical growth and maturation—abilities and skill.
Socialisation—expressive behaviour—effort contagion.
Intellectual growth: movement preferences; aesthetic
appreciation.

Development of movement concepts

Conceptual growth.
Development of concepts of motion: body: force: time: space
Movement experience and aesthetic education.

4

TOPIC AREAS - DEVELOPED

TOPIC AREAS-DEVELOPED

ACQUISITION OF SKILL

4 A

by J.C. EVANS

The purpose of this section is to focus attention on the many variables which affect the learning and performance of perceptual-motor skills. Overviews are provided by Cratty (1964); Knapp (1963); Fitts & Posner (1967); Welford (1968); Whiting (1969); Morris & Whiting (1971).

A.1. Theoretical frameworks for the study of perceptual-motor skill

This section attempts to identify the theoretical frameworks currently employed in the study of perceptual-motor skill. Theoretical models fall into three broad categories: 1. *Communications Models:* based on the coding, translation, transmission and storage of data; 2. *Control System Models:* the regulation of behaviour through the interplay of input, output, feedback and noise variables; cybernetic theory with particular emphasis on self-regulating closed-loop systems; 3. *Adaptive System Models:* based on the existence of hierarchical processes and analogous to computer operations with low and high order programmes or routines.

The first two categories suggest steretyped concepts, their characterisitics remaining unchanged in dynamic situations. Adaptive models are characterised by processes providing variability in input, modification of programmes to achieve optimum performance levels and a criterion measure. All models may involve continuous or discrete systems (Fitts, 1964; Smith, 1966; Welford, 1968).

The organisation, pattern and phases of skill learning stress the importance of task continuity, coherence and complexity, the changes in the factor structure of skills and the importance of abilities at different stages of learning (Woodworth, 1958; Fitts & Posner, 1967; Lawther, 1968; Fleishman, 1954; Gagné & Fleishman, 1959).

The application of learning theory to the acquisition of skill centres on the relevance of S—R association and more generalised learning effects through cognitive sets (Fitts, 1964; Lawther, 1968). The effects of S—R compatability and the relevance of statistical decision processes are central issues (Welford, 1968; Fitts & Posner, 1967). Information theory provides another means of measuring

51

skill and comparing skills, utilising the concepts of rate of information transmission and channel capacity (Fitts & Posner, 1967).

Individual differences in ability patterns, the development of within-task abilities and the importance of genetic/environmental influence are still open questions (Jones, 1966; Fleishman, 1966).

The relevance of 'body concept' is discussed in section I.7.

A.2. Factors affecting the acquisition and performance of skill

The advantages and disadvantages of various forms of display are discussed by Poulton (1966). Present knowledge as to whether learning involves discrete, serial or continuous patterns identifies conflicting evidence which is not resolved (Fitts, 1964; Craik, 1948; Miller, Galanter & Pribram, 1960).

Work and rest variables in learning and performance have received considerable attention. Hull's theories of reactive and conditioned inhibition and Kimble's concept of the 'critical level' are very relevant (Hull, 1943; Kimble, 1949).

The relative merits of whole and part learning appear to rest on the characteristics of the learner and the type of task and what is considered to be a 'whole' (Seagoe, 1936; Niemeyer, 1958; Gates, 1953). Current evidence is summarised by Lawther (1968).

The elimination of errors and the speeding up of the learning process by guidance techniques is discussed in Holding (1965) but there is little evidence in the area of gross motor skill to justify any clear conclusions.

The influence of initial performance is very important with particular reference to the ingraining of errors (Kay, 1951; Von Wright, 1957). The relative value of passive and active practice in learning seems to be affected by display-control compatibility and the simplicity or complexity of the skill (Macrae & Holding, 1965; 1966).

The value of mental practice is uncertain as tests are not very reliable. There is a suggestion that it is more beneficial at higher levels of skill learning (Clark, 1960) or in tasks involving a stable visual field (Cratty & Densmore, 1963; Ulrich, 1967).

The effectiveness of informing a subject of his performance depends on a number of variables. Holding (1965) suggests a classification showing different kinds of knowledge of results K.R. Bilodeau (1966) argues the merits of terminology and reviews four major considerations:
(1) temporal placement of KR, (2) frequency of KR, (3) additional KR,
(4) transformations of KR. Lawther (1968) gives a general survey of
findings and relevant references.

The transfer of the effects of training remains a central concern in the acquisition of skills. Theoretical explanations centre on either the importance of general elements, specific factors or concepts of transposition. Component analysis has been used in emphasising inter-task component relationships rather than gross transfer effects but this has not been developed in motor-skill learning at present.

An important principle suggested by recent research is that inter-task facilitation is produced by intra-task interference. A detailed discussion of these developments is given in Battig (1966); a general summary of relevant research in motor-learning is given in Cratty (1964).

Motivation is a key consideration in the acquisition of skill. The concept of activation or arousal in behaviour relates to a neuro/physiological dimension which is related to the behavioural dimension of 'drive'.

It is suggested that behavioural effects may arise through physiological activity or conversely, deprivation or the manipulation of motivating conditions can have physiological effects (Hull, 1943; 1952; Spence, 1958; Lindsley, 1957). The effect of an individual's expectations, i.e. level of aspiration, on future performance is of some importance. It is suggested that the aspiration level is affected by past experience (Gould, 1939), nature of the task, stage of maturity (Anderson & Brandt, 1939; Allport 1920) and parental influence, (D'Andrade & Rosen, 1959; Little & Cohen, 1950).

The effects of competition seem to depend upon the personality of the performer (Grossack, 1954), and the nature of the task (Shaw, 1958). Eysenck's theories of the three dimensions of personality give clear indications of the relationships between personality traits and motivation to action as well as performance levels. The work of Selye (1956) and studies of the effects of anxiety, stress and tension on performance are particularly important. Studies of group interaction, co-operation, leadership patterns and communication reveal significant findings which can affect the measurement, performance and learning of skills (Triplett, 1897; Gates, 1924; Abel, 1936; Carter & Nixon, 1949; Carter et al, 1951; Roseborough, 1953; Cratty & Sage, 1963).

The maintenance of vigilance under various motivating conditions and the effects of habituation, expectancy, fatigue and loading on performance covers a wide field of experimental data. Important source texts are Broadbent (1958); Mackworth (1950); Welford (1968). The terms stress and emotion are used interchangeably in the literature. Selye (1956) interprets stress as a disturbance of normal homeostasis. The work of Johnson (1956); Fleishman (1958); Ulrich (1957); Chase (1932); Hartrick (1960); Bayton & Conley (1957); Strong (1963) shows how deprivation, satiation, success, failure and anxiety affect performance.

The answer to how motor-skill learning is retained is still uncertain. Evidence suggests that retention involves the transformation and re-coding of information (Bartlett 1932). Henry (1960) postulates a memory-drum theory and Bilodean (1964) stresses the importance and influence of initial experience; the retention of *errors* is particularly relevant Kay (1951); Von Wright, (1957). For a neurophysiological theory, see Hebb (1949).

A summary of work on short-term memory (Posner, 1963; 1967) stresses information capacity, role of attention, rehearsal and practice and the possible existence of a peripheral storage and a central short-term storage, (Sperling, 1967).

The ability to react to various forms of stimuli is well documented. For basic references see Woodworth & Schlosberg (1963) and for recent research Teichner (1954); Whiting (1969). It is suggested that the inability of an individual to react quickly to successive stimuli depends partially upon the psychological refractory period (Welford 1952; 1968). Other causes of increased reaction time include lack of anticipation and peripheral interference. Davis (1957) and Creamer (1963) oppose this view.

A.3. Perceptual aspects of skill

This section identifies important perceptual processes.

The structure and importance of visual, auditory, kinaesthetic and balance senses are treated in detail by Howard & Templeton (1966) and the role of kinaesthesis in perceptual-motor learning in Fleishman & Rich (1963). The function of proprioception in more complex learning tasks is of particular interest (Bahrick, Noble & Fitts, 1954).

Attention can be seen as an intervening variable in the response mechanisms of observation and motor action (Wyckoff, 1952). Problems of inferring observation responses from eye movements exist, as looking and seeing are different (Mackworth & Mackworth, 1958; Sanders, 1963). The role of selective perception (Horn, 1966) and learning to ignore irrelevant information (Rabbit, 1967) together with the ability to discriminate stimuli are important features in the study of stimulus-response situations (Welford, 1968; Adams, 1966).

The existence of innate perceptual abilities has been suggested by Poole (1969) and Jenson (1967). Neonate studies by Fantz (1961) on responses to various arrangements of schematic facial features, Gibson & Walk's (1960) 'visual-cliff' experiment, backed by evidence from Hershenson et al. (1965) and Stechler (1964), have shown that some degree of perception is unlearned. Kessen et al. (1956) suggest the existence of neuro-physiological coding mechanisms similar in concept to the contour operations that Hubel & Wiesel (1962) found in the visual system of cats. Fleishman's (1967) interpretation of abilities as capacities for utilising different kinds of information and recent models of attention by Triesman (1969) and Norman (1968), incorporating templates to identify input information, and Piaget's (1952) development of schemata, provide a functional approach in which the interaction of genetic and experiential factors in the development of more complex skills can be understood.

Witkin et al. (1954, 1962) have developed a theory of personality based on inter-subject differences in perceiving figure-ground relationships; Messick & Damarin (1964) and Konstadt & Forman (1965) have dealt with the implication of perceptual 'types' for social interaction. Research inter-relating the theories of Witkin et al. (1962); Eysenck (1967); Cattell (1946) and Parnell (1958) have produced conflicting results. The present state of research into the relationship between personality and perceptual mode is best summarised by Vernon (1968) who states that 'the relationship . . . . . . is by no means clear nor well established.'

A.4. Decision making aspects of skill

After the findings of Craik (1948) which indicated that man performs as an intermittent-correction servo system, considerable evidence has accumulated to support the hypothesis that central limitations exist as to the rate at which information can be processed. These limitations have been expressed in terms of the size of transformations required between stimulus and response (Fitts & Posner, 1967). A linear relationship between reaction time and information gain led to the formulation of Hick's law (Hick, 1952). The slope of this relationship has been shown to be dependent on the degree of compatibility between stimulus and responce (Fitts & Deininger, 1954) and the effects of practice (Mowbray & Rhoades, 1959) with a slope reduction to near zero being obtained under highly compatible conditions and prolonged practice (Fitts & Posner, 1967; Smith, 1968). The ergonomic aspects of control-display relationships and population stereotypes are summarised in Murrell (1965).

The avoidance of anticipatory tendencies in reaction time studies is an important feature in the study of human movement, especially in highly paced tasks. The effects of stimulus coherency (temporal, spatial, event) and their subsequent effect on decision times have been investigated by Klemmer (1957), Noble & Trumbo (1967); Hyman (1953); Adams & Xhignesse (1960); Adams & Creamer (1962b); Hale (1967, 1968). Poulton (1952) distinguished between receptor and perceptual anticipation. Adams & Xhignesse (1960) in attempting to identify the processes by which S is able to anticipate events, suggests two possibilities. Extrinsic information from a dynamically changing environment and intrinsic information from a proprioceptive time base (see also Hutt, 1969.)

A study of serial choice reaction time tasks has shown that corrections to responses are faster than accurate responses (Rabbitt, 1966; Rabbitt & Phillips, 1967). The modification of a response seems to demand a different explanation in terms of the single-channel hypothesis (Welford, 1968). Models of skill behaviour accounting for error detection and control relevant to this topic have been postulated by Chase (1965) and Holst (1954).

A.5 Motor aspects of skill

The nature/nurture controversy has generated considerable debate and experiment in relation to cognitive development, but only recently has a systematic attempt been made to identify primary perceptual-motor abilities in certain tasks (Fleishman, 1967). The relationship between intelligence and motor learning is equivocal although it does seem that there is a point below which intelligence becomes important in learning motor skills (Oxendine (1968)). Motor educability, formerly a popular area of research, has proved to be of little value. There is increasing evidence to support the concept of specificity of skill (Henry, 1958; Jones & Seashore, 1944; Cumbee et al, 1957; Tyler, 1956; Marteniuk, 1969).

Amongst the factors which have been indentified as affecting the level of proficiency in athletics skills are reaction time (Olsen, 1956; Youngen, 1959; Knapp, 1961; Whiting & Hutt, 1971) and body-type (Tanner, 1964), although their contribution depends on the nature of the activity (Knapp, 1963). The role of certain abilities in selected skilled performers has been discussed by Evans (1967).

Complex behaviour appears to rest on the utilisation of highly developed hierarchies of skills and the efficient organisation of input and output information (Bryan & Harter, 1899; Pew, 1966; Fitts, 1964; Miller et al, 1960). Fitts distinguishes between executive programmes and subroutines and Miller et al's (1960) interpretation is in terms of behavioural units (TOTE) which are linked in a hierarchical structure for the execution of the overall plan.

In accounting for motor control, several investigators postulate an error sensitive device (Chase, 1965; Holst, 1954; Ruch, 1951, Gibbs, 1970). Ruch and Gibbs suggest that the locus of such a processing function is the cerebellum. Holst postulated that only when there is a mismatch between the motor 'command' and the reafferent feedback, need the higher brain centres be activated to produce a perception. This possibly over-simplfied view nevertheless provides an attractive explanation to account for the 'automatic' execution of many perceptual-motor skills where proprioceptive feedback seems to be monitored at a sub-conscious level. Ruch (1951) in accounting for the time tension pattern of muscular contractions involved in the sequential nature of movements proposed a reverberating circuit of neurons whereby nerve impulses give off one or more impulses per circuit, the delay characteristics giving a temporal patterning to the original cortical discharge.

The ability to 'time' responses is an important feature of skilled behaviour. Conrad (1956) distinguishes between paced tasks in which the performer has various amounts of control on the nature of the display. Adams & Xhignesse (1960) subdivide tasks into those where the performer learns to 'time' his actions using environmental cues and those in which the environment is stable. In the latter category, it is postulated that the delay of proprioceptive feedback in reverberating neural circuits can provide internal cues to aid timing (Adams & Xhignesse, 1960; Adams & Creamer, 1962b). More recently, Schmidt & Christina (1969) and Hutt (1969) have provided evidence to show how proprioceptive feedback from ongoing movements can aid the timing of responses. The ability to selectively attend to appropriate visual, auditory or proprioceptive cues becomes a central issue in the acquisitition of good timing. The implications of 'movement sense' and 'kinaes-thetic awareness' in relation to timing are discussed by Hutt (1971).

In striking moving objects, the moment at which the response is initiated is affected by the movement time of the effector. Schmidt (1969) has shown that increasing movement time decreases timing accuracy. He has also shown that in programmed movement, the performer may correct previous response errors by holding movement time constant and adjusting the starting time of his response.

Anatomical, physiological and psychological factors limit the rate at which information can be processed and responses executed. An undue emphasis on speed usually produces a reduction in accuracy or conversely greater precision will reduce speed of execution (Fitts, 1965). Performers can adopt speed or accuracy sets (Hale, 1969) and when no undue emphasis is given to these factors they seem to adopt an optimum rate of information transmission (Fitts, 1966). Leonard & Newman (1965) suggest that an emphasis on accuracy is only important during the first phase of learning a sensori-motor skill, i.e. when S-R relationships are being established.

The results of Hale (1969) revealed deficiencies in both serial classification and statistical decision models of choice reaction time (Smith, 1968) and indicates a need for more data on speed-error trade off. Vickers (1970) has proposed an accumulator model in which the speed-error trade off phenomenon is accounted for in terms of levels of arousal.

The relationship between personality and speed versus accuracy sets proposed by Eysenck (1967) was not substantiated in a study by Whiting & Hutt (1972).

References

BARTLETT F.C. (1932). "Remembering". Cambridge: University Press.

BROADBENT D.E. (1958). "Perception and Communication". London: Pergamon.

CRATTY B.J. (1964). "Movement Behaviour and Motor Learning". Philadelphia: Lea & Febiger.

EYSENCK H.J. (1967). "The Biological Basis of Personality". Springfield: Thomas.

FITTS P.M. & POSNER M.I. (1967). "Human Performance". Belmont: Brooks/Cole.

HEBB D.O. (1949). "The Organisation of Behaviour". New York: Wiley.

HOLDING D.H. (1965). "Principles of Training". Oxford: Pergamon.

HOWARD I.P. & TEMPLETON W.B. (1966). "Human Spatial Orientation". London: Wiley.

HULL C.L. (1943). "Principles of Behaviour". New York: Appleton-Century-Crofts.

KNAPP B. (1963). "Skill in Sport". London: Routledge & Kegan Paul.

MORRIS P.R. & WHITING H.T.A. (1971). "Motor Impairment and Compensatory Education". London: Bell.

OXENDINE J.B. (1968). "Psychology of Motor Learning". New York: Appleton-Century-Crofts.

PIAGET J. (1952). "The Origins of Intelligence in Children". New York: International University Press.

VERNON M.D. (1968). "The Psychology of Perception". Harmondsworth: Penguin.

WELFORD A.T. (1968). "Fundamentals of Skill". London: Methuen.

WHITING H.T.A. (1969). "Acquiring Ball Skill: A psychological interpretation". London: Bell.

WOODWORTH R.S. (1958). "Dynamics of Behaviour". London: Methuen.

WOODWORTH R.S. & SCHLOSBERG H. (1963). "Experimental Psychology". London: Methuen.

Specific References

ABEL L.B. (1936). The effects of shift in motivation upon the learning of a sensory-motor task. *Arch. Psychol.,* **29,** 1-57.

ADAMS J.A. (1966). Some mechanisms of motor responding: An examination of attention. In E.A. Bilodean (Ed.) "Acquisition of Skill". New York: Academic Press.

ADAMS J.A. & XHIGNESSE L.V. (1960). Some determinants of two dimensional visual tracking behaviour, *J. Exp. Psychol.,* **60,** 391-403.

ADAMS J.A. & CREAMER L.R. (1962a). Anticipatory timing of continuous and discrete responses, *J. Exp. Psychol.,* **63,** 84-90.

ADAMS J.A. & CREAMER L.R. (1962b). Proprioception variables as determiners of anticipatory timing behaviour. *Human Factors,* **4,** 217-222.

ALLPORT F.H. (1920). Influence of group upon association and thought. *J. Exp. Psychol.,* **3,** 159-182.

ANDERSON H.H. & BRANDT H.F. (1939). Study of motivation involving self-announced goals of fifth grade children of the concept of level of aspiration. *J. Soc. Psychol.,* **10,** 209-232.

BAHRICK H.P., NOBLE M.E. & FITTS P.M. (1954). Extra-task performance as a measure of learning a primary task. *J. Exp. Psychol.,* **48,** 292-302.

BATTIG W.F. (1966). Facilitation and Interference. In E.A. Bilodeau (Ed.) "Acquisition of Skill". New York: Academic Press.

BAYTON J.A. & CONLEY H.W. (1957). Duration of success background and the effects of failure upon performance, *J. Gen. Pschol.,* **LVI,** 179-85.

BILODEAU E.A., JONES M.B. & LEVY C.M. (1964). Long-term memory as a function of retention time and repeated recalling. *J. Gen. Psychol.,* **67,** 303-309.

BILODEAU I.McD. (1966). Information feedback. In E.A. Bilodeau (Ed.) "Acquisition of Skill". New York: Academic Press.

BRYAN W.L. & HARTER N. (1899). Studies on the telegraphic language: the acquisition of a variety of habits. *Psychol. Rev.,* **6,** 345-375.

CARTER L.F., HAYTHORN W. & HOWELL M. (1951). A further investigation of the criteria of leadership. *J. Ab. Soc. Psychol.,* **46,** 589-595.

CARTER L.F. & NIXON M. (1949). An investigation of the relationship between four criteria of leadership ability for three different tasks. *J. Psychol.,* **27,** 245-261.

CATTELL R.B. (1946). "The Description and Measurement of Personality". New York: Yorkers (World Book).

CHASE L. (1932). Motivation of young children. *The University of Iowa Studies in Child Welfare,* V **3,** 9-119.

CHASE R.A. (1965). An information-flow model of the organisation of motor activity. Part II: Sampling, central processing, and utilization of sensory information. *J. Nerv. Ment. Dis.,* **140,** 334-350.

CLARK L.V. (1960). Effect of mental practice on the development of a certain motor skill. *Res. Quart.,* XXXI : **4,** Pt. I, 560-69.

CONRAD R. (1956). The timing of signals in skill. *J. Exp. Psychol.,* 51, 365-370.

CRAIK K.J.W. (1948). Theory of the human operator in control systems: II. Man as an element in a control system. *Brit. J. Psychol.,* **38,** 142-148.

CRATTY B.J. & DENSMORE A.E. (1963). Activity during rest and learning a gross movement task. *Percept. Mot. Skills,* **17,** 250.

CRATTY B.J. & SAGE J.N. (1963). The effects of primary and secondary group interaction upon improvement in a complex motor task. Unpublished study, University of California, Los Angeles.

CREAMER L.R. (1963). Event uncertainty, psychological refractory period and human data processing. *J. Exp. Psychol.,* **66,** 187-194.

CUMBEE F.Z., MEYER M. & PATERSON G. (1957). Factorial analysis of motor co-ordination variables for third and fourth grade girls. *Res. Quart.,* **28,** 100-108.

D'ANDRADE R. & ROSEN B.C. (1959). The psychological origins of achievement motivation. *Sociometry,* **22,** 185-218.

DAVIS R. (1957). The human operator as a single channel information system. *Quart. J. Exp. Psychol.,* **9,** 119-129.

EVANS J.C. (1967). A comparative study of the performance of above-average athletes, gymnasts and games players on tests of selected basic motor abilities. In J. Wyn Owen (Ed.) *Research in Physical Education,* **Vol. 1,** No. 2, The Physical Education Association.

FANTZ R.L. (1961). The origin of form perception. *Sci. Amer.,* **204,** 66.

FITTS P.M. (1964). Skill Learning. In A.W. Melton (Ed.) "Categories of Human Learning". New York: Academic Press.

FITTS P.M. (1966). Cognitive aspects of information processing: Set for speed versus accuracy. *J. Exp. Psychol.,* **71,** 849-857.

FITTS P.M. & DEININGER R.L. (1954). S—R compatability: correspondence among paired elements within stimulus and response codes. *J. Exp. Psychol.,* **48,** 483-492.

FLEISHMAN E.A. (1954). Dimensional analysis of psychomotor abilities. *J. Exp. Psychol.,* **48,** 437-454.

FLEISHMAN E.A. (1958). A relationship between incentive motivation and ability level in psychomotor performance. *J. Exp. Psychol.,* LVI : **1,** 78-81.

FLEISHMAN E.A. (1966). Comments on Professor Jones' Paper. In E.A. Bilodeau (Ed.) "Acquisition of Skill". New York: Academic Press.

FLEISHMAN E.A. (1967). Individual differences and motor learning. In R.M. Gagné (Ed.) "Individual Differences". Ohio: Merrill.

FLEISHMAN E.A. & RICH S. (1963). Role of kinaesthetic and spatial visual abilities in perceptual motor learning. *J. Exp. Psychol.*, **66**, 6-11.

GATES A.I. (1953). "Educational Psychology". New York: Macmillan.

GATES G. (1924). The effect of an audience upon performance. *J. Ab. Soc. Psychol.*, **18**, 334-344.

GIBBS C.B. (1970). Servo-control systems in organisms and the transfer of skill. In D. Legge (Ed.) "Skills". Harmondsworth: Penguin.

GIBSON E.J. & WALK R.D. (1960). The "visual cliff". *Sci.Amer.*, **202**, 64-71.

GOULD R. (1939). An experimental analysis of "level of aspiration". *Gener. Psychol. Monog.*, **21**, 1-115.

GROSSACK M.M. (1954). Some effects of co-operation and competition upon small group behaviour. *J. Ab. Soc. Psychol.*, **49**, 341-348.

HALE D.J. (1967). Sequential effects in a two-choice serial reaction task. *Quart. J. Exp. Psychol.*, **19**, 133-141.

HALE D.J. (1968). The relation of correct and error responses in a serial choice reaction task. *Psychon. Sci.*, **13**, 299-300.

HALE D.J. (1969). Speed-error trade off in a three-choice serial reaction task. *J. Exp. Psychol.*, **81**, 428-435.

HARTRICK F.J. (1960). The effects of various incentives on performance in an endurance exercise. Unpublished M.Sc. thesis, Pennsylvania State Univ.

HENRY F.M. (1958). Specificity vs. general in learning motor skills. *Ann. Proc. Coll. Phys. Educ. Ass.*, **61**, 126-128.

HENRY F. (1960). Increased response latency for complicated movement, and a 'memory drum' theory of neuromotor reaction. *Res. Quart.*, **31**, 448-457.

HERSHENSON M., MUSINGER H. & HESSEN W. (1965). Preferences for shapes of intermediate variability in the newborn. *Human Science*, **144**, 315-317.

HICK W.E. (1952). On the rate of gain of information. *Quart. J. Exp. Psychol.*, **4**, 11-26.

HOLST E. von. (1954). Relations between the central nervous system and the peripheral organs. *Brit. J. Animal. Beh.*, **2**, 89-94.

HORN G. (1966). Physiological and psychological aspects of selective perception. In D.S. Lehrman, R.A. Hinde & E. Shaw (Eds.) "Advances in the Study of Behaviour, Vol. 1". New York: Academic Press.

HUBEL D.H. WIESEL T.N. (1962). Receptive fields, binocular interaction and functional architecture in the cat's visual cortex. *J. Physiol.*, **160**, 106-154.

HULL C.L. (1952). "A Behaviour System". New Haven: Yale University.

HUTT J.W.R. (1969). The role of proprioception in the anticipation and timing of simple motor responses. Unpub. M.Sc. dissertation, Univ. of Aston in Birmingham.

HUTT J.W.R. (1971). "Proprioception and Timing: some findings and speculation" (in press).

HYMAN R. (1953). Stimulus information as a determiner of reaction time. *J. Exp. Psychol.*, **45**, 188-196.

JENSEN A.R. (1967). Estimation of the limits of heritability of traits by comparison of monozygotic and dizygotic twins. *Proc. Nat. Acad. Sci.,* **58,** 149-156.

JOHNSON B.L. (1956). Influence of pubertal development on responses to motivated exercise. *Res. Quart.,* XXVII, **2,** 182-193.

JONES M.B. (1966). Individual Differences. In E.A. Bilodeau (Ed.) "Acquisition of Skill". New York: Academic Press.

JONES H.E. & SEASHORE R.H. (1944). The development of fine motor and mechanical abilities. "43rd Handbook Pt. 1. National Society for the Study of Education". Chicago: Univ. Chicago Press.

KAY H. (1951). Learning of a serial task by different age groups. *Quart. J. Exp. Psychol.,* **3,** 166-183.

KESSEN N., HAITH M.M. & SALAPATEK P. (1956). The ocular orientation of newborn infants to visual contours. Referred to in R.N. Haber (Ed.) (1968). "Contemporary Theory and Research in Visual Perception". New York: Holt, Rinehart and Winston.

KIMBLE G.A. (1949). An experimental test of a two factor theory of inhibition, *J. Exp. Psychol.,* **39:** 15-23.

KLEMMER E.T. (1957). Simple reaction time as a function of time uncertainty. *J. Exp. Psychol.,* **54,** 195-200.

KNAPP B. (1961). Simple reaction time of selected top class sportsmen and research students, *Res. Quart.,* **32,** 409-411.

KONSTADT N & FORMAN E. (1965). Field dependence and external directness. *J. Pers. Soc. Psychol.,* **1,** 490-493.

LAWTHER J.D. (1968). "The Learning of Physical Skills". New Jersey: Prentice-Hall.

LEONARD J.A. & NEWMAN R.C. (1965). On the acquisition and maintenance of high speed and high accuracy in a keyboard task. *Ergonomics,* **8,** 281-303.

LINDSLEY D.B. (1957). Psychophysiology and motivation. In M.R. Jones (Ed.) *Nebraska Symposium on Motivation,* **5,** 44-105. University of Nebraska Press.

LITTLE S.W. & COHEN L.D. (1950). Goal setting behaviour of asthmatic children and of their mothers for them. *J. Pers.,* **19.** 376-389.

MACKWORTH N.H. (1950). Researches on the measurement of human performance. "Medical Research Council Special Report Series No. 268". London: H.M.S.O.

MACKWORTH J.F. & MACKWORTH N.H. (1958). Eye fixations recorded on changing visual scenes by the television eye-marker. *J. Opt. Soc. Amer.,* **48,** 439-445.

MACRAE A.W. & HOLDING D.H. (1965). Guided practice in direct and reversed serial tracking. *Ergonomics,* **8,** 487-492.

MACRAE A.W. & HOLDING D.H. (1966). Transfer of training after guidance or practice. *Quart. J. Exp. Psychol.,* **18,** 327-333.

MARTENIUK R.G. (1969). Generality and specificity of learning and performance on two similar speed tasks. *Res. Quart.,* **40,** 518-522.

MESSICK S. & DAMARIN F. (1964). Cognitive styles and memory for faces. *J. Ab. Soc. Psychol.,* **69,** 313-318.

MILLER G.A., GALLANTER E. & PRIBRAM K.H. (1960). "Plans and the

Structure of Behaviour". New York: Holt.

MOWBRAY G.H. & RHOADES M.U. (1959). On the reduction of choice reaction times with practice. *Quart. J. Exp. Psychol.*, **11**, 16-23.

MURRELL K.F.H. (1965). "Ergonomics". London: Chapman and Hall.

NIEMEYER R.K. (1958). Part versus whole methods and massed versus distributed practice in the learning of selected large muscle activities. *62nd Proceedings of the College of Physical Education Association,* New York.

NOBLE M. & TRUMBO D (1967). The organisation of skilled response. *Organ. Beh. Human Perf.,* **2**, 1-25.

NORMAN D.A. (1968). Towards a theory of memory and attention. *Psychol. Rev.,* **75**, 522-536.

OLSEN E.A. (1956). Relationship between psychological capacities and success in college athletics. *Res. Quart.,* **27**, 79-89.

PARNELL R.W. (1958). "Behaviour and Physique". London: Arnold.

PEW R.W. (1966). Acquisition of hierarchical control over the temporal organisation of a skill. *J. Exp. Psychol.,* **71**, 764-771.

POOLE H. (1969). Reconstructuring the perceptual world of African children. *Teacher Education in New Countries,* **10**, 2, 165-172.

POSNER M.I. (1963). Immediate memory in sequential tasks. *Psychol. Bull.,* **60**, 333-349.

POSNER M.I. (1967). Short term memory system in human information processing. In A.F. Sanders (Ed.) "Attention and Performance". Amsterdam: North Holland Pub. Co.

POULTON E.C. (1952). Perceptual anticipation in tracking with two-pointer and one-pointer displays, *Brit. J. Psychol.,* **43**, 222-229.

POULTON E.C. (1966). Tracking behaviour. In E.A. Bilodeau (Ed.) "Acquisition of Skill". New York: Academic Press.

RABBITT P.M. (1967). Learning to ignore irrelevant information. *Am. J. Psychol.,* **80**, 1-13.

RABBITT P.M.A. (1966). Errors and error correction in choice response tasks. *J. Exp. Psychol.,* **71**, 264-272.

RABBITT P.M.A. & PHILLIPS S. (1967). Error-detection and correct latencies as a function of S-R compatability. *Quart. J. Exp. Psychol.,* **19**, 37-42.

ROSEBOROUGH M.E. (1953). Experimental studies of small groups. *Psychol. Bull.,* **50**, 275-303.

RUCH T.C. (1951). Motor systems. In S.S. Stevens (Ed.) "Handbook of Experimental Psychology". New York: Wiley.

SANDERS A.F. (1963). The selective process in the functional visual field. *Soesterberg, The Netherlands: Report of the Institute for Perception R.V.O.— T.N.O.*

SCHMIDT R.A. (1969). Movement time as a determiner of timing accuracy. *J. Exp. Psychol.,* **79**, 43-47.

SCHMIDT R.A. & CHRISTINA R.W. (1969). Proprioception as a mediator in the timing of motor responses. *J. Exp. Psychol.,* **81**, 303-307.

SEAGOE M.V. (1936). Qualitative wholes: A re-evaluation of the whole part problem. *J. Ed. Psychol.,* **27**, 537-545.

SELYE H. (1956). "The Stress of Life". New York: McGraw-Hill.

SHAW M.E. (1958). Some motivational factors in co-operation and competition. *J. Pers.*, **26**, 155-169.

SMITH E.A. (1968). Choice reaction time. *Psychol. Bull.*, **67**, 79-11.

SMITH K.U. (1966). Cybernetic theory and analysis of learning. In E.A. Bilodeau (Ed.) "Acquisition of Skill". New York: Academic Press.

SPENCE K.W. (1958). A theory of emotionally based drive and its relations to performance in simple learning situations. *Amer. Psychol.*, **13**, 131-141.

SPERLING G. (1967). Successive approximation to a model for short term meory. In A.F. Sanders (Ed.) "Attention and Performance". Amersterdam: North-Holland Pub. Co.

STECHLER G. (1964). The effect of medication during labour on newborn attention. *Science,* **14**.

STRONG C.H. (1963). Motivation related to performance of physical fitness tests. *Res. Quart.,* XXXIV: **4**, 497-507.

TANNER J.M. (1964). "The Physique of the Olympic Athlete". London: Allen & Unwin.

TEICHNER W.H. (1954). Recent studies of simple reaction time. *Psych. Bull.,* **51**.

TREISMAN A.M. (1969). Strategies and models of selective attention. *Psychol. Rev.,* **76**, 282-299.

TRIPLETT N. (1897-8). The dynamogenic factors in pace making and competition, *Am. J. Psychol.,* **9**, 507-533.

TYLER L.E. (1956). "The Psychology of Human Differences". New York: Appleton-Century-Crofts.

ULRICH E. (1967). Some experiments in the function of mental training in the acquisition of motor skills. *Ergonomics,* **10**, 411.

ULRICH C. (1957). Measurement of stress evidenced by college women in situations involving competition. *Res. Quart.,* XXVIII, **2**, 160-72.

VICKERS D. (1970). Evidence for an accumulator model of psychophysical discrimination. *Ergonomics,* **13**, 37-58.

WELFORD A.T. (1952). The psychological refractory period and timing of high speed performances: a review and theory. *Brit. J. Psychol.,* **43**, 2-19.

WHITING H.T.A. (1969). "Acquiring Ball Skill: a psychological interpretation". London: Bell.

WHITING H.T.A. & HUTT J.W.R. (1972). The effects of personality and ability on speed of decisions regarding the directional aspects of ball flight (in press).

WITKIN H.A., GOODENOUGH D.R., KARP S.A., DYKE R.B. & FATERSON D.R. (1962). "Psychological Differentiation". New York: Wiley.

WITKIN H.A., LEWIS H.B., MACHOVER K., MEISSNER P.B. & WAPNER S. (1954). "Personality Through Perception". New York: Harper.

WRIGHT J.M. von (1957). A note on the role of 'guidance' in learning. *Brit. J. Psychol.,* **48**, 133-137.

WYCKOFF L.B. (1952). The role of observing responses in discrimination learning. Part I. *Psychol. Rev.,* **59**, 431-442.

YOUNGEN L. (1959). A comparison of reaction and movement times of women athletes and non-athletes. *Res. Quart.,* **30**, 349-355.

TOPIC AREAS-DEVELOPED

MOVEMENT IMPAIRMENT

by H.T.A. WHITING

4 B

General

Movement impairment refers to the inadequacy of an individual's *physical* responses to the everyday demands of his environment. As such, it is a condition that manifests itself in performances which are subnormal or whose efficiency has been hampered in some way. These responses reflect inadequate attempts to perform those motor skills which can be regarded as being either essential or culturally desirable (Morris & Whiting, 1971). An *impairment continuum* might be proposed from *gross* impairment at the one extreme—which necessitates clinical treatment—to those *apparently* minor forms of impairment at the other which might more appropriately be classified as 'clumsiness' and which are often accepted as 'within the norm'. Considerable medical attention has of necessity been given to the more gross forms of impairment, but, until comparatively recent years, the 'clumsy child' had received little formal attention or experimental investigation. While gross forms of impairment such as ataxia, chorea, spasticity etc. must remain a clinical diagnostic and treatment problem, an understanding of the nature of such impairments and a concern for rehabilitation and compensatory education procedures may well fall within the bounds of interest of the human movement specialist.

The more minor forms of impairment (motor impairment—clumsiness) have received more detailed attention from educationists and psychologists rather than from the médical profession. To some extent, this has been brought about by postulated links between such forms of impairment and intellectual and social development (Morris & Whiting, 1971). Because of such wider implications, the major emphasis in this area of human movement study will be on *motor impairment* (Stott 1966).

B.1. Psychiatric Diagnosis

Observation of pathological movement behaviour in psychiatric classification and diagnosis is not a new idea. Such procedures were implicit in the work of

Kanner (1944) and Breuer & Freud (1968). More recently, Nathan (1967) has provided a systematised approach based on flow-diagram analysis. Other useful work in this area has been reported by Eysenck (1961).

B.2. Gross Movement Disabilities

Clearly, where conditions such as hemiplegia, diplegia and other related gross forms of paralysis exist, movement and skilled performance is going to be impaired. These conditions may form an interest to the movement specialist, but it is likely that such an interest would be the prerogative of a person medically biassed. Such disabilities will include:-

Rigidity—an aberration in state of muscle tone—normally *hyper*tonia

Tremor—postural or action—involuntary movements produced during rest or in the performance of voluntary acts.

Choreiform Movements—involuntary movements during rest or action—of a more purposeful kind—generally rapid and jerky.

Athetoid Movements—involuntary movements during rest or action—a slow movement often involving alternating hyperextension and flexion of the extremities.

Dystonic Movements—slow, writhing and involuntary—often involving bizarre movements.

Hemiballism—involuntary movements of the entire limb.

Chorea—an acute toxic disorder of the nervous system usually due to acute rheumatism occurring in early childhood and adolescence —characterised by involuntary movements often involving bizarre movement of the tongue and mouth.

Ataxia—a disturbance in equilibrium—righting reflex diminished and sense of position in space disturbed.

Apraxia—loss of ability to perform purposeful movements in the absence of paralysis. *Motor* ataxia—'a loss of the kinaesthetic meaning of an act'. (House & Pansky 1960).

Spasticity (Cerebral palsy)—a brain disorder affecting motor ability because of the resultant rigidity in particular muscle groups.

For specific aetiology and further description, the reader is referred to standard medical texts or to neuroanatomical works such as House & Pansky (1960).

B.3. Concept of Motor Impairment

Motor impairment—*clumsiness* syndrome (Stott 1966). Characteristic of children both of school and pre-school age who have difficulty in acquiring and performing even the simplest motor skills (Morris & Whiting 1971). The indadequacy of their responses may cause them to be labelled 'awkward' or 'clumsy'. This inability manifests itself particularly when such children are engaged in activities of a practical nature such as handicraft, painting, handwriting and games.

Disabilities of this kind may come under the heading of 'normal' particularly in those environments where motor impairment is common. Generally speaking, these forms of impairment are not sufficiently apparent as to merit clinical attention although increasingly the syndrome is being recognised in child guidance centres, physical education classes and schools for the educationally sub-normal.

B.4 Motor Impairment—a misnomer?

The use of the term 'motor' in the phrase 'motor impairment' can be misleading in that it gives the impression of a disability which is primarily concerned with the effector (output/motor) side of performance. While the syndrome manifests itself as an inability to carry out appropriate motor actions, the phrase 'motor impairment' can be a limitation when possible *explanations* of the inadequacy are being sought. The deficiency may lie in one or more parts of a whole network of processes involved in the input of information to the system, decision-making on the basis of such information and the formulating of an executive response. For this reason, a more acceptable and descriptive term would be *perceptual-motor impairment* which draws attention to the important relationship which exists between the input and output sides of performance. Such an analysis becomes more understandable if an information-processing model of skilled performance is adopted (Welford 1968; Whiting 1969).

B.5. Motor Impairment and Information Processing

A model of the kind proposed, treats the human organism as an information-processing system in which the primary subsystems are concerned with the reception of information (sense organs); the organisation and interpretation of such information (perceptual mechanisms) decision-making on the basis of such information (translatory mechanisms) and the initiation of a motor response (effector mechanisms). Such a model can be applied to skills of all kinds (verbal, motor, social, intellectual) and it is likely that a deficiency in any of the subsystems will affect the acquisition and performance of a wide range of such skills. Such deficits may be structural (e.g. brain-damage) or functional (e.g. a failure to establish appropriate mediating mechanisms) (Morris & Whiting 1971).

This kind of model has already proved its heuristic value and focused attention on the interaction betwen the organism and environment involved in any skilled behaviour.

B.6. Aetiology of Motor Impairment

As stated in B.2, the primary aetiological considerations under this heading are those involving neurological sequelae. On a broad basis, these can be categorised as *brain damage* and *brain dysfunction* with an appreciation that the two are obviously related. The impaired perceptual-motor performances observed in *known* cases of brain-damage has in the past led investigators to generalise to other instances of perceptual-motor impairment where brain-damage is less obvious.

Brain-damage per se represents a structural change brought about by physical causes, haemorrhage, toxic reactions and anoxia. However, a failure to detect signs of organic damage does not necessarily imply the absence of what has been termed *brain dysfunction*. The implication here, is of a cerebral disorganisation in a *neurophysiological* rather than a neuroanatomical sense. The concept has been discussed in the context of impairment by Morris & Whiting (1971); Clarke (1966); Stott (1966); Gubbay et al (1965); Walton et al (1962); Ford (1959); and Fish (1961).

A deficiency in skilled behaviour may be the result of a *breakdown* in already established skill performance, or an inherent inability to carry out successfully the skill which is demanded at a sufficiently adaptive level. Although major emphasis in this context will centre around the latter category, it should be borne in mind that temporary breakdowns can be brought about by toxic influences such as alcohol, drugs and mystical experience which may become permanent deficits if such behaviour persists. The aetiology of such inherent disabilities centres around genetic predisposition, brain-damage, an impoverished environment or combinations of two or more of these factors. Such aetiological considerations can best be illustrated by focusing attention on prenatal, paranatal and postnatal factors.

B.7. Prenatal Factors

Prenatal effects are the resultant of genetic influences, injurious conditions in utero or combinations of the two. Amongst the major factors in this respect are:-

Prenatal anoxia—inadequate oxygen supply to the developing foetus—multiple causality. (Courville 1953; Potter 1952; Ingalls 1950; Workany 1950).

Virus Infection—passed from the mother via the placenta to the foetus e.g. pneumonia, colds, polio, encephalitis, rubella, chickenpox, smallpox, mumps, measles and influenza.

Rhesus Factor—(Weiner 1946; Zeuler 1950)

Toxaemias of Pregnancy—a poisoned state of the blood detrimental to health of mother and foetus—notably pre-eclampsia and eclampsia leading to placental insufficiency and impairment of nutrition supply to the foetus.

Cigarette Smoking—(Buncher 1969; Scott-Russell 1969; Smithells & Morgan 1970).

Emotional Disturbance—The idea that psychological factors or events producing emotional stress in the mother can harm the unborn child still meets with some resistance. Yet, the marked increases in the incidence of certain types of human malformation in wartime and other times of stress lends some support to the contention (Eichmann & Gesenius 1952; Stott 1962b; Klebanov 1948).

Social Class—The high incidence of impairment of all kinds amongst children whose parents are classified in the lower social classes leads to a consideration of the factors which might be operating in such a social setting (Fairweather & Illsley 1960).

B.8. Paranatal Effects

Prematurity—(Dunsdon 1952; Lillienfeld et al 1953; Greenspan et al 1953; Drillien 1957; Donald 1958).

Placenta Praevia—placenta situated in the lower rather than the upper segment of the uterus.

Precipitate Labour—unusually rapid labour due to too frequent or too powerful contractions of the uterus or to a lower uterine segment that dilates with unusual ease.

Pelvic deformity—in the mother-due to disease, congenital effects or abnormalities of the spine.

Anomalies in Presentation—Breech presentations—prevention of the head engaging in the pelvic outlet.

Forceps Delivery

Caesarian section—risks from anaesthesia to mother and maternal asphyxia.

Twin pregnancies—(Potter 1952; Greenspan et al 1953).

B.9. Effects of pre and para natal damage on motor ability

May lead to the gross abnormalities listed in B.2. In addition, Knobloch & Pasamanick (1959) have postulated a continuum of reproductive casualty extending from foetal deaths, through a descending gradient of brain-damage resulting in such abnormalities as those in B.2. but at the other end of the continuum relatively minor forms of impairment which are not immediately clinically obvious. Support for such a continuum comes from Lillienfeld et al (1955); Rogers et al (1955); Gorah et al (1965).

B.10. Postnatal Factors

The primary areas around which present interest centres are sensory/perceptual and movement deprivation. Although some workers (e.g. Bowlby 1951) talk about maternal deprivation as a separate category, it is likely that this may reduce to sensory/perceptual deprivation (See Schaffer 1965; 1966, and Schaffer & Emerson 1968 for possible reasons).

It is clear from examination of evidence from some of the more extreme studies, that gross deprivation of one kind or another can result in impaired performance which at some later stage will give rise to maladaptive behaviour. This applies not only to those skills in which learning plays a major role, but also those skills which have perhaps inaptly been termed 'maturational skills'.

Useful studies in this area have been reviewed by Newton & Levine (1968); Schulz (1965). Whiting et al (1969a, 1969b, 1969c, 1970) have considered the problem in relation to motor impairment.

B.11. Motor Impairment and Intellectual Development

Motor ability and early cognitive development—the significance of movement for early cognitive development is being increasingly realised although it is by no means a novel relationship (See for example Welton 1912). Some of the evidence for the effects of movement experience on early cognitive development was given in B.10. Piaget (1952) in particular has described a sensori-motor period related to the development of intelligence in children. The concept 'significance of movement for cognitive development' has been discussed further by Whiting (1967).

Motor and Mental ability—the degree of relationship between motor and mental ability is difficult to establish and may depend upon such factors as age, environment and I.Q. Results of experimental work carried out have in the main been conflicting. Studies range from those at the turn of the century (Bagley 1900; Wisler 1901) to more recent studies such as those by Garfield (1923); Johnson (1932); McCloy (1934); Ray (1940); Espenchade (1940); Ismail & Gruber (1967) to select but a few at random. There would appear to be more agreement amongst workers in this field of a significant positive relationship between motor and mental ability if the complete range of mental ability is taken into account. Mentally retarded subjects in the main would appear to demonstrate less motor competence (Malpass 1961; Fish 1961; Sloan 1948; Whiting et al 1969a).

The ability concept—Fleishman (1967) differentiates between 'ability' and 'skill'. The latter representing the level of proficiency in a specific task or limited group of tasks. Abilities enter into and are necessary for the performance of skills. Certain abilities are more basic in the sense that they are related to performance on diverse tasks. The implication of this dichotomy, is that skilled performance depends upon abilities which are present before embarking on a task together with habits and subskills which are peculiar to and acquired within the task itself. The idea that basic abilities place limits on later skill proficiency is of fundamental importance in relation to motor and intellectual performance (Morris & Whiting 1971). If in fact, brain-damage is present, or children have undergone sensory/perceptual deprivation, movement deprivation or combinations of these, it might reasonably be supposed that the possibility of their developing the appropriate abilities to a normal level will be severely restricted and subsequent skilled behaviour will be adversely affected.

B.12. Motor Impairment and the Socialisation Process

Possible links between behaviour disorders and motor-impairment have been suggested by Knobloch & Pasamanick (1959); Pasamanick et al (1956); Rogers et al (1955); Corah et al (1965). Interpretations of why such a linkage should exist may be accounted for by such considerations as:-

i. A 'general' inability to acquire skill which affects social as well as other skills because of damage or dysfunction in the subsystems concerned with skill performance. In particular, the concept of selective attention/perception merits attention (Morris & Whiting, 1971).

ii. The interaction between mother and infant which results in tension patterns in the mother. (Barker & Wright 1955; Richardson 1964; Fairweather & Illsley 1960; Casler 1968; Prechtl 1963; Stott 1962a; Hewett et al 1970).

iii. The 'status significance' that competence in physical skills carries at particular stages of childhood and the effect this might have on the development of the individual's self-esteem (Jones & Bayley 1950; Hewett et al 1970; Morris & Whiting 1971).

Motor impairment and maladjustment—Stott (1959; 1962a; 1964) perhaps more than any other worker has linked motor-impairment with maladjustment and delinquency. He hypothesises a factor of congenital impairment. Support for this connection comes from the work of Prechtl (1961); Drillien (1964); Bamber (1966); Whiting et al (1970).

Brain-damage, personality and socialisation—Blakemore (1968) has accumulated evidence to the effect that groups of brain-damaged subjects are slow to acquire conditioned responses when compared with control subjects but this does not necessarily apply to *all* brain-damaged subjects. This idea can be related to Eysenck's (1967) personality theory. Anderson & Hanvik (1950) have produced evidence for differential effects on personality change dependent upon the site of particular brain lesions.

B.13 Body Concept and Motor Impairment

i. *Terminology*—problems of terminology make this a difficult field for evaluation although in the present context undoubtedly a valuable one. The following terms for example have been utilised by different writers for the same or related concepts:-

Body-schema—(Schilder 1935; Freedman 1961; Piaget 1952; Frostig & Horne 1964; Fisher 1966)

Body-image—(Fenichel 1945; Ritchie-Russell 1958; Fisher & Cleveland 1958; Wright 1960; Frostig & Horne 1964; McKellar 1965; Dibiase & Hjelle 1968).

Body-awareness—(Morison 1969)

Body-concept—(Witkin et al 1962; Frostig & Horne 1964)

Body-sense—(Allport 1955).

Body-experience—(Jourard 1967)

ii. *Development of the body-concept*—an original 'global' impression of the body gives way to an awareness of parts of the body, the way in which they interrelate in structure and function and their potential for displacement within the environment. That is, in Witkin et al's (1962) terms towards *differentiation* of inner structure and function and towards an appreciation of spatial concepts (Whiting et al 1973). The progress in ontological development is from a relatively field-dependent (Witkin et al 1962) mode of perceiving to a relatively field independent mode (see I.7) which is paralleled by progress towards a more sophisticated body-concept. Goodenough & Eagle (1963) and Karp & Konstadt (1963) have shown that young children are relatively field-dependent. Witkin (1967) has produced some evidence that differences in the global-articulated style dimension reflects differences in socialisation procedures. Dyk & Witkin (1965) indicate that women are more field-dependent and have a less sophisticated body-concept than men. Wober (1966) proposes the term 'sensotype' for the pattern of relative importance of the different senses by which a child learns to perceive the world and in which patterns his abilities lie. Witkin (1965) suggests an increase in psychological differentiation up to about seventeen years of age and then a plateau into young adulthood.

iii. *Body concept and compensatory education*—Kephart (1960) in particular has emphasised the relationship between the development of body-image and the development of lateral and directional awareness and feedback in relation to learning disabilities. Hill et al (1967) in their investigation into the effect of a systematic programme of exercises on the development of retarded children's awareness of right-left directionality propose that programmes for training retarded children should include activities which give them many experiences in orientating their attention to the position of their own bodies in space relative to that of other objects as well as directing children to make responses with specified body parts. Similar procedures are recognised by Tansley (1967); Stephenson & Robertson (1965) and Frostig & Horne (1964).

The interrelationship of body-image and compensatory education is not a particularly new idea. It was implicit in the early work of Schilder (1935) and Alexander (1957). Other useful work in this area has been carried out by Painter (1964); Holden (1962); Hobbs (1966). A useful overview has been provided by Morris & Whiting (1971).

iv. *Assessing Body-concept*—assessment of body-concept (or related concepts) has in the main been based on projective techniques amongst which the most frequently used have been:-

Draw-a-man Test (Witkin et al 1962)

Barrier Index—a Rorschach technique (Fisher 1964a)

Semantic differential—Fisher (1964b) More operationally defined tests have been used by Epstein (1957) (Finger apposition test) and Stone (1968) (dynamic tests of body-concept). Jones (1970) has carried out pilot studies on dynamic tests of body-concept.

B.14. Assessing Motor-impairment

A postulated link between brain-damage and motor-impairment has in the past led to the conclusion that a test of brain-damage would be the most appropriate means of diagnosis. While such tests might be a useful procedure with reasonably gross impairments, it is doubtful if they are sufficiently sensitive for cases of minimal motor impairment which can be attributed to brain-damage. The most useful procedures to date would seem to be:-

i. Teacher/parent assessments—observation of motor performances. School records/intelligence tests/medical histories/home environment.

ii. Medical examination—physical, neurological, electroencephalograph examination, motor impersistence tests (Garfield 1964; Fisher 1956; Joynt, Benton & Fogel 1962; Prechtl & Stemmer 1962; Rutter et al 1966)

iii. Specifically designed motor ability tests

Oseretsky tests (1923; 1929)

Yarmolenko's test (1933)

Lincoln/Oseretsky revision (Sloan 1955)

Vineland adaptation of the Oseretsky test (Cassell 1949)

General test of motor impairment (Stott 1966)

iv. Psychometric tests:-

a.   Tests used to identify brain-damage e.g. Bender-Gestalt test (1938)—

(Koppitz 1964; Pascal & Suttell 1957; Thweatt 1963; Miller et al 1963) Memory for Designs Test—(Graham & Kendall 1960; Garrett, Price & Deabler 1957; Clarke et al 1968).

b.   Tests of body-concept—See E.13d
c.   Tests of specific factors of motor performance—e.g. manual dexterity, speed/accuracy—spiral-maze test (Gibson 1964; Whiting et al 1969c).
d.   Intelligence tests e.g. Wechsler (1949) intelligence test for children
e.   Tests of perceptual abilities e.g. developmental test of visual perception (Frostig et al 1961; 1963).

B.15. Motor Impairment & Compensatory Education

On a broad basis compensatory education procedures can be classified as *Direct* (in which more stereotyped movement patterns are practised by the retardates— Kephart 1960; Cratty 1969; Oliver 1963; Oliver & Keogh 1967, 1968; Oliver 1955) and *Indirect* (freedom for experimenting with a wide range of movement patterns—(Reading Research Foundation Chicago: Argy 1965; Tansley & Gulliford 1960; Sherborne 1965; Bruce 1969; Allen 1970).

Other workers have based their compensatory education procedures upon the recapitulation of developmental sequences—(Frostig & Horne 1964; Cruickshank et al 1961; Tansley & Gulliford 1960; Kephart 1960; Sutphin 1964; Ausubel 1967; Kamii & Radin 1967; Gallagher 1964; Wedell 1964).

A further procedure has been to pay particular attention to certain general characteristics of the syndrome such as:-

Distractability—(Gessell & Amatrude 1941; Schaffer 1958; Cratty 1969; Floyer 1955; Cruickshank et al 1961; Cruickshank & Dolphin 1951; Francis-Williams 1964).

Disinhibition—(Strauss & Kephart 1955; Cruickshank et al 1961).

Perseveration—(Werner 1941; Strauss & Lehtinen 1948; Cardwell 1956; Tansley & Gulliford 1960; Cratty 1969.

Perceptual-motor functions—(Strauss & Lehtinen 1948; Frostig et al 1961; 1963; Cratty 1969).

Development of body-awareness—See E.13c—(Frostig & Horne 1964; Jakeman 1967; Cratty 1969; Frostig 1968; Kephart 1960).

Visual perception—(Frostig et al 1961, 1963; Cratty 1969.

Source Texts

BAMBER J. (1966). Motor impairment and delinquency. Unpublished M.A. thesis, University of Glasgow.

CLARKE P.R.F. (1966). "The Nature and Consequences of Brain Lesions in Children and Adults". Proceedings of course held by the English division of processional psychologists. London: Brit. Psych. Soc.

CLARKE T.A., JOHNSON G.A., MORRIS P.R. & PAGE M. (1968). Motor impairment—a study of clumsy children. Unpublished dissertation, Institute of Education, University of Leeds.

CRATTY B.J. (1969). "Motor Activity and the Education of Retardates". Philadelphia; Lea & Febiger.

EYSENCK H.J. (Ed.) (1961). "Handbook of Abnormal Psychology". New York: Basic Books.

FLEISHMAN E.A. (1967). Individual differences and motor learning. In R.M. Gagné (Ed.) "Individual Differences". Ohio: Merrill.

FROSTIG M. (1963). "Development Test of Visual Perception". California: Consulting Psychologists Press.

ISMAIL A.H. & GRUBER J.J. (1967). "Motor Aptitude and Intellectual Performance". Ohio: Merrill.

KEPHART N.C. (1960). "The Slow-learner in the Classroom". Ohio: Merrill.

MALPASS L.F. (1961). Motor skills in mental deficiency. In E. Ellis (Ed.) "Handbook of Mental Deficiency". New York: McGraw-Hill.

MORRIS P.R. & WHITING H.T.A. (1971). "Motor Impairment and Compensatory Education" London: Bell.

NEWTON G. & LEVINE S. (Eds.) (1968). "Early Experience and Behaviour". Springfield: Thomas.

SCHULZ D.P. (1965). "Sensory Restriction". New York: Academic Press.

STOTT D.H. (1959). "Unsettled Children and their Families". London: University Press.

WELFORD A.T. (1968). "Fundamentals of Skill". London: Methuen.

WITKIN H.A., DYK R.B., FATERSON M.F. & KARP S.A. (1962). "Psychological Differentiation". New York: Wiley.

Journal Overviews

FAIRWEATHER D.V. & ILLSLEY R. (1960). Obstetric and social origins of mentally handicapped children. *Brit. J. Prev. Soc. Med.,* **14,** 149-159.

GARFIELD J.C. (1964). Motor impersistence in normal and brain-damaged children. *Neurology,* **14,** 623-630.

OSERETSKY N. (1929). A group method of examining the motor functions of children and adolescents. *Z. Kinderforsch.,* **35,** 352-372.

STOTT D.H. (1962a). Evidence for a congenital factor in maladjustment and delinquency. *Am.J. Psychol.,* **118,** 781-794.

STOTT D.H. (1966). A general test of motor impairment for children. *Dev. Med. Child. Neur.,* **8,** 523-531.

WALTON J.H., ELLIS E. & COURT S.D.N. (1962). Clumsy children: development apraxia and agnosia. *Brain* **85**, 603.

Specific References

ALEXANDER F.M. (1957). "The Use of the Self". London: Re-educational Publicat.

ALLEN W. (1970). An investigation into the motor-impaired child with a view to devising a remedial programme based on Rudolf Laban's principles of movement. Unpublished dissertation. Institute of Education, University of Leeds.

ALLPORT G.W. (1955). "Becoming". New Haven: Yale Univ. Press.

ANDERSON A.L. & HANVIK L.J. (1950). The psychometric localisation of brain lesions: the differential effect of frontal and parietal lesions on M.M.P.I. profiles. *J. Clin. Psychol.,* **6**, 177-180.

ARGY W.P. (1965). Montessori versus orthodox. *Rehabilitation Literature,* **26**, 10.

AUSUBEL D.P. (1967). How reversible are the cognitive and maturational effects of cultural deprivation? In A.H. Passow, M. Goldberg, & A.J. Tannebaum (Eds.) "Education of the Disadvantaged". New York: Holt, Rinehart and Winston.

BAGLEY W.C. (1900). On the correlation of mental and motor ability in schoolchildren. *Am. J. Psychol.,* **12**, 193-205.

BAMBER J. (1966). Motor impairment & delinquency. Unpublished M.A. thesis, University of Glasgow.

BARKER R.G. & WRIGHT H.F. (1955). "Midwest and its Children: the psychological ecology of an American town". New York: Row and Peterson.

BENDER A.L. (1938). A visual-motor Gestalt test. *Res. Monog.* New York: Amer Orthopsychiatric Assn.

BLAKEMORE C.B. (1968). Personality and brain damage. In H.J. Eysenck (Ed.) "The Biological Basis of Personality". Springfield: Thomas.

BOWLBY J. (1951). "Maternal Care and Mental Health". Geneva: W.H.O. monograph series.

BREUER J. & FREUD S. (1968). "Studies in Hysteria: The standard edition of the complete psychological works of Sigmund Freud: Vol. II (1893-1895)". London: Hogarth.

BRUCE V.R. (1969). "Awakening the Slower Mind". London: Pergamon.

BUNCHER C.R. (1969). Cigarette smoking and duration of pregnancy. *Am. J. Obstet. Gynec.,* **103**, 942-946.

CARDWELL V.E. (1956). "Cerebral Palsy: Advances in understanding and care". New York: North River Press.

CASLER L. (1968). Perceptual deprivation in institutional settings. In G. Newton & S. Levine (Eds.) "Early Experience and Behaviour". Springfield: Thomas.

CASSELL R. (1949). Vineland adaption of the Oseretsky Tests. *Training School Bulletin* Supp. Vol. **43**, 3-4.

CLARKE P.R.F. (1966). "The Nature and Consequences of Brain Lesions in Children and Adults". Proceedings of a course held by the English division of professional psychologists. London: British Psychological Society.

CLARKE T.A., JOHNSON G.A., MORRIS P.R. & PAGE M. (1968). Motor impairment: a study of clumsy children. Unpublished dissertation, Institute of Education, University of Leeds.

CORAH N.L., ANTHONY E.J., PAINTER P., STERN J.A. & THURSTON D.L. (1965). Effects of perinatal anoxia after seven years. *Psychol. Monog., 79*, 3, 1-33.

COURVILLE C.B. (1953). "Contributions to the Study of Cerebral Palsy". Los Angeles: San Lucas Press.

CRATTY B.J. (1969). "Motor Activity and the Education of Retardates". Philadelphia: Lea & Febiger.

CRUICKSHANK W.M. & DOLPHIN J.E. (1951). Educational implications of psychological studies of cerebral palsied children. *Except. Child., 18*, 1-8.

CRUICKSHANK W.M., BENTZEN F.A., RATZEBURG F.H. & TANNHAUSER M.T. (1961). "A Teaching Method for Brain-injured and Hyperactive Children". Syracuse: University Press.

DIBIASE W.J. & HJELLE L.A. (1968). Body-image stereotypes and body-type preferences among male college students. *Percept. Motor Skills, 27*, 1143-1146.

DONALD I. (1958). *Scottish Med. J., 3*, 151.

DRILLIEN C.N. (1957). *J. Obst. Gyn. Brit. Emp., 61*, 161.

DRILLIEN C.N. (1964). "Growth and Development of Prematurely Born Infants". Edinburgh: Livingston.

DUNSDON M.I. (1952). "The Education of Cerebral-palsied Children". London: Newnes.

DYK R.B. & WITKIN H.A. (1965). Family experiences related to the development of differentiation in children. *Child. Dev., 30*, 1, 21-55.

EICHMANN R. & GESENIUS W. (1952). *Arch. Gynac., 181*, 186.

EPSTEIN L. (1957). The relationship of certain aspects of the body-image to the perception of the upright. Unpublished Doctoral dissertation, New York University.

ESPENCHADE A. (1940). Motor performance in adolescence. Monog. Soc. Res. *Child Dev., 5*.

EYSENCK H.J. (Ed.) (1961). "Handbok of Abnormal Psychology". New York: Basic Books.

EYSENCK H.J. (1967). "The Biological Basis of Personality." Springfield: Thomas.

FAIRWEATHER D.V. & ILLSLEY R. (1960). Obstetric and social origins of mentally handicapped children. *Brit. J. Prev. Soc. Med., 14*, 149-159.

FENICHEL O. (1945). "The Psychoanalytic Theory of Neurosis". New York: Nortan.

FISH B. (1961). The study of motor development in infancy and its relation to psychological functioning. *Amer. J. Psychiat. 17*, 1113-1118.

FISHER M. (1956). Left hemiplegia and motor impersistence. *J. Nerv. Ment. Dis., 123*, 201-218.

FISHER S. (1964a). Body awareness and selective memory for body versus non-body references. *J. Pers.,* **32,** 138-144.

FISHER S. (1964b). Power orientation and concept of self height in men: preliminary note. *Percept. Motor Skills,* **18,** 732.

FISHER S. (1966). Body attention patterns and personality defences. *Psychol. Monogr.,* **80,** 9.

FISHER S. & CLEVELAND R.L. (1958). "Body-image and Personality". Princeton: Van Nostrand.

FLEISHMAN E.A. (1967). Individual differences and motor learning. In R.M. Gagné (Ed.) "Individual Differences". Ohio: Merrill.

FLOYER E.B. (1955). "A Psychological Study of a City's Cerebral Palsied Children." London: British Council for the Welfare of Spastics.

FORD F.R. (1959). "Diseases of the Nervous System in Infancy, Childhood and Adolescence". Springfield: Thomas.

FREEDMAN S.J. (1961). Sensory deprivation: facts in search of a theory. *J. Nerv. Ment. Dis.,* **132,** 17-21.

FRANCES-WILLIAMS J. (1964). "Understanding and Helping the Distractible Child". London: Spastics Society.

FROSTIG M., LEFEVER D.W. & WHITTLESEY D.R.B. (1961). A developmental test of visual perception for everyday normal and neurologically handicapped children. *Percept. Motor Skills,* **12,** 383-389.

FROSTIG M. (1963). "Developmental Test of Visual Perception". California: Consulting Psychologists' Press.

FROSTIG M. & HORNE D. (1964). "The Frostig Program for the Development of Visual Perception". Chicago: Follet.

FROSTIG M. (1968).Sensory-motor development. *Special Education,* **57,** 2, 18-20.

GALLAGHER J.J. (1964). "The Tutoring of Brain-injured Mentally Retarded Children". Springfield: Thomas.

GARFIELD E. (1923). The measurement of motor ability. *Arch. Psychol.,* **62.**

GARFIELD J.C. (1964). Motor impersistence in normal and brain-damaged children. *Neurology,* **14,** 623-630.

GARRETT E.S., PRICE A.C. & DEABLER H.L. (1957). Diagnostic testing for cortical brain impairment. *Arch. Neurol. Psychiat.,* **77,** 223-225.

GESELL A. & AMATRUDE C.S. (1941). "Developmental Diagnosis". New York: Hoeber.

GIBSON H.B. (1964). The spiral maze: A psychomotor test with implications for the study of delinquency. *Brit. J. Psychol.,* **54,** 219-225.

GOODENOUGH D.R. & EAGLE C.J.A. (1963). A modification of the embedded figures test for use with young children. *J. Gen. Psychol.,* **103,** 67-74.

GRAHAM F.A. & KENDALL S. (1960). Memory for designs. *Percept. Motor Skills,* **11,** 147-188.

GREENSPAN H., LEON D. & WEAVER A. (1953). Clinical approach to the aetiology of cerebral palsy. *Arch. Phys. Med. Rehab.,* **34,** 478-485.

GUBBAY S.S., ELLIS E., WALTON J.N. & COURT S.D.M. (1965). A study of apraxic and agnosic defects in 21 children. *J. Neur.,* **88.**

HEWETT S., NEWSON J. & NEWSON E. (1970). "The Family and the Handicapped Child". London: Allen & Unwin.

HILL S.D. et al (1967). Relation of training in motor activity to development of right-left directionality in mentally-retarded children. *Percept. Motor Skills,* **24,** 363-366.

HOBBS N. (1966). Helping disturbed children: psychological and ecological strategies. *Amer. Psychol.,* **21,** 12, 1105-1115.

HOLDEN R.H. (1962). Changes in the body-image of physically handicapped children due to summer day camp experience. *Merrill Palmer Quart.,* 8, 19-26.

HOUSE E.L. & PANSKY B. (1960). "A Functional Approach to Neuroanatomy". New York: McGraw-Hill.

INGALLS T.H. (1950). Anoxia as a cause of foetal death and congenital defects in the mouse. *Am. J. Dis. Child.,* **80,** 34-45.

ISMAIL A.H. & GRUBER J.J. (1967). "Motor Aptitude and Intellectual Performance". Ohio: Merrill.

JAKEMAN D. (1967). The Marianne Frostig approach. *Forward Trends,* **11,** 3, 99-100.

JOHNSON G.B. (1932). Physical skill test for sectioning classes into homogeneous groups. *Res. Quart.,* **3,** 128-138.

JONES M.G. (1970). Perception, personality and movement characteristics of women students of physical education. Unpublished M.Ed. dissertation, University of Leicester.

JONES M.C. B BAYLEY N. (1950). Physical maturing among boys related to behaviour *J. Educ. Psychol.,* **41,** 129-148.

JOURARD S. (1967). Out of touch—body-taboo. *New Society,* **9.**

JOYNT R.J., BENTON A.L. & FOGEL M.L. (1962). Behavioural and pathological correlation of motor impersistence. *Neurology,* **12,** 876-881.

KAMII C.K. & RADIN N.L. (1967). A framework for a pre-school curriculum based on some Piagetian concepts. *J. Creative Beh.,* **1,** 3, 314-323.

KANNER L. (1944). Early infantile autism. *J. Paediatrics,* **25,** 211-217.

KARP S.A. & KONSTADT N. (1963). "Manual for the Children's Embedded Figures Test. Cognitive tests". Brooklyn: Authors (P.O. Box 4, Vanderveer Station 11210).

KEPHART N.C. (1960). "The Slow-learner in the Classroom". Ohio: Merrill.

KLEBANOV D. (1948). Hunger und Physiche Ereringe ab ovar und keinshadigungen. *Geburtsch und Frauenheilk,* **812,** 7-8.

KNOBLOCH H. & PASAMANICK B. (1959). Geographic and seasonal variation in birth rates. *Pub. Health Report* (U.S.A.), **74,** 4, 285-289.

KOPPITZ E.M. (1958). The Bender Gestalt Test and learning disturbances in young children. *J. Clin. Psychol.,* **14,** 292-295.

LILLIENFELD D. & ABRAHAM M. (1953). "Mass Study of Reproduction Wastage in Prematurity Congenital Malformation". New York: Assoc. for the Aid of Crippled Children.

LILLIENFELD A.M., PASAMANICK B. & ROGERS M. (1955). Relationship between pregnancy experience and the development of certain neuropsychiatric disorders in childhood. *Am. J. Pub. Health,* **45,** 637-643.

MALPASS L.F. (1961). Motor skills in mental deficiency. In E. Ellis (Ed.) "Handbook of Mental Deficiency". New York: McGraw-Hill.

McCLOY C.H. (1934). The measurement of general motor capacity and general motor ability. *Res. Quart.,* **5,** 46-61.

McKELLAR P. (1965). Thinking, remembering and imagining. In J.G. Howells (Ed.). "Modern Perspectives in Child Psychiatry". Edinburgh: Oliver & Boyd.

MILLER L.C., LOWENFELD R., LINDER R. & TURNER J. (1963). Reliability of Koppitz' scoring system for the Bender Gestalt. *J. Clin. Psychol.,* **19,** 2111.

MORISON R. (1969). "A Movement Approach to Educational Gymnastics". London: Dent.

MORRIS P.R. & WHITING H.T.A. (1971). "Motor Impairment and Compensatory Education". London: Bell.

NATHAN P.E. (1967). "Cues, Decisions & Diagnosis". New York: Academic Press.

NEWTON G. & LEVINE S. (Eds.) (1968). "Early Experience and Behaviour". Springfield: Thomas.

OLIVER J.N. (1955). Physical education for educationally sub-normal children. *Educat. Rev.,* **8,** 122-136.

OLIVER J.N. (1963). The physical education of E.S.N. children. *Forward Trends,* **7,** 3, 87-90.

OLIVER J.N. & KEOGH J.F. (1967). Helping the physically awkward. *Special Education,* **56,** 22-26.

OLIVER J.N. & KEOGH J.F. (1968). A clinical study of physically awkward E.S.N. boys. *Res. Quart.,* **39,** 301-307.

OSERETSKY N. (1923). Metric scale for studying the motor capacity of children. Published in Russian. Referred to in Rudolph Lassner, Annotated bibliography of the Oseretsky test of motor proficiency. *J. Consult. Psychol. (1948),* **12,** 37-47.

OSERETSKY N. (1929). A group method of examing the motor functions of children and adolescents. *Z. Kinderforsch.,* **35,** 352-372.

PAINTER G.B. (1964). The effect of rhythmic and sensory-motor activity programs on perceptual-motor-spatial abilities of kindergarten children. Unpublished M.S. dissertation, University of Illinois.

PASAMANICK B., ROGERS M.E. & LILLIENFELD A.M. (1956). Pregnancy experience and development of behaviour disorder in children. *Am. J. Psychiat.,* **112,** 613-618.

PASCALL C.R. & SUTTELL B.S. (1957). "The Bender-Gestalt Test". New York: Grune & Stratton.

PIAGET J. (1952). "The Origins of Intelligence in Children". New York: International Universities Press.

POTTER E. (1952). "Pathology of the Foetus and the Newborn". Chicago: Year Book Pub. Inc.

PRECHTL H.F.R. (1961). Neurological sequelae of pre-natal and paranatal complications. In B. Foss (Ed.). "Determinants of Infant Behaviour 1". London: Methuen.

PRECHTL H.F.R. & STEMMER C.J. (1962). The choreiform syndrome in children. *Dev. Med. Child. Neur.,* **8,** 149-159.

RAY H.C. (1940). Interrelationship of mental and physical abilities and achievement of high-school boys. *Res. Quart.,* **11,** 129-141.

RICHARDSON S.A. (1964). The social environment and individual functioning. In H.G. Birch (Ed.). "Brain Damage in Children: the biological and social aspects". New York: Williams & Wilkins.

RITCHIE-RUSSELL W. (1958). Disturbance of the body-image. *Cerebral Palsy Bull.,* **4,** 7-9.

ROGERS M., LILLIENFELD A.M. & PASAMANICK B. (1955). Prenatal and paranatal factors in the development of childhood behaviour disorders. *Acta Psychiat. et Neur. Scand.* Supplement No. **101.**

RUTTER M., GRAHAM P. & BIRCH D. (1966). Intercorrelation between the choreiform syndrome, reading disability and psychiatric disorders. *Dev. Med. Child. Neur.* **8,** 149-159.

SCHAFFER H.R. (1958). Objective observation of personality development in early infancy. *Brit. J. Med. Psychol.,* **31,** 174.

SCHAFFER H.R. (1965). Changes in developmental quotient under two conditions of maternal separation. *Brit. J. Soc. Clin. Psychol.,* **4,** 39-46.

SCHAFFER H.R. (1966). Activity level as a constitutional determinant of infantile reaction to deprivation *Child Dev.,* **37,** 3, 592-602.

SCHAFFER H.R. & EMERSON P.E. (1968). The effects of experimentally administered stimulation on developmental quotients of infants. *Brit. J. Soc. Clin. Psychol.,* **7,** 61-67.

SCHILDER P. (1935). "The Image and Appearance of the Human Body". London: Kegan Paul.

SCHULZ D.P. (1965). "Sensory Restriction". New York: Academic Press.

SCOTT-RUSSELL C. (1969). Another hazard of smoking. *New Scientist,* Jan. 9th.

SHERBOURNE V. (1965). "Movement for Mentally Handicapped Children". Bristol: Dept. of students of the National Assoc. of Mental Health.

SLOAN W. (1948). "Lincoln Adaptation of the Oseretsky Scale". Illinois: Lincoln.

SLOAN W. (1955). The Lincoln/Oseretsky motor development scale. *Gen. Psychol. Monog.,* **51,** 183-252.

SMITHELLS R.W. & MORGAN D.M. (1970). Transmission of drugs by the placenta and breasts. *Pract.,* **204,** 14-19.

STEPHENSON E. & ROBERTSON J. (1965). Normal child development and handicapped children. In J.G. Howells (Ed.). "Modern Perspectives in Child Psychology". Edinburgh: Oliver & Boyd.

STONE R.E. (1968). Relation between the perception and reproduction of body postures. *Res. Quart.,* **39,** 3, 721-727.

STOTT D.H. (1959). "Unsettled Children and Their Families". London: Univ. Press.

STOTT D.H. (1962a). Evidence for a congenital factor in maladjustment and delinquency. *Am. J. Psychol.,* **118,** 781-794.

STOTT D.H. (1962b). 'Mongolism Related to Early Shock in Pregnancy'. Proceedings of the London Conference on the scientific study of mental deficiency. Dagenham: May & Baker.

STOTT D.H. (1964). Why maladjustment? *New Society,* Dec. 10th.

STOTT D.H. (1966). A general test of motor-impairment for children. *Dev. Med. Child Neur.,* **8,** 523-531.

STRAUSS A.A. & LEHTINEN L.C. (1948). "Psychopathology and Education of the Brain-injured Child". New York: Grune & Stratton.

STRAUSS A.A. & KEPHART N.C. (1955). "Psychopathology and Education of the Brain-injured Child". New York: Grune & Stratton.

SUTPHIN F.E. (1964). "A Perceptual Testing-planning Handbook for First Grade Teachers". New York: Boyd.

TANSLEY A.E. & GULLIFORD R. (1960). "The Education of Slow-learning Children". London: Routledge & Kegan Paul.

TANSLEY A.E. (1967). The education of neurologically abnormal children. *Times Educ. Suppl.,* Jan. 20th.

THWEATT R.C. (1963). Prediction of school learning disabilities through the use of the Bender-Gestalt test. *J. Clin. Psychol.,* **19,** 216-217.

WALTON J.H., ELLIS E. & COURT S.D.N. (1962). Clumsy children: developmental apraxia and agnosia. *Brain,* **85,** 603.

WECHSLER D. (1949). "Wechsler Intelligence Scale for Children—Manual". New York: Psych. Corp.

WEDELL K. (1964). "Some Aspects of Perceptual-Motor Development in Young Children". London: The Spastics Society.

WEINER H. (1946). Preventative aspects of rhesus incompatability. *New York State Health News,* **26,** 9-14.

WELTON J. (1912). "The Psychology of Education". London: MacMillan.

WERNER H. (1941). Psychological approaches investigating differences in learning distractability. *Am. J. Ment. Def.,* **47,** 269.

WHITING H.T.A. (1969). "Acquiring Ball Skill: a psychological interpretation". *Remedial Gym. Rec. Therapy,* **45,** 14-19.

WHITING H.T.A., CLARKE T.A. & MORRIS P.R. (1969a). A clinical validation of the Stott Test of motor impairment. *Brit. J. Soc. Clin. Psychol.,* **8,** 270-274.

WHITING H.T.A., JOHNSON G.F. & PAGE M. (1969b). Factor analytic study of motor impairment at the ten-year age level in normal and E.S.N. populations. Unpublished paper, Physical Education Dept. University of Leeds.

WHITING H.T.A. (1969). "Acquiring Ball Skill: a psychological interpretation". London: Bell.

WHITING H.T.A., JOHNSON G.F. & PAGE M. (1969c). The Gibson Spiral Maze as a possible screening device for minimal brain damage. *Brit. J. Soc. Clin. Psychol.,* **8,** 164-168.

WHITING H.T.A., DAVIES J.G., GIBSON J.M., LUMLEY R., SUTCLIFFE R.S.E. & MORRIS P.R. (1970). Motor impairment in an approved school population. Unpublished paper, Physical Education Dept., University of Leeds.

WHITING H.T.A., HARDMAN K., HENDRY L.B. & JONES M.G. (1973). "Personality and Performance in Physical Education and Sport". London: Kimpton (in press).

WISLER C. (1901). The correlation of mental and physical tests. *Psychol. Rev. Monog. Supp.*

WITKIN H.A., DYK R.B., FATERSON M.F. & KARP S.A. (1962). "Psychological Differentiation". New York: Wiley.

WITKIN H.A. (1965). Psychological differentiation and forms of pathology. *J. Abn. Psychol., 70, 5.*

WITKIN H.A. (1967). A cognitive style approach to cross-cultural research. *Int. J. Psychol., 2, 4,* 232-250.

WOBER M. (1966). Sensotypes. *J. Soc. Psychol., 70,* 181-189.

WORKANY J. (1950). Congenital malformation. In M. Nelson (Ed.). "Textbook of Paediatrics". Philadelphia: Saunders.

WRIGHT B.A. (1960). "Physical Disability—a Psychological Approach". New York: Harper & Row.

YARMOLENKO A. (1933). The motor sphere of school-age children. *J. Genet. Psychol., 42,* 298-316.

ZEULZER W.W. (1950). Kernicterus. Aetiological study based on analysis of 55 cases. *Paediatrics 6,* 452-474.

TOPIC AREAS-DEVELOPED

CENTRAL & PERIPHERAL NERVOUS FUNCTION

by J.D. BROOKE

This section covers control of human movement and all other sections fall back to it. Much however is still unknown. Objective statements can be made about isolated function and simple reflex behaviours. Inferences about more complex formulations are made by studying behavioural effects of treatments and relating these to anatomical, ablative, electrical and chemical intrusion experiments, frequently on animals. There is a high need for knowledge at molecular, physiological and behavioural levels on the control of *human* movement.

Schade & Ford (1965) provides an introduction to basic neurology, Davies (1967) to anatomy and Campbell (1965) a clear analysis of many known central nervous system (CNS) interactions. Moruzzi et al (1963) in volume one of the constantly expanding series on progress in brain research provide a more complex set of papers and Luria (1966) and Hess (1957) fall into this category. Brain and behaviour are the orientation for Delafresnaye (1954); Brazier (1960; 1961), Fessard et al (1961) Russell (1966) and Butter (1968). The three volumes on neurophysiology in the American Handbook of Physiology, Field (1960) have an extensive coverage and Sherrington's 1906 series of lectures (1947) is still worthwhile reading. Loufbourrow (1960) deals directly with applications to sport and exercise. This is an exciting, and, because of its complexity, frightening field of study and research. Attention in this review is not to a comprehensive coverage but to a focus on particularly pertinent aspects and, where possible, an attempt to sketch a ground plan.

C.1.

There is observed traditionally a division between peripheral and central nervous systems. The division of the peripheral system can be summarised as:

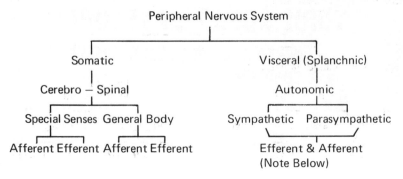

Note that although initially the autonomic was conceived as an efferent system it is common now to include the relevant afferents from the organs involved, see C.3. and C.5. The motor system is involved in the efferent aspects of the peripheral system. Schade & Ford (1965), Davies (1967) and Green (1968) develop the complexities around the diagram.

The Divisions of the CNS are

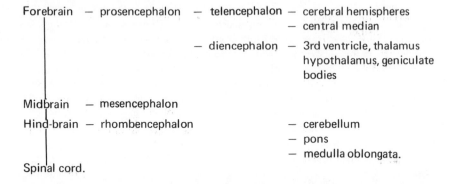

The cerebral hemispheres are often classified by the 50 areas first described by Brodmann (1909). Care must be taken at present to differentiate between anatomical classification and function exhibited, for the two are not necessarily synonymous.

C.2. Basic Neural Function.

There are three basic functions of all nerves.

1. *Excitability.* With adequate stimulus the resting electrical state of a nerve is disturbed sufficiently to propagate an electrical impulse.

2. *Conductivity.* If the stimulus is strong enough to propagate the impulse, amplitude and conduction speed are independent of stimulus strength (All or None Law).

3. *Transmission.* Impulses are always transmitted to another neuron, muscle or gland by chemically bridging the gap—synapse for neurones—between the two. At motor end-plates the chemical is acetylcholine. Schade & Ford (1965) expand greatly on the electrical, chemical, anatomical and physiological matters evolving from these three basic functions.

C.3. Peripheral Afferent Function.

All afferent nerves enter the spinal cord at posterior nerve roots (Bell-Magendie Law) 1. Those responsible for information on the position of joints in space (from joint capsule receptors) and accurate touch sensations travel in the central lemniscal cord path, relay in the medulla and the thalamus and arrive at the appropriate lobes of the sensory cortex. 2. Those responsible for proprioception from muscle contraction (from muscle spindle receptor and muscle tendon Golgi receptors) relay at the posterior root and enter the spino-cerebellar cord tracts to run direct to the cerebellum. 3. Those responsible for pain, temperature and crude sensation of touch relay at the posterior root, travel in the central spinothalamic cord path to relay in the thalamus for transmission to the sensory cortex. Mixing of impulses occurs in the latter, more diffuse, tract.

The specific sensory receptors for these sensations were at first thought to be identified anatomically by von Frey (1894) but the extension of this theory to function was broken by Sinclair, Weddell & Zander (1952) and Weddell et al (1961; 1962) who propose an alternative, that the temporal and spatial pattern of impulses entering the CNS may be the peripheral factor determining some aspects of sensory modality (Sherrick, 1966). For pain, Dzidzishvili (1965) suggested from a literature survey that a reception system was 'inherent in all organs and to all encapsulated receptors of any kind'. Keele & Smith (1962) and Soulairac et al (1968) are useful texts for the study of pain and Brooke (1971) contains a number of references to the neurophysiology of pain. There are behavioural studies of pain in physical performers, e.g. Borg (1962); Ryan & Kovack (1966); Brooke et al (1967; 1972). The use of nerve fibre diameter as classification (Erlanger & Gasser, 1937), has not been effective across the whole sensory modality range, although some relation is found with the perception of e.g. pain, (Collins et al, 1960).

The neurophysiology of the special senses of smell, taste, sight and sound is a topic well reviewed in physiology texts such as Green (1968) and Rosenblith (1961) or in more detail Field (1965). See also for *vision* Mishkin (1966) and Teuber (1965), for *sound* Rasmussen & Windle (1960), for *smell* de Vries & Stuiver (1961) and for *taste* Zotterman (1959).

There are visceral afferents in the vagus for cardio-respiratory regulation as described in C.5. below and there are some sensory fibres associated with the sympathetic system. They are carried over the normal somatic afferent pathways with cells of origin in posterior root ganglia.

It must be noted that perception is not dependent merely on peripheral afferent stimulation. From entry into the CNS (cord) the transmission of sensory information is subjected to modification by descending branches of the central integrating mechanisms (Mountcastle, 1961) and this is one of the major determinants of conscious perception. C.6. below considers some of these processes.

C.4. Motor System.

Adjustment of the whole body to the external environmental load is obtained through the metabolic work system. The effector for this movement behaviour is skeletal muscle, the support troops cardio-ventilatory-digestive musculature. Somatic efferent flow to skeletal muscle is two-fold. The primary innervation is from the cortex areas 4 and 6 (motor areas) through the corticospinal tract (pyramidal), with cross over of fibres at medulla level, down the lateral columns to the anterior horn cell of the cord (there is probably an internuncial neurone before the anterior horn). The final common path from each anterior horn cell allows transmission to one lower motor neurone of A alpha fibre (18.20μ diameter, which supplies a number of muscle fibres, a motor unit, the smallest contractile block. The secondary innervation is via small anterior horn cells which supply (by A gamma fibres 2-8 μ diameter) intrafusal muscle fibres in the muscle spindles resonsible for proprioceptive feedback. This gamma system is controlled by an extrapyramidal system from the reticular formation, vestibular nuclei, tectum, olive and red nucleus. The two systems allow motor reflex feedback. By the alpha system the force of contraction of the muscle is dependent on the impulse frequencies in the lower motor neurones and the number of motor units involved. The gamma system causes contraction of the intrafusal fibres which generate, at the muscle spindle sensory ending, impulses to the spinal cord. By reflex arc these fire anterior horn cells to the alpha system, which, by muscle contraction, decrease the muscle spindle activity as muscle spindles lie parallel to contractile fibres. Differing levels of 'setting' of the gamma system thus by reflex arc raise or lower maximum contractile force. This appears to be implicated in determining the degree of physiologically available strength that is permitted to be used by central psychological mechanisms (Jung & Hassler, 1960). Damage to the pyramidal (alpha) system leads to flaccid paralysis: interruption of the gamma system results in loss of movement associated with increased resistance to movement i.e. spasticity.

While there are no inhibitory nerves supplying the voluntary muscle fibres there are inhibitory as well as excitatory descending systems to the anterior horn cell and the sum of these negative and positive effects determines the resulting lower motor neurone impulse. Separate parts of the cerebellum play an important part in posture, balance and voluntary movement but the neural interactions are cortico-cerebellar-cortico, extra pyramidal-cerebellar-extra pyramidal and spino-cerebellar as described above, C.3. There are no direct cerebellar-spinal tracts (Snider, 1958).

This account of skeletal muscle innervation is much simplified see e.g. Campbell (1965) and the Handbooks of Physiology, particularly Jung & Hassler (1960) for more detail and research references to the various aspects.

The reflex arc is the functional basis of the central nervous system. Unconditioned ones are inherited and considered in this section. Conditioned ones (Pavlovian) are not inherited and are not considered here. The simplest reflex is the two neurone, one synapse arc at spinal level. Stimulation of muscle spindle stretch receptors initiates impulses to posterior roots, synapse is made with anterior horn cells back to the same muscle and efferent flow elicits muscle contraction. A development on these short reflexes are the longer spinal reflexes where the posterior-anterior synapse is separated by transmission up or down the cord through inter-nuncial neurones, so eliciting reaction at other body sites. As described above for muscle spindles, these simple responses are most frequently under the control of higher centres, but because many can be initiated when this control is absent they are termed spinal reflexes.

It can be demonstrated in the cat with brain transected just above the medulla, that the hindbrain is involved in that essential of CNS functions, integration, for more complex reflexes of postural adjustment still can be elicited, e.g. see Hardy (1934). As transections at successively higher levels are made (while still cutting out the involvement of the cerebral cortex) increasing complexity of reflex response can be demonstrated with the involvement of the midbrain. The archicerebellum and palaeocerebellum are the highest end points for reflex paths (Campbell, 1965). The former directs reflex control of body balance and the latter motor response to proprioceptive information from joint and muscle receptors, involving pyramidal and extra-pyramidal tracts. In the mature adult the forebrain exerts an inhibitory influence on reflex response. The lack of this inhibition can be seen in the sucking and grasping reflexes (of phylogenic origin) of the infant and of the adult with cerebral cortex frontal lobe damage (Teitelbaum, 1967). Loufbourrow (1960) provides a useful chapter on this whole aspect of neuro-muscular integration within the sports context. Sections C.6. and C.7. will indicate some of the complexities involved in the higher integration at telencephalonic and diencephalonic levels.

C.5. Maintenance of Homeostasis.

The adaptation of the internal medium of the body to the external environmental load is under neural and neuro-humoral control Anand (1967). Again all the mechanisms are not understood.

The lung ventilation, cardiac output and circulatory distribution on the one hand and the level of the Pco_2 and Po_2 on the other are interdependent, constantly cycling, with adjustments being made by feedback. The afferent information initiating neural efferents to these metabolic sites is complex. Baroreceptors sense blood pressure at the aortic arch, carotid sinus and carotid arteries and chemoreceptors sense Po_2 and Pco_2 at the carotid body and aortic body and transmit information via the vagus nerve (X) and glossopharangeal nerve (1X) to the CNS complexes globally termed the vasomotor centre (Heymans & Neil, 1958) and the respiratory centre in the medulla of the hind brain.

Increased metabolic function is initiated by increased afferent sympathetic flow and associated adrenal release of catecholamines and reduced metabolic function is initiated by increased afferent vagal and decreased sympathetic flow. Other factors affect the response of the vasomotor and respiratory centres, see J.1. and J.4., e.g. higher centre activity (particularly for the respiratory centre), the significance of the sensory afferent flow (Brooke, 1970), direct effect on the centres of low Po_2 (Dell 1964) and high or low Pco_2 [probably by altering the $[cH^+]$ of the cerbrospinal fluid (Mitchell et al, 1963)], lung stretch receptors and joint proprioceptors (Flandrois et al, 1967) (respiratory centre), metabolic disturbances altering the chemical function of motor organs and the interaction of activity in one centre upon the activation of the other (Cotes, 1968). This is a complex field, see Caro (1966); Heymans & Neil (1958); Bell et al (1965); Porter (1970) and the Handbooks of Physiology Sections 2 and 3, e.g. Dejours (1965). Also note Milhorn et al (1965) and Pribram (1968) for mathematical and systems theory analyses of respiratory control. The activating and integrating systems discussed in C.7. are much involved in the overall maintenance of homeostasis.

The salt-water balance is maintained at the kidney under control of the antidiuretic hormone from the adenohypophysis under the control of the hypothalamus, which receives information on the osmotic pressure of extracellular fluid and blood from osmoreceptors in the internal carotid artery (Giebisch, 1962). Feeding and body temperature are regulated each by two control centres in the hypothalamus one for each end of each scale. The processes involved are complex. (See for feeding Anand, 1967 and for temperature, Society for Experimental Biology, 1964). The level of the appetite is also affected by the effective blood glucose level, the difference between the arterial and venous blood sugar levels (Mayer, 1953). The blood sugar is controlled by insulin which facilitates cellular uptake of glucose. Insulin is secreted in the pancreatic islets of Langerhans and is used in the treatment of diabetes mellitus. Protein metabolism is controlled by the hormone cortisol produced by the adrenal cortex. It modifies metabolism under physical stress by facilitating the transfer to carbohydrate metabolism and is regulated by the adrenocorticotrophic hormone (ACTH) release via the hypothalamus and the neurohypophysis. The release of the catecholamine hormones of the adrenal medulla, adrenaline and nor-adrenaline, also prepares the organism for physical or mental load (Cannon, 1915) with action on heart, circulation, bronchioles, digestion and mobilization of liver glycogen. The adrenal medulla is innervated by and often augments the action of the sympathetic nervous system. Other hormones controlled by the hypothalmic-anterior pituitary (adenohypophysis) link are the growth hormone complex for overall body growth, the three gonadotrophic hormones, and thyrotrophic hormone which controls the release of thyroxine from the thyroid gland. Again this is a complex study, see, Beach (1940), Spence (1953) and Creese (1963).

The acid-base state of the body is regulated by a number of mechanisms. Respiration in partial anaerobiosis, one of the possible causes of exhaustion with physical movement (Brooke et al (1972) , raises blood lactates and $[cH^+]$; ventilation reduces Pco_2: lactates are oxidised in the liver: in the intervening space of time some of the hydrogen ions are buffered by the blood bases, see J.3. Changes in acid-base state due to respiration are termed respiratory acidosis or alkalosis. As the regulation of bicarbonate and the secretion of hydrogen ions lies in the kidney, so variability in this control or the ingestion of acids or alkalis will lead to metabolic acidosis or alkalosis. Further information on this detailed area of study is contained in Mills (1963); Milne (1957); Pitts (1963) and Robinson (1954).

The metabolic condition of the brain itself is adjusted to attain its own homeostasis. With alterations in blood gases, temperature and nutrients, marked changes in CNS function occur e.g. see Dell (1964). A number of workers assume the existence of a blood brain barrier e.g. Mitchell (1966); Dell (1964) but it has not yet been clearly identified Tschirgi (1962), only assumed e.g. due to the different rates of change of Pco_2 and $[cH^+]$ between cerebro-spinal fluid and arterial blood. Adequate potentials for oxygen and blood sugar are essential for CNS cells. The brain burns almost exclusively carbohydrate and hypoglycaemia is characterised by CNS malfunction. This condition can be seen in otherwise healthy human beings who exercise vigorously for several hours or who inadvertantly go without food and continue gross motor behaviour, as in emergencies on expeditions. See N.2. Also with arrested circulation for more than two to three minutes cerebral anoxia leads to pathological changes in CNS cells (hence the urgency in starting cardiac massage following cardiac arrest).

C.6. Arousal, Emotion and Conditioning.

Useful texts are Granit (1955); Jasper et al (1957); Loufbourrow (1960); Rosenblith (1961); and Evans & Mulholland (1969). Brooke (1971) lists two hundred references to arousal and exercise tolerance. Meldman (1970) draws together research under concepts related to diseases of attention. There is at present indadequate anatomical definition behind the physiological information in these fields. Three areas of research work can be identified.

1. Sensory impulses via collaterals to the brain stem reticular activating system are associated with *generalised arousal* (sleep-awake-active-hyperactive); Moruzzi & Magoun (1949); Lindsley (1957); Fuster (1958); Anokhin (1961); Narikashvili (1963) and Wilson & Radloff (1967). There are afferent and efferent connections with cerebral cortex, main sensory paths down to the first synapses and emotional centres. There occur diffuse activating impulses over sensory cortical areas facilitating consciousness. The inverted U curve of skill performance and arousal is associated with this general arousal system.

2. Ascending impulses through sensory collaterals to cortical sensory areas through thalamic reticular nuclei appear directed to *more specific shifts* in the comparative dominance of perception of the sensory modes (Anokhin, 1947; Rothballer, 1955; Sharpless & Jasper, 1956; Lindsley, 1957; Jasper, 1963). Sokolov (1960) proposes a useful model.

3. *Specific modifications* of sensory flow occur as distal as the first sensory synapse (Granit & Kaada, 1953; Hagbarth, 1960 and Hernandez–Peon, 1956 1966). Inhibition of the afferent transmission from peripheral sensory receptors was observed in recordings made at the second sensory neurone when attention was altered. It has been suggested that the selection of information for behaviour e.g. for skilled human movement, can be adequately accounted for by the addition to the information from these areas of the actions of the limbic system (the hippocampus, orbito-frontal and temporal lobes surrounding the Sylvian fissure, the cingulate gyrus, some of the thalamic and hypothalamic nuclei and the basal ganglia and amygdaloid nucleus), see Kaada et al (1949) and Olds (1958). This would allow the use of additional information on recent memories via the hippocampus and visceral emotional traces via the amygdala, hypothalamus and neurohypophysis. Note that Grastyan et al (1959) suggest that the hippocampus modulates during different phases of learning the orienting response (the over-ride response to novel or high amplitude stimuli or those of great meaning).

With regard to learning and the activating systems, Girden et al (1936) found that removal of the forebrain stopped discrimination between different conditioned stimuli in the dog. However simple conditioning still occurred and it is possible that the limbic system and reticular activating system maintained this integration at the lower brain level.

This very active area of work is beginning to uncover a biological filtering system where the focus of the highest central control is directed, through inhibition and facilitation of input at various neural levels, to particular physical events in the environment of the organism so that stimuli from other physical sources evoke little or no conscious perception. Brooke, Hamley and Stone (1972) demonstrated such a mechanism in subjects whose perceptual fields narrowed under the distress of movement exhaustion due to inadequate oxygen potential. Peripheral features of the environment, perceived during homeostasis and exercise, were lost at exhaustion in exercise.

C.7. Learning and Memory.

From ancient times sensation and intelligence have been associated with the cerebral cortex (Sherrington 1933). This is still the case although the role of lower brain filters in the selection of input now is perceived, see C.6. and the cerebrum is held by some authors to be primarily an evolutionary data bank on input (Pribram, 1963). In reviewing the role of the cerebral cortex it is clear that there is very inadequate functional research and incomplete anatomical definition behind the human capacity that can be observed (Pribram, 1958). The complexity is great and most direct investigation has to be on animals. Campbell (1965) provides a useful review and Weiner & Schade (1963) a stimulating study on brain models.

There are two interconnected cerebral hemispheres, one more dominant than the other. Within each, each of the four cerebral lobes has its own association area, with the parietal lobe receiving stimuli from somesthetic and taste sensations, the occipital-vision and the temporal sound. The frontal is discussed below. Most of the sensory input passes through the common integrative area, the angular gyrus. By inter-connecting fibres in the association areas there are complex linkages between the various sensory and motor areas. This complexity increases as study is made of higher animals in the phylogenic scale. For the extreme complexity in surface anatomy and in neuronal layers see Davies (1967).

Two primary characteristics of CNS function are inhibition and integration of information up to cerebral level. Many CNS areas are involved, reticular formation (above C.6.) thalamus, Fessard et al (1964) and, according to Klosovskii & Kosmorskoya (1963) vitally as a junction point between cerebral areas, inhibitory areas 8, 24 and the band in area 4, the angular gyrus, the limbic system (above C.6.) association areas and other sites. Sperry (1964) points to the importance of inter-hemispheric transfer of information and reviews the function of the forebrain commissure, the corpus callosum.

It is clear that the old concept of a reflex arc through simple anatomical pathways is not tenable to account for the learning associated with cerebral function (Lashley, 1942). Campbell (1965) discusses the neurological aetiology of different types of learning. As discussed above, C.5., the phylogenically older reflexes associated with central grey function in ontogenically young humans, by maturity appear to be under cerebral grey inhibition. Also some simple 'learning' by conditioning in animals still can be obtained after ablation of the cerebrum. But for the integrated learning and behaviour characterising the human the inspecting and co-ordinating role of the cerebral cortex is essential (Campbell, 1965). Not enough is known.

One physiological experimental technique that has allowed some indication of cerebral states is electro-encephalography, eeg., the recording of the electrical activity of the brain by scalp electrodes. There is much individual variability but generally:- 1. with inattention 10 c/s alpha rhythm predominates, 2. in attention, faster asynchronous activity occurs, 3. beta rhythm of > 14 c/s is associated with emotional tension, 4. theta rhythm 4-7 c/s is associated with emotional frustration, particularly in psychopathic individuals and 5. delta rhythm 1-3.5 c/s is usually found only in adults in deep sleep but is common in children before age 12. Its persistence in consciousness after that stage is associated with immaturity. There are in addition a number of other identifiable characteristics of the electrical activity but in the interpretation of eeg's much remains to be done. It is noteworthy that Morrell & Jasper (1956) could trace in the eeg similar changes to those observed behaviourally during classical and discriminatory conditioning, and that the 'conditioned' electrical activity always preceeded the behavioural conditioned response. Brazier (1962) provides an introductionary review to electroencephalography: see also Walter (1953; 1959). Little is reported about the eeg in exercise.

Much work is being directed to study information storage in the cerebrum, a good deal of it at molecular levels. Psychologists identify short and long term memory (seconds or minutes versus years of retention). It may be that the former can be attributed to the reverberation of incoming signals through neural circuits in the cerebral layers i.e. primary, secondary, tertiary cells etc. The long term memory may be accounted for by modifications being made at the CNS synapses due to the repeated passage of particular signal patterns. See Hebb (1951) for a psychological interpretation. This modification may be by physical growth of neuronal components, by chemical changes in the nerve cells and/or by changes in the composition of the neuroglia, the special connective tissue of the CNS. In activated nerve cells the concentration of ribonucleic acid is known to increase (Hydén, 1959) and that in surrounding neuroglia to decrease. At a more detailed level the relative concentrations of particular bases in the nuclear RNA alters in such activated neurons (Hydén & Egyhazi, 1962). It is probable that a particular pattern of protein molecular blocks is formed at active neurones and that these represent the long term memory trace or 'engram' (Smith, 1962). The reverberation suggested for short term memory would facilitate this chemical printing. The immediate memory loss following concussion could also be accounted for in this theory, for the physical blow stopping short term reverberation would terminate the printing and so early memories would return on recovery but most recent traces would be lost. Again this is a difficult field to attempt synthesis in at present, e.g. Dingman & Sporn (1964) launch strong criticism at the RNA theories, preferring consideration of lipids and they set up research criteria that must be met to authenticate the constituents of the engram. It must also be noted that the integrating and inhibiting functions of activating systems such as those described in C.6. are intimately involved in learning and memory.

For a large area of the frontal lobes it is difficult to identify function either as a sensory area or as a site for motor discharge. The areas 9, 10, 11 & 12 of the frontal lobes connect profusely with other areas of the brain, including association fibres to the other three lobes and to the diencephalon. For relief of chronic depression and intractable pain the operation of frontal lobotomy, severing the white matter connection with the diencephalon, in addition to providing relief from the excessive emotional feelings for the patient and insight for the student, also routinely involves impairment of the ability to memorize. The basic intelligence is not normally affected. It is suggested that the frontal lobes are involved with the limbic system in some types of information storage. It also appears that the uncal region of the temporal lobe is involved in long term memory. There appears to be no anatomical or physiological evidence for a separate unconscious mind nor in content does this differentiation occur qualitatively according to Campbell (1965).

In summary, much is unknown. Behaviourally the central nervous system is composed of ascending levels of competence and intricacy from simple spinal reflex arcs, through the primitive integrative arcs of the hindbrain, the more sophisticated but still largely automatic reflexes of the mid-brain to the activating emotive and visceral-directing synthesising of the diencephalon or 'old' brain and eventually the final co-ordinating, inhibiting, integrating and storing functions of the telencephalon or 'new' brain that results in the assertion of reason upon the lower animal characteristics of the previous level. With advancing technology removing the need for high physical power ability in occupational tasks, it is clear that research on nervous function from behavioural to molecular levels is necessary for understanding in particular (a) the demands of new occupational loads (b) the relationship between human movement and central nervous homeostasis both in the education of the developing child and the leisure pursuits of the mature and (c) the interaction between CNS function, human movement and clinical abnormality.

Source Texts

BRAZIER M.A.B. (Ed.) (1960). "The Central Nervous System and Behaviour". New York: Macy.

BRAZIER M.A.B. (Ed.) (1961). "Brain and Behaviour Vol. 1". Washington: Amer. Inst. Biol. Sci.

BUTTER C.M. (1968). "Neuropsychology: the study of brain and behaviour". California: Brooks/Cole.

CAMPBELL H.J. (1965). "Correlative Physiology of the Nervous System". London: Academic Press.

DELAFRESNEYE J.F. (Ed.) (1954). "Brain Mechanisms and Consciousness". Oxford: Blackwell.

FESSARD A. et al (Eds.) (1961). "Brain Mechanisms and Learning". Oxford: Blackwell.

FIELD J. (Ed.) (1960). "Handbook of Physiology. Section 1 Neurophysiology Vols. 1. 11. 111". Washington: Amer. Physiol. Soc.

DAVIES D.V. (Ed.) (1967). "Grays Anatomy" (34th Ed.) London: Longmans.

EVANS C.R. & MULHOLLAND T.B. (Eds.) (1969). "Attention in Neurophysiology". London: Butterworth.

HESS W.R. (1957). "The Functional Organization of the Diencephalon". New York: Grune & Stratton.

LURIA A.R. (1966). "Higher Cortical Functions in Man". New York: Basic Books.

MORUZZI G. et al (Eds.) (1963). "Progress in Brain Research Vol. 1.—Brain Mechanisms". New York: Elsevier.

RUSSELL R.W. (Ed.) (1966). "Frontiers in Physiological Psychology". London: Academic Press.

SCHADE J.P. & FORD D.H. (1965). "Basic Neurology". London: Elsevier.

SHERRINGTON C. (1947). "The Integrative Action of the Nervous System". New Haven: Yale University Press.

Specific References

ANAND B.K. (1967). In C.F. Code (Ed.) "Handbook of Physiology Section 5 Alimentary Canal. Vol. 1".Washington: Amer. Physoil. Soc.

ANOKHIN P.K. (1947). Teoriya funkstional 'noi sistemy kak osnova' dlya ponimaniya kompensatornykh protsessor organisma. *Uchevye Zapiski Moskovskgv Gosundanstvernnogo Universiteta* Im: M.V. Lamonosova, **2, 3**: Moskva, 32-41.

ANOKHIN P.K. (1961). The multiple ascending influences of the subcortical centres on the cerebral cortex. In: M.A.B. Brazier (Ed.). "Brain and Behaviour Vol. 1". Washington: Amer. Inst. Biol. Sci.

BEACH F.A. (1940). Effects of cortical lesions upon the copulatory behaviour of male rats. *J. comp. Psychol.,* **29,** 193-239.

BELL G.H., DAVIDSON J.N. & SCARBOROUGH H. (1965). "Textbook of Physiology & Biochemistry". London: Livingstone.

BORG G.A.V. (1962). "Physiological Performance and Perceived Exertion". Sweden: Lund.

BRAZIER M.A.B. (1962). The analysis of brain waves. *Sci. Amer.* **6.**

BRODMANN K. (1909). "Vergleichende Localizationslehre der Grosshirnrinde in Ihren Prinzipien Dargestellt auf Grund der Zellenbanes". Leipzig.

BROOKE J.D. (1970). The use of lung function measures to predict work abilities within a sports group. In J.D. Brooke (Ed.). "Lung Function and Work Capacity". Salford: The University.

BROOKE J.D. (1971). "Handbook of References on Human Movement. No. 3. Central Nervous System Filtering of Afferent Impulses, Pain and Exercise". Eccles: Worthwhile Designs.

BROOKE J.D., COOPER D., HAMLEY E.J. & SAVILLE B. (1967). Electromyographical analysis of noxious ischaemic work. *Bull. B.A.S.M.,* **3,** 1, 26.

BROOKE J.D., HAMLEY E.J. & STONE P.T. (1972). Disturbance of attention and metabolic homeostasis during performance of an exhausting physical activity. In H.T.A. Whiting (Ed.) "Readings in Sports Psychology". London Kimpton.

CANNON W.B. (1915). "Bodily Changes in Pain, Hunger, Fear and Rage", New York: Appleton.

CARO C.G. (Ed.) (1966). "Advances in Respiratory Physiology". London: Arnold.

COLLINS W.F. et al (1960). Relation of peripheral nerve fibre size and sensation in man. *Arch. Neurol.,* **3**, 381-385.

COTES J.E. (1968). "Lung Function (2nd Ed.)". Oxford: Blackwell.

CREESE R. (Ed.) (1963). "Recent Advances in Physiology". London: Churchill.

DEJOURS P. (1965). Control of respiration. In W.O. Fenn & H. Rahn (Eds.). "Handbook of Physiology. Section 3 Respiration Vol. 1". Washington: Amer. Physiol. Soc.

DELL P. (1964). Reticular homeostasis and critical reactivity. In G. Moruzzi et al (Eds.). "Progress in brain research: Vol. 1. Brain Mechanisms". New York: Elsevier.

DINGMAN W. & SPORN M.B. (1964). The incorporation of 8-azaquanine into rat brain DNA and its effect on maze learning by the rat: An inquiry into the biochemical bases of memory. *J. Psychiat. Res.,* **1**, 1-11.

DZIDZISHVILI N.N. (1965). O bolevoc retseptsii. *Zhurnal Vysshei Nervnoi Deiatel nosti,* **15**, 6, 1026-1035.

ERLANGER J. & GASSER H.S. (1937). "Electrical Signs of Nervous Activity". Philadelphia: The University.

FESSARD A.D. & FESSARD A. (1964). Thalamic integrations and their consequences at the telencephalic levels. In G. Moruzzi et al (Eds.). "Progress in Brain Research. Vol. 1. Brain Mechanisms". New York: Elsevier.

FLANDROIS R. et al (1967). Limbs mechanoreceptors inducing the reflex hyperpnea of exercise. *Resp. Physiol.,* **2**, 335-343.

FREY M. von (1894). *Ber. d. kgl. sachs. Ges. d. Wiss.,* 283.

FUSTER J.M. (1958). Effects of stimulation of brain stem on tachistoscopic perception. *Science,* **127**, 150.

GIEBISCH G. (1962). Kidney, water and electrolyte metabolism. *Ann. Rev. Physiol.,* **24**, 357.

GIRDEN E. et al (1936). Conditioned responses in decorticate dog to acoustic, thermal and tactile stimulation. *J. Comp. Psychol.,* **21**, 367-385.

GRANIT R. (1955). "Receptors and Sensory Perception". New Haven: Yale University Press.

GRANIT R. & KAADA B.R. (1953). Influence of stimulation of central nervous structures on muscle spindles in cats. *Acta. Physiol. Scand.,* **27**, 130.

GREEN J.H. (1968). "An Introduction to Human Physiology (2nd Ed.)" London: Oxford University Press.

GRASTYAN E. et al (1959). Hippocampal electrical activity during the development of conditioned reflexes. *EEG. Clin. Neurophysiol.,* **11**, 409-430.

HAGBARTH K.E. (1960). Centrifugal mechanisms of sensory control. *Ergeb. Biol.,* **22**, 47-66.

HARDY M. (1934). Observations on the innervation of the masculi sacculi in man. *Anat. Rec.,* **59**, 403-417.

HEBB D.O. (1951). The role of neurological ideas in psychology. *J. Personal.,* **20**, 39-55.

HERNANDEZ-PEON R. (1966). Physiological mechanisms in attention. In R.W. Russell (Ed.). "Frontiers in Physiological Psychology". New York: Academic Press.

HERNANDEZ-PEON R., SCHERRER H. & JOUVET M. (1956). Modifications of electrical activity in cochlear nucleus during attention in unanaesthetised cats. *Science,* **123,** 331-332.

HEYMANS C. & NEIL E. (1958). "Reflexogenic Areas of the Cardiovascular System". London: Churchill.

HYDÉN H. (1959). Biochemical changes in glial cells and nerve cells at varying activity. In O. Hoffman-Ostenhoff (Ed.). "Biochemistry of the Central Nervous System. Vol. 3". London: Pergamon.

HYDÉN H. & EGYHAZI H. (1962). Nuclear RNA changes of nerve cells during a learning experience in rats. *Proc. Nat. Acad. Sci.,* **48,** 1366-1375.

JASPER H.H. et al (Eds.) (1957). "Reticular Formation of the Brain". London: Churchill.

JASPER H.H. (1963). Electrical responses in sensory systems. In G. Morruzi et al (Eds.). "Progress in Brain Research Vol. 1. Brain Mechanisms". London: Elsevier.

JUNG R. & HASSLER R. (1960). The extrapyramidal motor system. In J. Field (Ed.). "Handbook of Physiology, Section 1. Neurophysiology Vol. 2". Washington: Amer. Physiol. Soc.

KAADA B.R., PRIBRAM K.H. & EPSTEIN J.A. (1949). *J. Neurophysiol.,* **12,** 347-356.

KEELE C.A. & SMITH R. (Eds.) (1962). "The Assessment of Pain in Man and Animals". London: Livingstone.

KLOSOVSKII B.N. & KOSMORSKOYA E.N. (1963). "Excitatory and Inhibitory States of the Brain". Jerusalem: Isreal Programme Sci. Trans.

LASHLEY K.S. (1942). The mechanism of vision. *J. Genet. Psychol.,* **60,** 197-221.

LINDSLEY D.B. (1957). The recticular system and perceptual discrimination. In H.H. Jasper et al (Eds.). "Reticular Formation of the Brain". London: Churchill.

LOUFBOURROW G.N. (1960). Neuromuscular integration. In W. Johnson (Ed.). "Science and Medicine of Exercise and Sports". New York: Harper.

MAYER J. (1953). *New Engl. J. Med.,* **249,** 13.

MELDMAN M.J. (1970). "Diseases of Attention and Perception". Oxford: Pergamon.

MILHORN H.T. et al (1965). A mathematical model of the human respiratory control system. *Biophysic. J.,* **5,** 27-47.

MILLS J.N. (1963). Mechanisms of renal homeostasis. In R. Creese (Ed.). "Recent Advances in Physiology". London: Churchill.

MILNE M.D. (1957). Renal control of acid-base balance. *Lect. Sci. Bas. Med.,* **5,** 404-420.

MISHKIN M. (1966). Visual mechanism beyond the striate cortex. In R.W. Russell (Ed.). "Frontiers in Physiological Psychology". London: Academic Press.

MITCHELL R.A. (1966). Cerebrospinal fluid and the regulation of respiration. In C.G. Caro (Ed.). "Advances in Respiratory Physiology". London: Arnold.

MITCHELL R.A. et al (1963). Respiratory responses mediated through superficial chemosensitive areas on the medulla. *J. Appl. Physiol.,* **18,** 523-533.

MORRELL F. & JASPER H.H. (1956). Electrographic studies of the formation of temporary connections in the brain. *EEG. Clin. Neurophysiol.*, 8, 201-215.

MORUZZI G. & MAGOUN H.W. (1949). Brain stem reticular formation and activation of the eeg. *EEG. Clin. Neurophysiol.*, 1, 455-473.

MOUNTCASTLE V.B. (1961). Duality of function in the somatic afferent system. In M.A.B. Brazier (Ed.). "Brain and Behaviour". Washington: Amer. Inst. Biol. Sci.

NARIKASHVILI S.P. (1963). Influence of non-specific impulses on the sensory cortex. In G. Moruzzi, et al (Eds.). "Progress in Brain Research Vol. 1. Brain Mechanisms". London: Elsevier.

OLDS J. (1958). Self-stimulation of the brain. *Science*, 127, 315-324.

PITTS R.F. (1963). "Physiology of the Kidney and Body Fluids". Chicago: Year Book Med. Pub.

PORTER R. (Ed.) (1970). "Breathing". London: Churchill.

PRIBRAM K.H. (1958). Neurocortical function in behaviour. In H.F. Harlow & C.N. Woolsey (Eds.). "Biological and Biochemical Bases of Behaviour". Wisconsin: The University Press.

PRIBRAM K.H. (1963). The new neurology: memory, novelty, thought and choice. In G.H. Glaser (Ed.). "EEG and Behaviour". London: Basic Books.

PRIBAN I.P. (1968). Control of respiration, self-adaptive. In "Encyclopaedia of Linguistics, Information and Control".

RASMUSSEN G.L. & WINDLE W.F. (Eds.) (1960). "Neurophysiological Mechanisms of the Auditory and Vestibular Systems". Springfield Thomas.

ROBINSON J.R. (1954). "Reflections on Renal Function". Oxford: Blackwell.

ROSENBLITH W. (Ed.) (1961). "Sensory Communication". New York: Wiley.

ROTHBALLER A.B. (1955). Studies on the adrenaline sensitive component of the reticular system. Ph.D. Thesis, McGill University.

RYAN E.D. & KOVACK C.R. (1966). Pain tolerance and athletic participation. *Percept. Mot. Skills.* 22, 383-390.

SHARPLESS S. & JASPER H. (1956). Habituation of the arousal reaction. *Brain*, 79, 655-680.

SHERRICK C.E. (1966). Somesthetic senses, *Ann. Rev. Psychol.*, 17, 309-336.

SHERRINGTON C.S. (1933). "The Brain and its Mechanism". Cambridge: University Press.

SINCLAIR D.C., WEDDELL G. & ZANDER E. (1952). The relationship of cutaneous sensibility to neurohistology in the human pinna. *J. Anat.*, 86, 402-411.

SMITH C.E. (1962). Is memory a matter of enzyme induction? *Science*, 138, 889-890.

SNIDER R.S. (1958). The cerebellum. *Sci. Amer.* 8.

SOCIETY FOR EXP. BIOL. (1964). "Homeostasis & Feedback Mechanisms". Cambridge: The University Press.

SOKOLOV E.N. (1960). Neuronal models and the orienting reflex. In M.A.B. Brazier (Ed.). "The Central Nervous System and Behaviour". New York: Macy.

SOULAIRAC A., CAHN J. & CHARPENTIER J. (Eds.) (1968). "Pain". London: Academic Press.

SPENCE A.W. (1953). "Clinical Endocrinology". London: Oxford University Press.

SPERRY R.W. (1964). The great cerebral commissure. *Sci. Amer.* **1**

TEITELBAUM P. (1967). "Physiological Psychology". New Jersey: Prentice Hall.

TEUBER H.L. (1965). Perception. In J. Field (Ed.). "Handbook of Physiology, Section 1, Neurophysiology Vol. 111". Washington: Amer. Physiol. Soc.

TSCHIRGI R.D. (1962). Blood-brain barrier—fact or fancy? In G.H. Glaser (Ed.). "EEG and Behaviour". London: Basic Books.

VRIES H.A. de & STUIVER M. (1961). The absolute sensitivity of the human sense of smell. In W. Rosenblith (Ed.). "Sensory Communication". New York: Wiley.

WALTER W.G. (1953). "The Living Brain". London: Duckworth.

WALTER W.G. (1959). Intrinsic rhythms of the brain. In J. Field, (Ed.). "Handbook of Physiology Section 1 Neurophysiology, Vol. 1". Washington: Amer. Physiol. Soc.

WEDDELL G. (1961). Receptors for somatic sensation. In M.A.B. Brazier (Ed.). "Brain and Behaviour". Washington: American Institute of Biological Science.

WEDDELL G. & MILLER S. (1962). Cutaneous sensibility. *Ann. Rev. Physiol.*, **24**, 199-222.

WEINER N. & SCHADE J.P. (Eds.) (1963). "Nerve, Brain and Memory Models". London: Elsevier.

WILSON G.T. & RADLOFF W.P. (1967). Degree of arousal and performance: effects of reticular stimulation on an operant task. *Psychonomic Sci;.* **7**, 1, 13-14.

ZOTTERMAN Y. (1959). Thermal sensations. In J. Field, (Ed.). "Handbook of Physiology Section 1 Neurophysiology Vol. 1". Washington: Amer. Physiol. Soc.

TOPIC AREAS - DEVELOPED

4 D

COMMUNICATION THROUGH MOVEMENT

by ELIZABETH MAULDON

General

Communication is a complex concept and when extended beyond language to include non-verbal behaviour, the arts and the applied sciences, its range embraces man's total activity.

This section is confined to an investigation of the part movement plays in inter-personal communication, whether information is conveyed verbally or non-verbally, deliberately or otherwise. Focus has been directed towards the tactile, auditory and visual channels of communication, since in the main it is through these three senses that information is emitted and received. The kinesthetic senses have not been specifically included since proprioceptive cues, stimulated mainly by actions of the body itself, generate information internally and as such are of an intra-personal nature. Information through the kinesthetic channels may well be integrated with other sensory cues and indeed it is assumed that no channel works in isolation. Although the three sub-divisions are used for convenience the transmission and reception of information involves the simultaneous use of both verbal and non-verbal sources.

It is primarily through language that men communicate thought, feelings and purposes but the crucial part that non-verbal cues play in social encounters has, in recent years, become a focus of attention for social scientists as well as psychiatrists and others concerned with the diagnosis of psychosomatic disorders.

Competency in social relationships is a crucial skill and knowledge of the processes involved in interaction would seem a vital tool for those, for example, in teaching, medicine or management, primarily involved as they are with effective communication.

In a study of social interaction as an aspect of movement behaviour it is of interest to note theories relating the social behaviour of non-human primates to that of man and although this section does not cover such an area, a survey of research findings together with a comprehensive bibliography of work in the field is provided by Carroll (1971a, 1971b) see also Section H. The following references will provide alternative or supplementary reading:

Chance (1962); Diebold (1967); Hebb & Thompson (1954); Morris (1967, 1969); Nissen (1951); Zuckerman (1932).

This section does not include a psychological view of the communicative processes but Davitz (1964); Newcomb, Turner & Converse (1952) and Parry (1967) give useful background information on this topic.

D.1. The tactile channel

Sensitivity to tactual stimuli is probably the most primitive sensory process and as such constitutes a basic communication system, (Frank 1960).

Argyle (1967, 1969) deals fully with the 'language' of bodily contact. Cultural and sex differences, physical proximity, social distance and eye-contact are also well documented. Work in this area has also been reported by Kinsey et al (1948, 1953). Jourard (1966), deals specifically with cultural comparisons in relation to accessibility of body zones and Hall (1959, 1966) highlights cross-cultural differences in degree of physical distance, position and degrees of touch. Williams (1963) and Porter, Argyle & Salter (1969) attempt to relate personality with degree of spatial proximity but find no consistent correlation.

The relationship of tactual experiences to self—structure and body-image, kinesthesis, body awareness and perceptual judgements is dealt with in Cratty (1964) and relevant references are provided. Hunt (1962) records the positive results of an experiment in the treatment of brain-damaged children involving the maintenance of continual body contact with each other in a play situation, which appeared to heighten the body-image and affect social behaviour. Cratty (1964) also refers to the contribution of 'manual guidance' in the learning of motor skills (see also Holding 1966). Brown (1965) studying the development of intelligence, reports evidence from Piaget and others of the relatedness of mental operations and concept formation to physical manipulation of objects. There is in fact a growing body of knowledge which correlates sensory experience with cognitive development. Whiting (1967) investigating sensory deprivation in the pre-school child has amassed the findings of a considerable number of workers in this field. Sensory deprivation—including that of tactual and kinesthetic experiences—in the young, has been shown to result in social inadequacy. Work in this field was originally carried out with monkeys (Harlow 1962), but later with institutionalised babies. Results are reported by Davis (1940, 1947); Ribble (1944); Spitz (1958, 1962); Bowlby (1960) and Morris & Whiting (1971).

A need to relearn the 'lost language of the body' is established by Schutz (1967) who expounds the rationale underlying the Esalen movement of which an important aspect is the emphasis on 're-embodying procedures' involving communication through touch.

D.2. The auditory channel

The capacity to communicate through language, as distinct from sound, evolved only in Man.

Theories regarding the origin or words and language are many. Langer (1942) speculates on the origins of speech and its function in communication and distinguishes between the 'vocal signs' of animals and speech as a human act of symbolic transformation; Langer (1962) extends her original thesis of speech as symbol-making and symbol-using activities. Hockett (1960) proposes that Man's ability to communicate by means of speech shares many features with communication systems of animals and develops his theory that speech actually arose from these more primitive origins. Redfern (1961a, 1961b, 1962a, 1962b, 1963, 1965) provides a valuable overview of several of the classic theories of the origin of speech, comments fully on Paget's 'gesture' theory, and makes explicit 'words as action', the association of speech and action, sound and 'effort' and sound and emotion. Davitz (1964) within an analysis of the psychology of non-verbal behaviour examines the role of oral communication in terms of inflexion, speed and tone of voice in the transmission of spoken language. Arunguren (1967) deliberately omitting reference to the genesis of language, subjects it to analysis in terms of 'transmitter', 'transmission' and 'receiver'. A comprehensive 'grammar of modern communication' is elicited including forms such as the non-linguistic, artistic, socio-economic and electronic.

Sociologically the importance of language in communication theory is fully documented. Argyle (1967, 1969) delineates such relevant areas as the universal structure of language, language in social interaction, the development of language in children, the paralinguistic or non-verbal element, and verbal behaviour as well as the link between speech and non-verbal signals and their alternative use. Bernstein (1961, 1965, 1970) proposes a connection between modes of speech which he terms 'formal' or 'public' (later to become known as 'elaborated' and 'restricted'), social class and educational performance. Robinson & Rackstraw (1967) support this hypothesis and Bernstein & Henderson (1969) enlarge the original theory. Of interest to those concerned with movement behaviour is the implied importance attached to the non-verbal cues accompanying each speech mode, and the relationship of the verbal and the gestural aspects of communication—an area in need of investigation. 'Rules of talk' and conventions in ritual interchange are proferred by Goffman (1967)—a unique contributor in the analysis of social interaction. Two original essays by Riesman on oral communication and tradition, and Carpenter and McLuhan on 'acoustic space' are found in Carpenter & McLuhan (1960).

Menuhin (1969) contends that sound is 'the supreme element in communication' and develops the theory that 'communication through music begins where words end.' Sound, whether vocal or instrumental, is produced by movements of the body and the association of sound with emotional mood and action originates in antiquity. The Greeks attributed to the Lydian, Dorian and Phyrigian rhythms a specific significance. Dalcroze (1921, 1930) based his musical theories on the relationship of movement to sound and the importance of the 'feel' of rhythms. Cage (1966) identifies sound *with* movement and his writings on 'music-dance composition' contain his revolutionary theories concerning the comparative nature of the two media. A comparison of corresponding features of music and movement has yet to be developed and literature dealing explicitly with the subject is scarce. Langer (1942) devotes a chapter to 'Significance in Music' and includes music as 'communicated feeling', 'unconsummated symbol' and 'language'. References can be found to 'vocal, instrumental and visual rhythms' (Leonard 1954a, 1954b); motor elements and rhythmic elements' (Chilkovsky 1958); 'rhythms and harmonies of movement and sound' (Laban 1958); 'movement-anatomy and musical acoustics' (Laban 1959) and 'music as audible movement and movement as inaudible sound' (Laban 1960). A recent addition to the literature on this subject is Bruce (1970).

D.3. The visual channel

Communication refers to the transmission of messages from one place to another and in the visual field this is occasioned by the use of signal, sign or symbol. Sign stimuli in the establishment of fixed action patterns in animals is well documented in Brown (1965). Distinction is made between signal, sign and symbol by Reid (1961) and between sign and symbol by Langer (1942) and Cassirer (1944); Metheny (1968) conceives of symbol as part of the human world of meaning.

Movement expression was studied extensively by Delsarte (1811-1871) and reported by Shawn (1954). The work of psychologists Allport & Vernon (1932) and Downey (1932) which aimed at determining relationships between personality and expressive movement, although naive in movement analysis, contributed to knowledge in this area. Correlation between movement behaviour and personality is implicit in Laban's 'effort' and 'spatial' theories and Ullman (1970) makes these explicit. Whereas psychologically the approach to the problem is unsophisticated the means of analysis is relatively sound. Lamb (1965), influenced by the work of Laban in this field, distinguishes between 'posture' and 'gesture' in the observation of movement behaviour and applies his observation strategies to personality assessment and recruitment procedures in industry (Lamb 1966, 1967).

One of the most expressive areas of the body is the face and Kendon, Osgood, Ekman & Friesen, and Birdwhistell have attempted to categorise face signals; Argyle (1967, 1969) summarises these results and develops his own theories of 'looking', referring to the 'kinesic dance', 'direction of gaze' and 'eye-contact' and the significance of these as stated in Goffman (1963) and Chance's 'cut-off' theory.

Goffman (1955, 1956, 1957, 1961, 1963a, 1963b) is a prolific writer on the subject of 'managing the self' in behavioural terms. His theory of social interaction as a 'performance' or 'information game' is a convincing one and his reference to gestural inconsistencies in expressions 'given' and 'given-off' augment Laban's concept of 'shadow movements' (Layson 1969). To some degree this theory also correlates with Bateson's (1956) concept of 'double-bind' (the contradiction of verbal utterances and non-verbal behaviour), Ekman & Friesen's (1967) 'leakage' theory, and Birdwhistell's (1960) 'meta-incongruency'.

Layson (1969) hypothesises other similarities between the theories of Goffman and Laban and selects areas of work of several notable social scientists which have relevance for movement study.

The emergence of leaders in both formal and informal groupings, personality correlates with leadership and leadership 'style' has attracted researchers in social psychology such as Merei, Bales and Kahn & Katz, and findings of these writers and others are collated in Newcomb, Turner & Converse (1952); Krech, Crutchfield & Ballachey (1962) and Argyle (1967, 1969). Weber's concept of 'charisma' has been developed within social theory and of relevance to movement study is the connection between Birdwhistell's (1960) 'kinesic leaders' and Lamb's (1965) 'charismatic movement leaders'.

Within the field of social psychology studies of disturbed behaviour resulting in disruption of communication include descriptions of abnormal movement patterns as symptomatic of mental disorder. Literature focusing on this area is summarised in Argyle (1969) and specific references include authoritative research findings on depression, schizophrenia, hysteria and autism. The latter, characterised by failure to communicate through speech, inferior auditory discrimination, preference for taste and touch, bizarre movement patterns and ritualised gestures (Savage 1968), should prove of particular interest since it would seem that non-verbal behaviour will prove the most rewarding area for diagnosis and treatment in such cases and therefore of significance for the student of human movement.

The specific function of movement within the visual arts as communicating media is a relatively undeveloped area, so too is the possible relatedness of kinesthetic sensitivity, visual perception and aesthetic connotation, and whereas there is speculation regarding these particular issues (Horst & Russell (1961) Logan (1964); Mettler (n.d.)) both areas lack substantive enquiry. Writers concerning themselves with art theory imply relationships (Cary (1958); Gabo (1962); Stanislavski (1963); de Sausmarez (1964); Champigneulle (1967); Menuhin (1969); Melnitz (1969) but there have been few authoritative attempts to make explicit associations. Dance is an exception and part of Section N is devoted to work in this area.

Explicit reference is however made in Leonard (1952a, 1952b, 1953); Robertson (1954, 1963); Pease (1959) and Hinley (1966). Students interested in developing this theme might discover starting points in these conjectures.

Laban (1950, 1960, 1966) views movement as an integral factor in all art forms and his theories relating to dance, drama and mime have been studied and implemented by experts in these arts.

Source Texts

ALLPORT G.W. & VERNON P.E. (1932). "Studies in Expressive Movement". London: Macmillan.

ARANGUREN J.L. (1967). "Human Communication". New York: World Univ. Library.

ARGYLE M. (1967). "The Psychology of Interpersonal Behaviour". Harmondsworth: Penguin.

ARGYLE M. (1969). "Social Interaction". London: Methuen.

BROWN R. (1965). "Social Psychology". Illinois: Free Press.

CASSIRER E. (1944). "An Essay on Man". Yale: Univ. Press.

DAVITZ J.R. et al (1964). "The Communication of Emotional Meaning". New York: McGraw Hill.

GOFFMAN E. (1956). "The Presentation of Self in Everyday Life". Edinburgh: University Press.

GOFFMAN E. (1961). "Asylums". New York: Anchor Books.

GOFFMAN E. (1963a). "Behaviour in Public Places". Illinois: Free Press.

GOFFMAN E. (1963b). "Stigma". New Jersey: Prentice-Hall.

GOFFMAN E. (1967). "Interaction Ritual". Chicago: Aldine.

HALL E.T. (1959). "The Silent Language". New York: Premier.

HALL E.T. (1966). "The Hidden Dimension". New York: Doubleday.

HEBB D.O. & THOMPSON W.R. (1954). The social significance of animal studies. In G. Lindzey (Ed.). "Handbook of Social Psychology". Vol. 1. Cambridge Mass: Addison-Wesley.

KRECH D., CRUTCHFIELD R.S. & BALLACHEY E.L. (1962). "Individual in Society". New York: McGraw Hill.

LAMB W. (1965). "Posture and Gesture". London: Duckworth.

NEWCOMB T.M., TURNER R.H. & CONVERSE P.E. (1952). "Social Psychology. A study of human interaction". London: Routledge & Kegan Paul.

PARRY J. (1967). "The Psychology of Human Communication". London: University Press.

Journal Overviews

CARROLL J. (1971a). A survey of some research findings from the social sciences that relate to movement behaviour. *Anstey Mon.,* 1 (Anstey College of P.E. England).

CARROLL J. (1971b). Perception and deception in social interaction. Anstey Papers No. 3. (Anstey College of P.E. England).

LAYSON J. (1969). An introduction to some aspects of sociology of physical education. In Conference Report (Ass. of Principals of Women's Coll. of P.E.) Physical Education: Aesthetic and social aspects.

SAVAGE V.A. (1968). Childhood Autism: a review of the literature with particular reference to the speech and language structure of the autistic child. *Brit. J. Disorders. Comm.,* 1, 75-88.

WHITING H.T.A. (1967). Significance of movement for early development. *Rem. Gym Rec. Ther.,* 45, 1479.

Specific References

(N.B. L.A.M.G.—Laban art of movement guild magazine).

BATESON G. et al. (1956). Towards a theory of schizophrenia. *Beh. Sc.,* **1,** 251-264.

BERNSTEIN B. (1961). Social class and linguistic development: a theory of social learning. In A.H. Halsey, J. Floud & C.A. Anderson (Eds.) "Education, Economy and Society". New York: Free Press.

BERNSTEIN B. (1965). Speech and the home. In *Nat. Assn for the Teachers of English Bulletin.* **Vol. 2** No. 2, 18-21.

BERNSTEIN B. & HENDERSON D. (1969). Social class differences in the relevance of language to socialisation. *Sociology,* **3,** 1-20.

BERNSTEIN B. & BRANDIS W. (1970). Social class differences in communication and control. In W. Brandis & D. Henderson (Eds.) "Social Class, Language and Communication".

BIRDWHISTELL R. (1960). Kinesics and communication. In E. Carpenter & M. McLuhan (Eds.) "Explorations in Communication". Boston: Beacon Press.

BOWLBY J. (1960). Grief and mourning in infancy and early childhood. In "The Psychoanalytic Study of the Child". New York: International University Press.

BRUCE V. (1970). "Movement in Silence and Sound". London: Bell.

CAGE J. (1966). "Silence". Massachusetts: Institute of Technology Press.

CARPENTER E. & MCLUHAN M. (Eds.) (1960). "Explorations in Communication." Boston: Beacon Press.

CARY J. (1958). "Art and Reality". New York: Anchor Books.

CHAMPIGNEUELLE B. (1967). "Rodin". London: Thames & Hudson.

CHANCE M.R.A. (1962). The interpretation of some agonistic postures: the role of 'cut-off' acts and postures. *Symp. Zool. Soc. Lond.,* **8,** 71-89.

CHILKOVSKY N. (1958). Labanotation for the ethnomusicologists. *L.A.M.G.* **21,** 34-39.

CRATTY B.J. (1964). "Movement Behaviour and Motor Learning". Philadelphia: Lea & Febiger.

DALCROZE E.J. (1921). "Rhythm Music and Education". London: Chatto & Windus.

DALCROZE E.J. (1930). "Eurhythmics, Art and Education". London: Chatto & Windus.

DAVIS K. (1940). Extreme social isolation of a child. *Am. J. Sociol.,* **45,** 554-565.

DAVIS K. (1947). Final note on a case of extreme social isolation. *Am. J. Sociol.,* **52,** 432-437.

DAVITZ J.R. (1964). "The Communication of Emotional Meaning". New York: McGraw Hill.

DIEBOLD A.R. (1967). Anthropology and the comparative psychology of communicative behaviour. In T.A. Sebeok (Ed.) "Animal Communication—Techniques of Study and Results of Research". Indiana: University Press.

DOWNEY J.E. (1932). "The Will Temperament and its Testing". New York: World Book Co.

EKMAN P. & FRIESEN W.V. (1967). Non-verbal behaviour in psychotherapy research. *Res. Psychother.,* **3.**

FRANK L.K. (1960). Tactile communication. In E. Carpenter & M. McLuhan (Eds.) "Explorations in Communication". Boston: Beacon Press.

GABO N. (1962). "Of Divers Arts". London: Faber & Faber.

GOFFMAN E. (1955). On face-work. *Psychiatry,* **18,** 213-231.

GOFFMAN E. (1957). Alienation from interaction. *Hum. Relat.,* **10,** 47-60.

HARLOW H.F. & M. (1962). Social deprivation in monkeys *Sci. Am.,* **207,** 5, 136-146.

HINLEY J. (1966). Space/Time/ Movement. A correlation of movement and art. *L.A.M.G.,* **36,** 10-13.

HOCKETT C.D. (1960). The origin of speech. *Sci. Am.*

HOLDING D.H. (1966). "Principles of Training". London: Pergamon.

HORST L. & RUSSELL C. (1961). "Modern Dance Forms in Relation to the Other Arts". San Francisco: Impulse.

HUNT V. (1962). Cerebral palsied youngsters and body-image problems. Unpub. report to the Faculty of Un. of Calif. Los Angeles.

JOURARD S.M. (1966). An exploratory study of body accessibility. *Brit. J. Soc. Clin. Psychol.,* **5,** 221-231.

KINSEY A.C. et al (1948). "Sexual Behaviour in the Human Male". Philadelphia: Saunders.

KINSEY A.C. et al (1953). "Sexual Behaviour in the Human Female". Philadelphia: Saunders.

LABAN R. (1950). "Mastery of Movement on the Stage". London: Macdonald & Evans.

LABAN R. (1958). The world of rhythm and harmony. *L.A.M.G.,* **20,** 6-9.

LABAN R. (1959). Dance as a discipline. *L.A.M.G.,* **22,** 33-40.

LABAN R. (1960). "Mastery of Movement". London: Macdonald & Evans.

LABAN R. (1966). "Choreutics". London: Macdonald & Evans.

LAMB W. (1965). "Posture and Gesture: An Introduction to the Study of Physical Behaviour". London: Duckworth.

LAMB W. (1966). Recruiting and assessing from the evidence of movement behaviour. *L.A.M.G.,* **37,** 16-20.

LAMB W. (1967). The work of Warren Lamb—B.B.C. T.V.'s 'Tomorrows World'. *L.A.M.G.,* **39,** 48-51.

LANGER S. (1942). "Philosophy in a New Key". New York: Mentor.

LANGER S. (1962). "Philosophical Sketches". New York: Mentor.

LEONARD M. (1952a). Architecture and dance. *L.A.M.G.* **8,** 21-25.

LEONARD M. (1952b). Architecture and dance. *L.A.M.G.* **9,** 24-30.

LEONARD M. (1953). Architecture and dance. *L.A.M.G.* **10,** 18-22.

LEONARD M. (1954a). Rhythm and dance. *L.A.M.G.,* **12,** 29-33.

LEONARD M. (1954b). Rhythm and dance. *L.A.M.G.,* **13.**

LEONARD M. (1955). Rhythm and dance. *L.A.M.G.,* **14,** 35-42.

LEONARD M. (1956). Rhythm and dance. *L.A.M.G.,* **17,** 45-51.

LOGAN G.A. (1964). Movement in art. *Quest Monog.,* **2,** 43-45.

MELNITZ W. (1969). The theatre and its continuing social function. "The Arts and Man". Paris: UNESCO.

MENUHIN Y. (1969). Music and the nature of its contribution to humanity. "The Arts and Man". Paris: UNESCO.

METHENY E. (1968). "Movement and Meaning". New York: McGraw Hill.
METTLER B. (n.d.) "Nine Articles on Dance". Boston: Mettler Studios.
MORRIS D. (1967). "The Naked Ape". London: Cape.
MORRIS D. (1969). "The Human Zoo". London: Cape.
MORRIS P.R. & WHITING H.T.A. (1971). "Motor Impairment and Compensatory Education." London: Bell.
NISSEN H.T.W. (1951). Social behaviour in primates. In C.P. Stone (Ed.) "Comparative Psychology". 3rd ed. New York: Prentice-Hall.
PEASE E. (1959). Stage movement. *L.A.M.G.,* **23,** 39-43.
PORTER E.R., ARGYLE M. & SALTER V. (1969). "The Role of Spatial Proximity in Social Interaction". (roneoed) Oxford Instit. of Exp. Psych.
REDFERN H.B. (1961a). Movement and speech. *L.A.M.G.* **26,** 32-38.
REDFERN H.B. (1961b). Movement and speech. *L.A.M.G.* **27,** 31-36.
REDFERN H.B. (1962a). Movement and speech. *L.A.M.G.* **28,** 40-48.
REDFERN H.B. (1962b). Movement and speech. *L.A.M.G.,* **29,** 28-37.
REDFERN H.B. (1963). Movement and speech. *L.A.M.G.,* **31,** 22-30.
REDFERN H.B. (1965). Movement and speech. *L.A.M.G.,* **34,** 38-42.
REID L.A. (1961). "Ways of Knowledge and Experience". London: Allen & Unwin.
RIBBLE M.A. (1944). Infantile experience in relation to personality development. In J.McV. Hunt (Ed.) "Personality and the Behaviour Disorders Vol. II". New York: Ronald.
ROBERTSON S.M. (1954). On modelling blindfold. *L.A.M.G.* **13,** 56-62.
ROBERTSON S.M. (1963). "Rosegarden and Labyrinth". London: Routledge & Kegan Paul.
ROBINSON P. & RACKSTRAW S.J. (1967). Variations in mother's answers to children's questions as a function of social class, verbal intelligence test scores and sex. *Sociology,* **1,** 259-276.
DE SAUSMAREZ M. (1964). "Basic Design: The Dynamics of Visual Form". London: Studio Vista.
SCHULTZ D.P. (1965). "Sensory Restriction". New York: Academic Press.
SCHUTZ J. (1967) "Joy". New York: Grove Press.
SHAWN T. (1954). "Every Little Movement." New York: Dance Horizons.
SPITZ R.A. (1958). Hospitalism: an enquiry into the genesis of psychiatric conditions in early childhood. In "The Psychoanalytic Study of the Child, Vol. 1". New York: International Univ. Press.
SPITZ R.A. (1962). Autoerotism re-examined. The role of early sexual behaviour patterns in personality formation. "The Psychoanalytic Study of the Child, Vol. XVII". New York: International Univ. Press.
STANISLAVSKI C. (1963). "Creating a Role". London: Geoffrey Bles.
ULLMAN L. (1970). Movement as an aid to the understanding and development of personality. In "Conference Report Movement, Dance, Drama". University of Hull.
WILLIAMS J.L. (1963). Personal space and its relation to extra-version-introversion. Unpub. M.A. thesis Univ. of Alberta.
ZUCKERMAN S. (1932). "The Social Life of Monkeys and Apes". London: Routledge & Kegan Paul.

4 E
TOPIC AREAS - DEVELOPED

AESTHETICS AND HUMAN MOVEMENT

by JUNE LAYSON

Introduction

a) *Aesthetics* is a relatively well-established branch of *philosophy* in which concepts such as 'the work of art', 'beauty' and 'the spectator response' are central. Saw & Osborne (1968) discuss the various definitions of aesthetics and concomitant theories within a historical continuum, and Osborne (1968a) traces the development of the major 'isms' (naturalism, romanticism, etc.) in the area. Wollheim (1968), focusing attention on the art object, examines a wide range of issues, while Reid (1969b) develops a single theme of 'embodiment' as the 'central defining characteristic of the aesthetic'.

    Studies in aesthetics are rarely concerned with human movement per se although some aestheticians do make direct references. Langer (1942, 1953, 1957, 1962) writes on dance as an art form, and Reid (1969a, 1970) considers the possible aesthetic content of physical activities, distinguishing these from dance as the art form of movement.

b) *Perception* is generally accepted as a vital element in aesthetics and Mace (1968) sees an inter-relationship between aesthetics and *psychology.* Individual psychologists, notably Arnheim (1956; 1967; 1969), have contributed to the study of aesthetics though most of the research is concerned with static two-dimensional forms; human movement receiving scant attention. Relevant psychological studies (behaviourist, gestalt and phenomenological) are referred to in sections E.1, E.2 and E.3.

c) *Cross-cultural differences* in movement behaviour (see Section J), art forms and perception are well documented though little or no work has been directed towards cultural and sub-cultural aspects of movement and aesthetics. A few anthropologists, sociologists and others, however, have studied cultural norms in one area of movement, usually dance, and have made the aesthetic content salient, though not important, but an overall picture has yet to emerge. Kealiinohomoku (1970) examines ballet as ethnic dance, Kurath (1960) outlines the field of dance ethnology and provides a comprehensive bibliography, (Kealiinohomoku & Gillis, 1970 annotate a Kurath bibliography), Boas (1944), Mead (1946), Rust (1969), Shipman (1968) and Waterman (1962) discuss dance within societies, and Mead (1928, 1930, 1950) shows awareness of aesthetic content in her dance descriptions. Daniels (1969) refers to Cozens' and Stumpf's notion of sport as art, giving this a cultural setting and Denney (1969) describes the aesthetic element in spectatorial forms of sport within the American culture. Schilder (1950) briefly notes cultural influences on the concept of the beauty of the body in rest and in motion and Smith (1968) refers to different cultural and sub-cultural values in the appreciation of movement.

d) *Physical education,* perhaps in the present context best termed an applied area of human movement studies (Abernathy & Waltz 1964), also is of relevance to the study of aesthetics and movement. Phenix (1964) delineates six 'realms of meaning' in a structure of knowledge subsuming physical education and dance under 'esthetics', the third realm. Anthony (1968) proposes that sport and physical education should become a means of aesthetic education and Carlisle (1969) argues that physical education, including dance, can only be justified in the school curriculum on aesthetic grounds, a point disputed by Adams (1969). Aesthetic education through movement, particularly dance, is implicit in statements on 'Objectives of Physical Education' (Schools Council, 1968) and 'The Concept of Physical Education' (Morgan et al 1970).

It is debatable whether physical education should include or be distinct from dance but the latter, considered in its own educational right, is seen as part of aesthetic education by Carroll & Lofthouse (1969), Layson (1970), Reid (1969a, 1970) and Russell (1965, 1969, 1970). Justifications of dance, particularly modern dance, as contributing mainly to aesthetic education are comparatively recent, Laban (1948) presenting a multiplicity of claims of which aesthetic education through dance is one. The foregoing texts are British and a difference of emphasis in dance in education in the U.S.A. should be noted, for American writers from H'Doubler (1940) onwards have invariably justified dance in education on aesthetic criteria.

Relevant physical education and dance texts are referred to in Section E.1 particularly, as well as in E.2 and E.3.

e) Apart from the studies in aesthetics, psychology, anthropology, sociology, physical education and dance in education which are relevant to the area 'aesthetics and human movement', pertinent information exists in *general philosophy, literature, works on dance in the theatre* and *writings on sport.*

Philosophers have often used the aesthetic aspect of movement, particularly dance, as a paradigm or allegory in the development and exposition of their theories; Duncan (1928) terms Neitzsche 'the first dancing philosopher' and Prall (1929) cites Plato, Spinoza and Ellis.

British literary works have yet to be examined systematically for their aesthetic and movement content. Priddin (1952) has collected and collated references to dance in French literature including the works of Voltaire, Noverre, Gautier, Mallarmé, Valéry and Bergson. Many of these extracts are discourses on the aesthetics of dance.

The literature on dance in the theatre is vast, varying in its scholarly content. General annotated bibliographies on dance are those of Beaumont (1929, 1966, 1968), Belknap (1948-1963), Forrester (1968) and Magriel (1936). Among modern American writers on dance in the theatre who take cognisance of an aesthetic content are Armitage (1966), Cohen (1966), Horst & Russell (1961), Humphrey (1959), Kirstein (1967), Martin (1933, 1936, 1939), Rochlein (1964), Sorell (1966) and Terry (1956). Articles in the annual Impulse magazines (1952-1970) are also useful.

The literature on sport as an art form outside the educational context is speculative rather than informative. Anthony (1968) gives a short bibliography; other references are more appropriately placed in Section E.1.

f) *Historically* it is possible to outline the development of the notion of an aesthetic content to movement, although much evidence is as such uncollated. Movement and dance historians tend to attribute aesthetic awareness to the work of certain movement theorists and in various dance styles, yet Osborne (1968a) asserts that 'formal aesthetics' is a recent phenomenon and that throughout man's history 'aesthetic concepts' have been implicit, 'the aesthetic function seldom, if ever, stood alone and autonomous'. Thus, historically, aesthetic concepts of movement may be inferred with hindsight rather than extracted from contemporary source materials.

Ancient Greek attitudes to aesthetics and the body are noted by Osborne (1968a) and to aesthetics and movement by Sachs (1937), H'Doubler (1940) and Hinks, Archbutt & Curl (1971). Lawler (1964), in a definitive text, examines the aesthetic element in Greek dance through the contemporary works of Plato, Aristotle, Athenaeus and Xenophon.

Weaver's (1721) theories of movement contain references to beauty of motion (Fletcher, 1960, credits him with 'aesthetic concepts') and Noverre (1803) also considers harmony of movement. In Delsarte's (Shawn, 1963) movement theories the aesthetic element is evident and Delsartean techniques gave rise to the 'aesthetic gymnastics' of Mackaye (Shawn 1963) and the 'harmonic gymnastics' of Stebbins (1886, 1893). The Eurhythmics of Jaques-Dalcroze (1921, 1930) has an aesthetic basis as does Bode's (1931) 'Expression Gymnastics'. Duncan, Ginner, Laban, Ling, Morris and Streicher (see Section N) all refer to the aesthetic aspects of movement

American views on the development of an aesthetic concept of movement can be found in Brown & Sommer (1969), Kirstein (1967) and Maynard (1965).

g) It is evident that the area of aesthetics and human movement is diffuse, lacking coherent structure. The following three sections on *'The Activity'*, *'The Performer'* and *'The Spectator'* are presented in an attempt to provide a unified framework within which existing studies on aesthetics and movement can be understood and future study undertaken. The sections are not arbitrary. Adams (1969) notes that participant and observer differences are relevant. Kaelin (1968) indicates that while his work is focussed on the 'sensitive viewer', Minkowski (1967) is interested in the way 'the event is performed'. Kaelin adds 'the middle ground is the event itself'.

E.1 The Activity

a) The intention in this section is to examine various *movement activities* from an *aesthetic viewpoint;* an expedient approach since activities exist only when performed. Two caveats are made in this context, firstly, the terms 'movement' and 'activity' do not exclude stillness, secondly, the phrase 'movement activities' encompasses all overt human behaviour. Stillness, both as the antithesis of movement and as part of movement, is a relevant concept here. Aldrich (1971) and Clark (1956) are concerned with the human body in stillness, particularly in the visual arts, but when, as Cohen (1966) reports, a 1957 Paul Taylor dance consisted entirely of Taylor standing motionless, the relationship is patent. Movement as human behaviour includes everyday activities of eating, walking as well as the highly complex, organised and distinctive activities of a game of football or a performance of a classical ballet.

Work in this area can be divided into two broad categories; firstly (b) attempts to provide an aesthetic account of movement in general, secondly (c) literature concerned with aesthetic accounts of separate movement activities.

b) (i) *Concept approach*—Some authorities have taken concepts of 'beauty' and 'grace' and examined their relationship to movement. Kirstein (1967) has an interesting section on grace in movement in which he refers to 17th and 18th century authors. Richter (1967) quotes the 19th century French aesthetician Guyau who considers that the movements of men both at work and play become beautiful when they contain 'strength, harmony, rhythm, order, grace and are expressive of an admirable inner life'. Lamb (1965) devotes a chapter to 'Beauty in physical behaviour' explaining beauty in manual labour, the act of walking, the athlete performing, the dancer dancing in terms of 'posture' and 'gesture' with 'shape/effort matching'. Streicher (1965) proposes that since graceful movement is 'inherent in all free living animals . . . it is or should be inherent in ourselves'. In contrast Mason & Ventre (1965), commenting on the objective nature of physical education, state that 'functional efficiency may conflict with graceful movement'; this is later qualified 'functionally applied power in a mechanically efficient movement is a subtle form of gracefulness'.

A criticism of the application of concepts of 'grace' and 'beauty' to movement in general is that two important variables, the *performer* and the *observer,* are often discounted. Lamb's (1965) thesis does allow for these factors and he also acknowledges a sex variable, discussing male notions of beauty in the movement of females.

Other writers start from an aesthetic concept or premise and proceed to an examination of the area of movement. Valéry (1960) considers that the aesthetic nature of dance qualifies it as an art form, while physical acts in normal life are non-aesthetic or 'merely transitional'. Henry (1965) distinguishes between those 'motor performances' of everyday life and 'other motor performances yielding aesthetic values'. Smith (1968) claims that 'all human movement, for whatever purpose it is performed, is intimately related to aesthetics'; and although she distinguishes between 'art forms and functional forms' (in the former the intention is to meet aesthetic criteria, in the latter aesthetic criteria may be met) she maintains that in their evaluation the 'same general principles of aesthetics can be employed'.

(ii) *Continuum/Spectrum approach*—Several attempts have been made to array movement activities according to their aesthetic content. In a description of his system Bode (1931) places gymnastics between 'controlled, goal aiming movement, sport' and 'irresible motion, the dance'. Hinks, Archbutt & Curl (1971) write 'naturally the dance occupies a central role in aesthetic aspects of movement although the aesthetic element in gymnastics, athletics and sport generally . . . is of vital importance'. Anthony (1968) classifies sport and physical education on the basis of an 'aesthetic element'. He considers that 'most sports have the straight-forward, uncomplicated objective of scoring goals . . . any aesthetic element is incidental to the main aim', that other activities, like gymnastics and diving, involve in part an aim which is aesthetic and that dance activities are entirely within the aesthetic realm. Anthony poses the question whether sports 'in close contact with nature' (climbing, ski-ing, etc.) offer aesthetic experiences. Reid (1969a, 1970) provides the most sophisticated and detailed spectrum account. He acknowledges that the subject matter of aesthetics is wider than the arts and then discusses those aspects of movement which have aesthetic elements but are not art forms. Reid illustrates the differences between art and games and sport through an analysis of rules, styles and techniques, examines the 'borderline cases' such as gymnastics and skating, and discusses dance as the art form of movement.

The continuum/spectrum approach seems promising, although as yet is cannot accommodate those movement activities which are relatively un-organised such as playing or walking (but presumably when the latter becomes competitive and subject to rules it can be considered, even though its aesthetic content might be impaired).

(iii) *Model approach*—In physical education efforts to clarify or delineate the field often result in the construction of models some of which include aesthetic aspects. Streicher (1970) has a simple two-fold division of 'acrobatic forms', e.g. vaulting, tumbling, diving, and 'beautiful forms' which express a 'spiritual content in movement' and include all types of dance. Mason & Ventre (1965) place 'aesthetically pleasing activities' which give 'insight into spiritual values' in a sub-division of teaching aims, but along with Streicher their account is open to criticism on the grounds that the aesthetic and the spiritual are not necessarily interdependent. Brown & Cassidy (1963) and Brown (1967) structure 'the knowledge of human movement', one of its subjects being 'expressive form, the art of movement'. The conceptual model of Kenyon (1968) characterises 'physical activity as a sociopsychological phenomenon' and is six-dimensional, the fourth sub-domain being 'physical activity as an aesthetic experience'.

(iv) *Phenomenological approach*—This is a sophisticated approach which is gaining ground in the U.S.A. yet remains undeveloped in Britain. Kaelin, Kleinman, Metheny and Sheets have made significant contributions to this area in which 'the experience of moving', 'the act' and 'the meaning of the act' are key concepts. Kaelin (1964; 1968) discusses several movement activities in order to make their aesthetic content salient. He considers diving, skating and ski-ing as 'abstractions of movement, particularly locomotion, which approach fine art' and he examines the aesthetic aspects of the sex act. Kleinman (1964) uses Sartre's three dimensions of the body as a means of understanding the significance of human movement and although Sheets (1966) is primarily concerned with dance, her phenomenological approach extends to movement generally. Metheny (1965; 1968) does not term herself a phenomenologist though her work on significance and meaning in movement appears to lean in this direction, similarly her writings on meaningful movement forms seem to incline towards the aesthetic.

(v) *Terminology*—It is useful to note certain terminological confusions which arise when the aesthetic content of movement is debated and this can be demonstrated in the various meanings attached to the phrase 'the art of movement'. To Laban (1948) party games, masquerades, dance, games, ceremonies, rituals and so on all form part of the art of movement; in this extended sense of the phrase the activity would seem to be of less importance than the attitude to movement the performer brings to his performance. Phenix (1964) uses the term 'the arts of movement' to encompass all physical education activity including dance and he sees these as 'the source of esthetic meanings in which the inner life of persons is objectified through significant dynamic forms using the human body as the instrument'. Reid (1969a) is, by contrast, simple in his definition of dance as 'the *art* form of movement'. (A more comprehensive discussion of terminology in this area is in Layson, 1970.)

c)   Various forms of *sport* are often claimed to be *art forms* particularly by participants, sports journalists and spectators. Beyond such claims few justifications exist. Gray (1966) compares sport with dance improvisation, the former being 'instant art, calling for remarkable mental and kinesthetic responses and decisions, occurring continuously within the boundaries and limitations of a set of rules', and the latter also involving 'spontaneity of movement and instant reaction to stimuli'. Thus, to Gray, sport is 'art on a different plane', for it is 'more spontaneous, more uncertain, perhaps more irregular' than a structured art and it is characteristically unpredictable to the end. Friedenberg (1967) sees sport as an 'applied art' similar to dance in that the body is the medium, similar to architecture in that rules or laws have to be obeyed, but in contrast to the extemporaneous sport of Gray, Friedenberg considers that *athletics* is among the most formal of the arts.

Those who write on the *aesthetic content of sport* tend to substantiate arguments. Kaelin (1968) analyses the structure of sports to explain their aesthetic content. He states: 'the rules of the game have been set up to maintain aesthetic quality' and although he considers that the outcome of the game is 'aesthetically irrelevant' he asserts that 'the desire to win is never aesthetically irrelevant'. Kaelin concludes that competitive sport 'reaches its aesthetic heights when the victor narrowly surpasses a worthy opponent'. The characteristic sports struggle is given a new facet by Dumazedier (1968) who regards it as 'aesthetic transposition of war'. In contrast to Kaelin and Dumazedier, Smith (1968) contends that methods of scoring in sport may hamper the aesthetic content. Reid (1970) answers the question 'are some sports arts?' in the affirmative in so far as a few sports (e.g. ice-skating) have 'artistic elements', but in games and sports generally he considers that 'the aesthetic enjoyment is parasitic upon the central games-purpose of the game'. Slusher's (1967) opinions on sport could be regarded as implicitly aesthetic though his ontological approach and conclusions are criticised by Broekhoff (1968). Rodionov (1966) also subscribes to the notion of an aesthetic element in sport but not in professional forms.

Although the aesthetic content in *diving* has been traditionally acknowledged, the claim that *aquatics* is an art form is comparatively recent. Rochman (1965) credits Carrell in 1940 with the first exploration of 'movement in water as a potential art form'. Gray (1966) considers synchronised swimming as an acquatic art, Maver (1968) notes its aesthetic elements and Haueter (1965) discusses composition in aquatics.

Gymnastics is one of the few activities that Reid (1970) accepts as 'incorporating artistic elements'. The International Gymnastics Federation (1970) guides female judges in their assessment of 'rhythm and beauty' and 'aesthetic beauty', but Gray (1966) suggests that gymnastics has not yet progressed beyond the 'aesthetic enjoyment of mere skill' although it is capable of development to an art form.

To many the structure and aim of *boxing* is aesthetically displeasing. Rodionov (1966) hypothesises that the competitive element, particularly in single combat sports, may mask aesthetic content, but he concludes that boxing has unique aesthetic qualities. James (1970) notes that the concept of boxing as the 'noble art of self-defence' is well over a century old.

Mace (1968) refers to *cricket, fencing* and *fishing* as activities often described as arts, however, the literature on these aspects is mainly conjectural and unsystematic.

Of all movement activities *dance* is the one universally accepted as an art form, although whether dance subsumes *mime* (Aldrich 1963, Duncan 1928, Langer 1953) or is itself part of the arts of movement alongside *music, drama* and the *visual arts* (Hinks, Archbutt & Curl, 1971) and *motion pictures* (Borodin 1945) is debatable. (Mawer 1932 and Lawson 1966 provide basic texts on mime, with bibliographies.) Dance is often typified as the 'mother of arts' (Ellis 1923, Collingwood 1938, Sachs 1937) while to Mettler (1947) it is the 'integrating factor among all arts'. Martin (1933) explains the reasons underlying the exclusion of dance from the fine arts, Aldrich (1963) terms dance an 'impure' or minor art' because it is a hybrid of music and sculpture and to Arnheim (1956) dance can only be a secondary art since it 'relies on the given form and functioning of the human body'. Dance is not easily categorised within art (various classifications are examined by Langer, 1953); perhaps this underlines its uniqueness, as suggested by Wollheim (1971) in his discussion on the differences in or between the arts.

As an art form dance is multiplex consisting of interacting elements of ballet, ballroom, ethnic, folk, historical, modern, stage, social, etc. Each 'sub-dance form' has its own literature of varying standards but as Hall (1967) notes 'the literature on the aesthetics of dance is sparse'. The bibliographies and books on dance in the theatre mentioned in the Introduction (e) provide background material for the study of *aesthetics* and dance, though the aesthetic nature of dance is mainly implied or only discussed briefly. Other authors whose work includes reference to aesthetics and dance are H'Doubler (1940), Johnston (1965), Mettler (n.d.), Sachs (1937), Sheets (1966), Vuillermoz (1958) and Wigman (1952). Although Laban's influence on dance in education and the theatre is generally acknowledged only a few of his published works (1950, 1960, 1966) could be said to contain material on aesthetics, these being mainly on composition.

There has been little application of the theory of art to dance or research into the subject in Britain. Research in the U.S.A. on the *philosophy* and *aesthetics of dance* is listed in A.A.H.P.E.R. (1968) and Pease (1964). Langer (1942, 1953, 1957, 1962) is the main contributor to the literature on the *aesthetics of dance*, particularly in the exposition of the concepts of 'self' and 'conceptual' expression and 'actual' and 'virtual' gesture. However, Langer stresses certain areas, such as the theory of expression, and an overall systematic and comprehensive view of the aesthetics of dance is not yet available. Kaelin (1964) and Reid (1970) are critical of some aspects of Langer's work and also contribute to the literature (Kaelin 1968 on the 'autosignificance' of dance,

Reid on themes such as 'programme' and 'pure' dance).

E.2. The Performer

a) In this section the focus of attention is the *performer's aesthetic experience of movement*. In this context 'performer' is used to indicate the person moving, the subject of the movement activity. In structured movement activities the performer tends to be denoted by that activity—an archer, a golfer, etc. In the art form(s) of movement the performer, as artist, may be creative, re-creative or interpretive (Reid 1961) and his role is crucial (see Wollheim's 1968, discussion of the physical-object hypothesis). As yet the 'performer dimension' in the study of aesthetics and movement is unstructured and lacks research. The sub-headings used indicate themes around which existing work and hypotheses can be clustered.

b) *Skill/Technique.* Literature on motor skill acquisition occasionally implies that when the skill is learned or habitual, opportunities arise for aesthetic appreciation, e.g. *after skill refinement, attention can turn to the aesthetic* (Mason & Ventre 1965), or beauty in skilled action is 'an outcome of . . . mastery of the physical properties of all elements in the situation', (Metheny 1965). It could be hypothesised that 'aesthetic insight' may occur earlier, even in random trials, being both concurrent and subsequent. If this is so, there would be important motivation and learning implications in some motor skills. The 'technique v. creativity' controversy in dance education may be of relevance here. Similarities and differences in the concept of technique in sport and art are commented upon by Reid (1969a).

c) *Kinaesthetic perception.* Some attempts have been made to explain the *performer's aesthetic experience* of his movement in terms of *kinaesthetic perception.* Mace (1968) comments on originality and creativity in movement from a behaviourist viewpoint. Using gestalt theory Arnheim (1969) examines the performer's experience as 'kinesthesia' and the notion of 'resonance of movement' would seem to be impressive upon the performer as well as the spectator. In addition, Arnheim (1967) discusses 'isomorphism' ('the structural similarity or correlated processes occurring in different media') from the performer's experience, although this is essentially a performer/spectator situation. The performer/spectator relationship is also vital in phenomenology. Such an approach is taken by both Kaelin (1964) who sees the experience of the performer as 'kinaesthetic reaction', and Kleinman (1964), who describes the performer's absorption in the act mainly in terms of Sartre's first dimension.

Schilder (1950) is particularly interested in the link between 'expressive movement' and 'body image', he considers that the performer deliberately 'dissolves or weakens' the 'rigid form of the postural model of the body' and he comments upon the inter-relation between 'muscular states' and 'psychic attitudes'. H'Doubler (1940) stresses the importance of 'strong motor imagery' in the performer's experience of movement. Metheny and Ellfeldt (1961) and Metheny (1965) structure the movement experiences perceived by the performer as 'kinecept', 'kinestruct' and 'kinesymbol'. Smith (1968) coins the term 'inner aisthetikos' for the performer's 'own proprioceptively perceived performance'.

d) *Individual differences.* Implicit in much of the literature is the idea that aesthetic movement experiences are private and/or peculiar to the performer (Mettler n.d., Reid 1961, Rodionov 1966). To Lange (1970) 'aesthetic pleasure in dance is derived mainly from the feeling of balanced "effort capacity" ', a reference to Laban's theory of effort. Mason & Ventre (1965) postulate that an individual's aesthetic movement appreciation may be retained even to the extent of becoming a 'source of solace in old age'. This point raises the relation of growth and de-generation to the aesthetic appreciation of movement, seemingly a promising area for research. Reid (1969a, 1970) considers the performer's aesthetic appreciation within the framework of the activity, so that while a dancer may 'contemplate' and assume the 'aesthetic attitude' during his performance, a games-player so doing may jeopardise his game—his appreciation would be retrospective. The act of composing a dance involves aesthetic appraisal. Composition as a process is increasingly well-documented (see bibliographies and recent publications on dance in the theatre and education) and insight into individual differences within this process can be gained from choreographers and dance-biographers (Cohen 1966; Gray 1969; Wigman 1952, 1966; Wooten 1962, 1965).

E.3 The Spectator

a) A further dimension is added to the study of aesthetics and human movement when the *spectator's aesthetic attitude* is considered. Neither the term 'spectator' nor any of its synonyms is entirely apposite since degrees of passivity or aloofness might be inferred. Most writers on this aspect claim that aesthetic appreciation of movement by a spectator is one of involvement in a situation in which the mover becomes the object of attention.

Literature on aesthetics pays considerable attention to the aesthetic attitude generally; Osborne (1970) presents a historical overview of the development of this and related notions. However, the particular problems arising from the observation of movement and the nature and manner of its appreciation have yet to be examined in detail. The sub-headings used provide possible points of entry in a preliminary study of the area.

b) *Aesthetic attitude.* Succinct references to the 'aesthetic situation', the 'aesthetic attitude' and the 'aesthetic object' in relation to sport are made by Reid (1970); other related concepts are those of 'aesthetic identification' (Berenson 1948), 'aesthetic experience' (Koestler 1964), 'kinaesthetic sympathy' (Martin 1933), 'bodily resonance' (Prall 1929) and 'psychical distance' (Reid 1961).

For information on specific aspects of the aesthetic attitude reference must be made to disparate disciplines. Straus (1970) examines the function of man's upright posture in aesthetic attitude and Schilder (1950) discusses the beauty of the body in motion in terms of body image. Much of Gombrich's (1962) work on visual perception and art is concerned with static two-dimensional forms, as are references to tests of aesthetic discrimination (Parry 1967) and means of aiding aesthetic appreciation (Machotka 1970), but these have implications for the study of the aesthetic appreciation of movement. Similarly, the readings edited by Hogg (1969) which examine the aesthetic response within a psychological and physiological context have a movement relevance, e.g. Gregson (1969) on 'aesthetic response typologies' and Anastasi (1969) on tests of 'artistic aptitude'. Arnheim (1956, 1967 and 1969) devotes considerable space to discussions on movement and the spectator in which his 'principle of isomorphism' and a schema of what might be called 'kinaesthetic empathy' are prominent. The suggestion made by Read (1931) that, as with colour-blindness, aesthetic appreciation could be impaired by defective shape and mass perception would seem prima facie to be of relevance to the consideration of the aesthetic attitude and movement.

c) *Spectator typologies.* Some authors are concerned with general and specific aspects of spectating in sport and dance (Anthony 1968, Impulse 1962, Kaelin 1964, 1968, Langer 1957, 1962) others develop ideas on different types of spectators. In a discussion of the kinaesthetic sensations of the spectator Metheny (1968) distinguishes between the perception of movement sensation through miniature movement patterns and the conception of movement sensation through a recognition of movement patterns; a theory which could serve to identify different kinds of spectators. In addition Metheny offers an explanation of spectator choice in sport. Reid (1970) proposes that spectators of games and sports are of three types, partisanship, knowledge and aesthetic sensibility being important considerations. The element of knowledge and experience in spectating is referred to by Mason & Ventre (1965), and Child (1969) discusses the correlation between aesthetic judgment and education and experience in art. Denney (1969) examines 'spectatorial forms' and comments on Hemingway's attempts to characterise various types of spectators and to establish a 'code for spectators'.

d) *Films and Television.* Since movement is an ideal subject for both these media their present and possible future impact on the act of spectating cannot be ignored. Visual recordings of movement enable a performer to become spectator of his own performance and in this situation opportunities for aesthetic appreciation exist which were perhaps denied by the demands of the activity.

Reid (1969a) has noted that aesthetic appreciation of movement is often enhanced when films are run at slow speed. Arnheim (1956) considers that 'a change of velocity may not only make expressive qualities perceivable but may also modify them qualitatively'; a phenomenon he explains in relation to gravity and resistance. A similar tentative solution in terms of Laban's (1948, 1960) concept of effort could be offered in which elements would be seen to change from 'contending' to 'yielding'. The relation of aesthetic appreciation to artificial means of changing the speed of movement (including the 'fast speed comic effect') appears to be a profitable area for research.

Denney (1969) is particularly interested in the overlap between 'spectator and participant roles', though some of the sports he characterises as distinctly participatory, e.g. rock climbing and skin-diving, would already seem to have been encroached upon by television, in Britain at least. The difference between the actual spectator and the television spectator is also discussed by Denney within the context of realism.

Since there are no source texts or overviews in this area references are not categorised.
A.A.H.P.E.R. (1968). *see American Association for Health, Physical Education and Recreation.*
ABERNATHY R. & WALTZ M.A. (1964). Towards a discipline: first steps first. *Quest* II.
ADAMS M. (1969). The concept of physical education II. *Proc. Phil. Ed. Soc.* III.
ALDRICH V.C. (1963). "The Philosophy of Art". Englewood Cliffs, N.J.: Prentice Hall.
ALDRICH V.C. (1971). Art and the human form. *J. Aesth. Art. Crit.,* **XXIX**, 3.
AMERICAN ASSOCIATION FOR HEALTH, PHYSICAL EDUCATION AND RECREATION. DANCE DIVISION (1968). "Research in Dance I".
ANASTASI A. (1969). Artistic aptitudes. In J. Hogg (Ed.) "Psychology and the Visual Arts". Harmondsworth: Penguin.
ANTHONY D.W.J. (1968). Sport and physical education as a means of aesthetic education. *Phys. Ed.,* **60,** 179.
ARMITAGE M. (1966). "Martha Graham". New York: Dance Horizons.

ARNHEIM R. (1956). "Art and Visual Perception. A Psychology of the Creative Eye". London: Faber & Faber.

ARNHEIM R. (1967). "Towards a Psychology of Art". London: Faber & Faber.

ARNHEIM R. (1969). The gestalt theory of expression. In J. Hogg (Ed.) "Psychology and the Visual Arts". Harmondsworth: Penguin.

BEAUMONT C.W. (1929). "A Bibliography of Dancing". New York: Blom.

BEAUMONT C. (Ed.) (1966). "A Bibliography of the Dance Collection of Doris Niles and Serge Leslie Part I, A-K.". London: Beaumont.

BEAUMONT C. (Ed.) (1968). "A Bibliography of the Dance Collection of Doris Niles and Serge Leslie Part II, L-Z". London: Beaumont.

BELKNAP S.Y. (1948-1963). "Guide to Dance Periodicals, Vols. 1-10". New York: Scarecrow Press.

BERENSON B. (1948). "Aesthetics and History". New York: Doubleday.

BOAS F. (1944). "The Function cf Dance in Human Society". New Yorl.: Boas School.

BODE R. (1931). "Expression—Gymnastics". London: Barnes.

BORODIN G. (1945). "This Thing Called Ballet". London: MacDonald.

BROEKHOFF J. (1968). Cues to reading. *Quest* **X.**

BROWN C. (1967). The structure of knowledge of physical education. *Quest* **IX.**

BROWN C. & CASSIDY R. (1963). "Theory in Physical Education". Philadelphia: Lea & Febiger.

BROWN M.C. & SOMMER B.K. (1969). "Movement Education: Its Evolution and a Modern Approach". London: Addison-Wesley.

CARLISLE R. (1969). The concept of physical education. *Proc. Phil. Ed. Soc.,* **III.**

CARROLL J. & LOFTHOUSE P. (1969). "Creative Dance for Boys". London: Macdonald & Evans.

CHILD I.L. (1969). Personality correlates of esthetic judgement in college students. In J. Hogg (Ed.) "Psychology and the Visual Arts". Harmondsworth: Penguin.

CLARK K. (1956). "The Nude". Harmondsworth: Penguin.

COHEN S.J. (Ed.) (1966). "Modern Dance—Seven Statements of Belief". Connecticut: Wesleyan University Press.

COLLINGWOOD R.C. (1938). "The Priciples of Art". Oxford: Clarendon Press.

DANIELS S. (1969). The study of sport as an element of the culture. In J.W. Loy & G.S. Kenyon (Eds.) "Sport, Culture and Society". London: Macmillan.

DENNEY R. (1969). The spectatorial forms. In J.W. Loy & G.S. Kenyon (Eds.) "Sport, Culture and Society". London: Macmillan.

DUMAZEDIER J. (1968). Some remarks on sociological problems in relation to physical education and sports. *Int. Rev. Sp. Soc.,* **3.**

DUNCAN I. (1928). "My Life". London: Gollancz.

ELLIS H. (1923). "The Dance of Life". London: Constable.

FLETCHER I.K. (1960). Ballet in England 1660-1740. In I.K. Fletcher, S.J. Cohen & R. Lansdale (Eds.) "Famed for Dance. Essays on the Theory and Practice of Theatrical Dancing in England 1660-1740". New York: New York Public Library.

FORRESTER F.S. (1968). "Ballet in England. A Bibliography and Survey c. 1700 to June 1966". London: The Library Association.

FRIEDENBERG E.Z. (1967). Foreword. In H.S. Slusher "Man, Sport and existence". Philadelphia: Lea & Febiger.

GOMBRICH E.H. (1962). "Art and Illusion. A Study in the Psychology of Pictorial Representation. (2nd Ed.)" London: Phaidon Press.

GRAY M. (1966). The physical educator as artist. *Quest* **VII.**

GRAY M. (Ed.) (1969). "Focus on Dance V". Washington: A.A.H.P.E.R.

GREGSON R.A.M. (1969). Aspects of the theoretical status of aesthetic response typologies. In J. Hogg (Ed.) "Psychology and the Visual Arts". Harmondsworth: Penguin.

HALL F. (1967). Book review. *Brit. J. Aesth., 7.*

HAUETER P. (1965). Approaches to composition in aquatics. In "Focus on Dance III". Washington: A.A.H.P.E.R.

H'DOUBLER M. (1940). "Dance, a Creative Art Experience". Wisconsin: University Press.

HENRY F. (1965). Physical education. An academic discipline. *Leaflet*, Jan.-Feb.

HINKS E.J., ARCHBUTT S.E. & CURL G.F. (1971). Movement studies. *Bull. Univ. London Inst. Ed., 23.*

HOGG J. (Ed.) (1969). "Psychology and the Visual Arts". Harmondsworth: Penguin.

HORST L. & RUSSELL C. (1961). "Modern Dance Forms in Relation to the Modern Arts". San Francisco: Impulse Publications.

HUMPHREY D. (1959). In B. Pollack (Ed.) "The Art of Making Dances". New York: Grove.

IMPULSE (1952-1970). Annual publication (now ceased). San Francisco: Impulse Publications.

IMPULSE (1968). "Dance, a Projection for the Future". San Francisco: Impulse Publications.

INTERNATIONAL GYMNASTICS FEDERATION (1970). "Code of Points". Women's Technical Executive Commission F.I.G.

JAMES D.N. (1970). Sport in the '70's: amateur boxing. *Spt. Rec.* Oct. C.C.P.R.

JAQUES-DALCROZE E. (1921). "Rhythm, Music and Education". London: Chatto & Windus.

JAQUES-DALCROZE E. (1930). "Eurhythmics, Art and Education". London: Chatto & Windus.

JOHNSTON J. (1965). The new American modern dance. In R. Kostelanetz (Ed.) "The New American Arts". London: Collier-MacMillan.

KAELIN E.F. (1964). Being in the body. "N.A.P.E.C.W. Report". Washington: A.A.H.P.E.R.

KAELIN E.F. (1968). The well-played game: notes toward an aesthetics of sport. *Quest* **X.**

KEALIINOHOMOKU J.W. (1970). An anthropologist looks at ballet. *Impulse.*

KEALIINOHOMOKU J.W. & GILLIS F.J. (1970). Special bibliography: Gertrude Prokosch Kurath. *Ethnomusicology,* **14,** 1.

KENYON G.S. (1968). A conceptual model for characterizing physical activity. *Res. Quart.* **39,** 1.

KIRSTEIN L. (1967). "Three Pamphlets Collected". New York: Dance Horizons.

KLEINMAN S. (1964). The significance of human movement: a phenomenological approach. "N.A.P.E.C.W. Report". Washington: A.A.H.P.E.R.

KOESTLER A. (1964). "The Act of Creation". London: Pan.

KURATH G.P. (1960). Panorama of dance ethnology. *Current Anthrop.* I, 3.

LABAN R. (1948). "Modern Educational Dance". London: Macdonald & Evans.

LABAN R. (1950). "The Mastery of Movement on the Stage". London: Macdonald & Evans.

LABAN R. (1960). "Mastery of Movement (2nd Ed.)". London: Macdonald & Evans.

LABAN R. (1966). "Choreutics". London: Macdonald & Evans.

LAMB W. (1965). "Posture and Gesture". London: Duckworth.

LANGE R. (1970). The nature of dance. *Laban Art of Movement Guild Magazine.* **44.**

LANGER S. (1942). "Philosophy in a New Key. A Study in the Symbolism of Reason, Rite and Art". New York: Mentor.

LANGER S. (1953). "Feeling and Form". London: Routledge & Kegan Paul.

LANGER S. (1957). "Problems of Art". New York: Scribner.

LANGER S. (1962). "Philosophical Sketches". Maryland: Johns Hopkins Press.

LAWLER L.B. (1964). "The Dance in Ancient Greece". London: Black.

LAWSON J. (1966). "Mime. The Theory and Practice of Expressive Gesture with a Description of its Historical Development". London: Pitman.

LAYSON J. (1970). The contribution of modern dance to education. Unpublished M. Ed. thesis University of Manchester.

MACE C.A. (1968). Psychology and aesthetics. In H. Osborne (Ed.) "Aesthetics in the Modern World". London: Thames & Hudson.

MACHOTKA P. (1970). Visual aesthetics and learning. *J. Aesth. Ed.* **4,** 3.

MAGRIEL P. (1936). "A Bibliography of Dancing. A List of Books and Articles on the Dance and Related Subjects". London: Blom.

MARTIN J. (1933). "The Modern Dance". New York: Dance Horizons.

MARTIN J. (1936). "America Dancing. The Background and Personalities of the Modern Dance". New York: Dance Horizons.

MARTIN J. (1939). "Introduction to the Dance". New York: Dance Horizons.

MASON M.G. & VENTRE A.G.L. (1965). "Elements of P.E. 1. Philosophical Aspects". Leeds: Thistle.

MAVER J.A. (1968). Movement in the Aquatic Medium. In H.M. Smith (Ed.) "Introduction to Human Movement". London: Addison-Wesley.

MAWER I. (1932). "The Art of Mime. Its History and Technique in Education and the Theatre". London: Methuen.

MAYNARD O.(1965)."American Modern Dancers.The Pioneers". Boston: Little, Brown.

MEAD M. (1928). "Coming of age in Samoa". Harmondsworth: Penguin.

MEAD M. (1930). "Growing up in New Guinea". Harmondsworth: Penguin.

MEAD M. (1946). Dance as an expression of culture patterns. In F. Boas (Ed.) "The Function of Dance in Human Society". New York: Boas School.

MEAD M. (1950). "Male and Female". Harmondsworth: Penguin.

METHENY E. (1965). "Connotations of Movement in Sport and Dance". Iowa: Brown.

METHENY E. (1968). "Movement and Meaning". London: McGraw Hill.
METHENY E. & ELLFELDT L. (1961). Dynamics of human performance. In L. Larson (Ed.) "Health and Fitness in the Modern World". Athletic Institute.
METTLER B. (n.d.). "Nine Articles on Dance". Massachusetts: Mettler Studios.
METTLER B. (1947). The relation of dance to the visual arts in *J. Aesth. Art Crit.,* **V**, 4.
MINKOWSKI E. (1967). Imagination? and Spontaneity (. . . spontaneous movement like this!) In N.M. Lawrence & D.J. O'Connor (Eds.) "Reading in Existential Phenomenology". N.J.: Prentice Hall.
MORGAN R.E. et al. (1970). The concept of physical education. *Brit. J. Phy. Ed.,* **1**, 4.
NOVERRE J.G. (1803). "Letters on Dancing and Ballets". New York: Dance Horizons.
OSBORNE H. (1968a). "Aesthetics and Art Theory. An Historical Introduction". London: Longmans.
OSBORNE H. (Ed.). (1968b). "Aesthetics in the Modern World". London: Thames & Hudson.
OSBORNE H. (1970). "The Art of Appreciation". London: Oxford University Press.
PARRY J. (1967). "The Psychology of Human Communication". London: University Press.
PEASE E. (Ed.) (1964). "Compilation of Dance Research 1901-1964". Washington: A.A.H.P.E.R.
PHENIX P.H. (1964). "Realms of Meaning". London: McGraw-Hill.
PRALL D.W. (1929). "Aesthetic Judgment". New York: Crowell.
PRIDDIN D. (1952). "The Art of the Dance in French Literature". London: Black.
READ H. (1931). "The Meaning of Art". Harmondsworth: Penguin.
REID L.A. (1961). "Ways of Knowledge and Experience". London: Allen & Unwin.
REID L.A. (1969a). Aesthetics and education. In Conference Report "Physical Education: Aesthetic and Social Aspects". Association of Principals of Women's Colleges of Physical Education.
REID L.A. (1969b). "Meaning in the Arts". London: Allen & Unwin.
REID L.A. (1970). Movement and meaning. *Laban Art of Movement Guild Magazine.* **45.**
RICHTER P.E. (1967). "Perspectives in Aesthetics". New York: Odyssey Press.
ROCHLEIN H. (1964). "Notes on Contemporary American Dance". Maryland: University Extension Press.
ROCHMAN R.P. (1965). Dance and aquatic art. A comparative analysis of two movement studies. In "Focus on Dance III". Washington: A.A.H.P.E.R.
RODIONOV A. (1966). Sport as a means of aesthetic education. In "Sport in Education and Recreation, Conference Report". London: Physical Educ. Assoc.
RUSSELL J. (1965). "Creative Dance in the Primary School". London: MacDonald & Evans.
RUSSELL J. (1969). "Creative Dance in the Secondary School". London: MacDonald & Evans.
RUSSELL J. (1970). Dance in education. *Bull. Phys. Ed.* **VIII,** 1.

RUST F. (1969). "Dance in Society". London: Rouledge & Kegan Paul.

SACHS C. (1937). "World History of the Dance". New York: Norton.

SAW R. & OSBORNE H. (1968). Aesthetics as a branch of philosophy. In H. Osborne (Ed.) "Aesthetics in the Modern World". London: Thames & Hudson.

SCHILDER P. (1950). "The Image and Appearance of the Human Body". New York: Wiley.

SCHOOLS COUNCIL (1968) Objectives of physical education. In *Dialogue,* 1.

SHAWN T. (1963). "Every Little Movement (2nd Ed.)". New York: Dance Horizons.

SHEETS M. (1966). "The Phenomenology of Dance". Wisconsin: University Press.

SHIPMAN M.D. (1968). "A Sociological Perspective of Dance". A.T.C.D.E.

SLUSHER H. (1967). "Man, Sport and Existence". Philadelphia: Lea & Febiger

SMITH H.M. (1968). Movement and aesthetics. In H.M. Smith (Ed.) "Introduction to Human Movement". London: Addison-Wesley.

SORELL W. (1966). "The Dance Has Many Faces". New York: Columbia University Press.

STEBBINS G. (1886). "Delsarte System of Dramatic Expression". New York: Werner.

STEBBINS G. (1893). "Dynamic Breathing and Harmonic Gestures. A Complete System of Physical, Aesthetic and Physical Culture". New York: Werner.

STRAUS E.W. (1970). Born to see. Bound to behold. Reflections on the function of upright posture in the aesthetic attitude. In S.F. Spicker (Ed.) "The Philosophy of the Body". Chicago: Quadrangle.

STREICHER M. (1965). Aesthetic experience of movement. In Conference Report Fifth International Congress I.A.P.E.S.G.W.

STREICHER M. (1970). "Reshaping Physical Education". Manchester: University Press.

TERRY W. (1956). "The Dance in America". London: Harper Row.

VALERY P. (1960). The dance from Degas, dance, drawings. In J. Mathews (Ed.) "The Collected Works of Paul Valéry, Vol. 12". New York: Pantheon.

VUILLERMOZ E. (1958). Modern ballet. In S.K. Langer (Ed.) "Reflections on Art". New York: Johns Hopkins Press.

WATERMAN R. (1962). Role of dance in human society. In "Focus on Dance II". Washington: A.A.H.P.E.R.

WEAVER J. (1721). "Anatomical and Mechanical Lectures Upon Dancing". London: Brotherton.

WIGMAN M. (1952). Composition in pure movement. In B. Ghiselin (Ed.) "The Creative Process". Califonia: University Press.

WIGMAN M. (1966). "The Language of Dance". London: Macdonald & Evans.

WOLLHEIM R. (1968). "Art and Its Objects—An Introduction to Aesthetics". London: Harper & Row.

WOLLHEIM R. (1971). Lecture in the series "Talking with Philosophers". B.B.C. 24th January. (In press).

WOOTEN B.J. (Ed.) (1962). "Focus on Dance II". Washington: A.A.H.P.E.R.

WOOTEN B.J. (Ed.) (1965). "Focus on Dance III". Washington: A.A.H.P.E.R.

TOPIC AREAS - DEVELOPED

HUMAN MOVEMENT AND THE VISUAL ARTS

by D.W. MASTERSON

General

The representation of movement may be considered an archetype of art. Movement signifies life, and its illusion has been sought since the origins of visual art. The painting and sculpture of all cultures have made the depiction of motion one of their principal concerns, and great thinkers on art have considered it one of the ways whereby we recognise vitality and relive it in our own bodies. Art which depicts movement provides one of the principal means of experiencing aesthetic pleasure and delight.

Human actions, mechanical energy and motion produced by natural or contrived forces have been motivating factors for art of this kind. Whilst the literature especially devoted to human movement as a *subject* of art objects is not extensive, that concerned with motion in the *production* or *appreciation* of art is more abundant. In the former instance Lord Clark (1960), Kepes (1965), Moholy-Nagy (1947), and Scharf (1968) have devoted some attention to this subject in works of a more general nature; in the latter case Arnheim (1956), Kepes (1965) and Gombrich (1960) provide the best overviews.

Aesthetic aspects of human movement have been described by Souriau (1889, (1956), Valéry (1964), Pouret (1965), Reid (1969) and numerous writers on dance (see section S). The beauty of human movement in sport has been considered by Bellugue (1963), Gebelwicz (1965), Ghose (1962), Hohler (1961), James (1963), Keller (1969) and Vialar (1963). The representation of sport in art has been described by Masterson (1968, 1971); Toynbee (1962, 1963) and Umminger (1969). Aesthetic content in Physical Education has been examined by Adams (1969), Anthony (1968), Carlisle (1969), Reid (1969) and Rodionov (1966) and the concept of sport as an art form, or relationships between art and sport, have been described by Elliot (1969), Fraysinnet (1960), Höhne (1969), Masterson (1970), Osterhoudt (1971), Umminger (1958) and Pouret (1970).

F.1. Art and the Idea of Movement

The image of movement in static art objects must be related to theories of perception.

To perceive movement is to experience change. The theory involves not only physiological optics and psychology, but also ideas about time, space, velocity, etc. The relationship of moving objects and the sensation of 'pure movement' perceived by the eye is described in Wertheimer's theory of phiphenomena (Boring 1942). This shows how apparent motion is created when two fixed light sources of similar shape are suitably timed so that one appears in a different place shortly after that preceding it has disappeared. Alpha, beta and gamma movement, and the effects of after-imagery, which is induced from observation of continual movement, are also phenomena involved in the visual perception of movement. All depend on retinal stimulation and have been exploited by various artistic methods and techniques. Wallach & O'Connell (1953) have investigated this form of perception, as have Oppenheimer (1935), Michotte (1946) and Duncker, whose experiments on relationships between an object and the framework in, or around which it moves, are reported in Ellis (1939).

It has been shown, however, that when motion is observed, retinal stimulation alone cannot account entirely for the sensation of movement and it has been proposed that a contributing factor to this form of perception is the kinesthetic sensation (Kepes, 1965). Gellerman (1933) has shown how muscular movement accompanies the identification of moving objects. Michotte (1946) has observed that the function of the body in kinesthesis, especially with regard to the appreciation of movement, is outside that area where the motion is observed against a 'background' which acts as a framework of reference. Rorschach (1942) and his followers have regarded visual perception in itself to be limited to static patterns and that any suggestion of movement is only by association. Arnheim (1956), on the other hand, believed visual perception to involve dynamics which arise from the conflict between physiological perception and the psychological reduction of external stimulii to their simplest pattern. The Gestalt theory of movement perception is best reviewed by Koffka (1935).

The function of time in the concept motion goes back to Aristotle who considered it 'the number of motion according to what is before and what is beyond'. The performing arts appear to occur more *within* time than immobile art objects, which exist outside it, but Arnheim (1956) has claimed that:

> ... the experience of happenings, from that of things, is not that it involves the perception of passing time, but that we witness an original sequence in which phases follow each other meaning-fully in a one dimensional order.

Human movement in drama, dance and sport may be taken, therefore, as a 'timeless whole', rather than a series of individual sequences. These time-space concepts,

which Arnheim applied to the performing arts, he also extended to art objects.
Friedlander (1942) has considered the role of space in human movement in his
description of how a moving body seems to create a volume which surrounds it. A
dancer widens the space in which he dances, and a runner, diver or gymnast
'suggests' the space he has left behind.

F.2. The Representation of Movement in Art Objects

Certain artistic techniques used to translate both the kinesthetic experience and
the optical signs of movement into picture images have been described by Kepes
(1965) and Moholy-Nagy (1947). Modifications made in the art work to the
'constant characteristics' of its subject have been employed for this purpose
throughout the history of art. These modifications include not only the addition of
attributes which the subject does not possess in reality, but also the distortion of
those which it does.

Attitude and gesture have been employed particularly in the figuration of human
movement. Arrested motion, and the agitation of figures have been used especially.
Positions of instability, or imbalance, imply states of transition or the process of
transformation, and therefore convey the idea of movement. Such attitudes or
configurations, have been used frequently in art. Repeated usage of these devices
caused some of them to become symbols for movement during certain periods of
art history (Clark, 1960; Frankfort, 1951; Richter, 1959).

Repetition implies rhythm which, in turn, suggests movement. But it can also
convey the passage of time when the image of the principal subject of the picture
appears in different attitudes in various parts of the same picture. This device is so
diverse as to have been used in works of art from Ancient Egypt and Assyria to
Futurist art of the twentieth century. Processional friezes also may be associated
with it.

By encompassing dynamic shapes, or compelling the eye to follow directions, line
has been used to convey the idea of movement. Form also has been modified to this
end by techniques which relate it to the inability of our vision to observe a moving
body as closely as one at rest. Rapid and sketchy handling tend to lessen reality and
impart motion to artistic form (Friedlander, 1942). To these devices may be added
others, such as the interplay of contrasts in light and shade, or the use of colour
vibrations, both of which were described by Popper (1968) and Moholy-Nagy
(1961). The use of multiple view points, designed to create the illusion of three
dimensional space in two dimensional painting, as if the spectator were moving
around the subject of the picture to experience its different aspects, was an
important principle of cubist painting.

In more recent years Kinetic Art has elaborated certain aspects of the dynamics of
motion propounded by the Futurist school as well as other developments in art,
such as those emanating from Gabo and Calder (Popper 1968). This art pays some
attention to human movement but places it more on spectator participation in
appreciation of the art object than on human movement as a subject of the work of

art. 'Pure movement', created by the use of optical phenomena, transformation of
the visual image, or electro-magnetic forces etc., receives a greater degree of
emphasis. A morphology of this art has been compiled by Rickey (1965).

F.3. Human Movement in Painting and Sculpture

Lord Clark (1960) devoted special attention to human movement in painting and
sculpture as an expression of energy. He has described how distortion has been used
for this purpose from as early as the sixth century B.C. The black-figure vase
painters of Ancient Greece lengthened the limbs and emphasised certain muscles of
the athletic figures they used as decorative subjects for domestic pottery. Later,
Italian artists adopted similar distortions. Pollajuolo, for example, overstated the
musculature of his men in combat, and Michelangelo exaggerated the physiques of
his athletes. Modern art uses the same device. To impart them with movement
Giacometti elongated his walking figures. Distortion has also been applied to facial
expression when it reflects physical action. Examples here are legion, but for those
which indicate athletic effort, reference may be made to the masks of athletes
sculpted by Tait-McKenzie (Hussey, 1929). The most celebrated athletic
sculpture—that of the fifth century B.C.—relied on neither of these latter means to
evoke the sensation of movement. The physiques of these figures were exaggerated
in no other way than their idealised proportions (Gardiner, 1930). Facial
expressions were never contorted, for classic sculpture of the 'severe' period was
characterised by its impassivity (Hauser, 1968). Movement was imparted instead
by the ways in which the works were posed. Ridder & Deonna (1968) have
described how technical developments in bronze production, and methods of
casting during the sixth century B.C., permitted greater freedom in the choice of
attitude and gesture. The works of Ageladas, Pythagoras, Pheidias, Polyclitus and
Myron could, therefore, be posed in positions that had hitherto been impossible.
Substantive sculpture was able to break away from the conventional static attitude
of the archaic Kouroi figures, movement could be introduced and the three
dimensional properties of sculpture exploited. Attitudes of arrested motion were
chosen. Hauser (1968) claimed that from these developments the 'pregnant
moment', seen especially in the Discobolus, was born. Where movement was
depicted it was incorporated into the ideal conception; only when fifth century
idealism waned, and more sensate characteristics appeared in Greek art, was its
representation more closely allied to the emotional content of the work of art.

F.4. Movement in Idealistic and Naturalistic Art

Sorokin (1962) has described the divergent and opposing characteristics of
Ideational and Sensate Art. Other writers prefer different terms for these styles.
Hauser (1968) used 'Idealistic' and 'Naturalism', Ehrenzweig (1967) chose
'Apollonian' and 'Dionysian'. On the one hand Ideal art interprets the world as

eternal, static, calm, serene and immovable; beauty is conceived as everlasting 'Being'. On the other, Sensate art describes existence as incessant change, or a state of 'Becoming', and the world as constantly undergoing a process of progressive evolution. Movement is more a characteristic of the latter than the former conception of art, and was, therefore, more predominant in the Baroque or Romantic styles where it was calculated to overwhelm the spectator by its direct emotional appeal (Murray & Murray 1959) than in, for example, the Neo-Classical style. Cubism, especially in its geometric Analytic and Hermetic phases, may be cited as a form of modern art evocative of the static world.

F.5. Human Movement and Sport in Modern Art

The use of multiple viewpoints in Cubist painting implied a degree of movement on the part of the artist and spectator, but in its 'Synthetic' phase subjects which featured movement were chosen by Braque, Picasso, L'Hote, Gleizes, etc. Subsequent developments saw cubist principles adopted in the decor for theatrical productions to complement the movement of dancers, and modified forms, such as those of Leger, Villon and Delaunay, were used to depict movement in space (Leger's 'Divers'), or sporting subjects (Villon's wrestlers, Laren's boxer, and Delaunay's rugby players and runners) (de Francia, 1970; Masterson, 1971). From its beginnings, Futurist art set out to depict motion by deliberately choosing subjects which reflected the 'universal dynamism'. These included dancers, footballers, swimmers, boxers and cyclists (Carrieri, 1963). Boccioni focused much of his attention on muscular dynamism and sport (Ballo, 1964) and the movement's concern with motion was described in the Futurist Manifesto (Balla, Boccioni, Carra, Russolo, & Severini, 1910). Other writers, especially Marinetti (1914), described the principles of the movement. Photography had a marked effect on this art and the analyses of Marey and Muybridge have been related to Futurist theory about the image of movement (Scharf, 1968).

The development of Purism and Suprematism was an evolution towards a formalism which sacrificed subject to style and form. Sporting motifs occurred occasionally in representative abstract art (de Stael's footballers (Cooper, 1961), Baumeister's games players, gymnasts and athletes (Grohmann, 1966) and Archipenko's boxers (Masterson, 1971).

Outside the comparatively small groups which introduced new artistic schools and movements to the first half of the twentieth century stood a majority which Dorival (1958) called Neo-Realist. Amongst these artists was André Dunoyer de Segonzac, who was awarded the Pierre de Coubertin Prize for his works devoted to modern sport (Berlioux, 1968).

Continental developments in art did not go unfelt in England. The effects of Cubism and Futurism were reflected in Vorticism. William Roberts, a painter associated with this school, in seeking subjects with a social content produced a plethora of sporting scenes (Alley, 1965). The 14th Olympiad brought exhibitions of art based on sport to London and the competition 'Football and the Fine Arts', arranged jointly by the Arts Council and the Football Association in 1954, attracted 1,710 entries which included submissions of football scenes by Lowry, Ayrton, Toynbee, etc. Today, Hill, Unsworth and Kershaw frequently select sport as a subject for their art.

The tradition of sport in American art has been continuous from Bingham to Rauschenberg. Described as America's greatest painter of the nineteenth century, (Evershed, 1965), Eakins specialised in scenes of rowing, baseball and boxing (Porter, 1959). In portraying the American scene in the 1920's Bellows discovered the picturesque in boxing, footballing, wrestling, polo, tennis and fishing. Brown, following in the footsteps of Tait McKenzie, also finds sportsmen and athletes a source of inspiration. Today, Rauschenberg uses images from the dance and various sports (Ashton 1965; Robertson 1964), whilst Wahol has included modern ballet and baseball in his screen prints. Art which seeks to portray popular culture frequently uses sport as a subject. Laing chooses dragsters and sky divers (Finch, 1968), Kanovitz uses basketball players and, outside the United States, Blake paints wrestlers, Paolozzi includes football players and van Bohemen portrays motor racing. Some artists have described their personal reactions to sport (Urbscheitt, 1968; Toynbee, 1961; 1962; 1963; Holme, 1960; Gray, 1967).

Contemporary art in the Soviet Union and Eastern Europe is dominated by the Socialist Realist style. Since the periods of recovery and reconstruction in the U.S.S.R. during the 1920's sport has been a popular subject and numerous artists have portrayed it (Bush & Zamoshkin et al., 1934). The most notable of these is Alexander Deineka, Vice-President of the Soviet Academy of Fine Arts and winner of the Lenin Prize in 1964 (Dobrov, 1966). His work includes respresentations of a wide variety of sports (Chegodayev, 1959; Maga, 1959; Nikiforov, 1937). The tradition established by Motovilov, Dormidontov, Schwartz and Chaikov forty years ago are being continued today by Ozlovsin, Veivezite, Nissky, Kalkinia, Yablonskaya and Manisev.

From the attempts made to forge a strong relationship between sport and art in the German Democratic Republic during the last decade (Hohne, 1969a; 1969b; Witt, 1965) such artists as Blume, Bondzin, Geyer and Zitzmann have emerged as masters of the sporting genre in the Realist style.

Source Texts

ARNHEIM R. (1956). "Art and Visual Perception". London: Faber & Faber.

BALLO G. (1964). "Boccioni". Milan: Ed. da'll Saggiatore'.

BELLUGUE P. (1963). A propos d'art de forme et de mouvement". Paris: Maloine, S.A.

BORING E.G. (1942). "Sensation and Perception in the History of Experimental Psychology". New York: Appleton-Century Crofts.

BUSH M. & ZAMOSHKIN A. (1934). Soviet pictorial art. In: "Painting, Sculpture and Graphic Art in the USSR". Moscow: Voks.

CARRIERI R. (1963). "Futurism". Milan: Ed. del Milioni.

CHEGODAYEV (1959). "Alexander Alexandrovich Deineka". Moscow.

CLARK LORD (1960). "The Nude". Harmondsworth: Pelican.

COOPER D. (1961). "Nicolas de Staël". London: Weidenfeld & Nicolson.

DORIVAL B. (1958). "Twentieth Century Painters: From Cubism to Abstract Art". Paris: Editions Pierre Tisne.

EHRENZWEIG A. (1967). "The Hidden Order of Art". London: Weidenfeld & Nicolson.

ELLIS W.D. (Ed.) (1939). "A Source Book of Gestalt Psychology". New York.

EVERSHED E. (1965). Impressionism and symbolism. In: "Larousse Encyclopaedia of Modern Art". London: Hamlyn.

FINCH C. (1968). "Pop Art: object and image". London: Studio Vista.

FRANKFORT H. (1951). "Arrest and Movement". London: Faber.

FRAYSINNET P. (1960). "Le Sport Parmi les Beaux Arts". Paris: Dargaud.

FRIEDLANDER M.J. (1942). "On Art and Connoisseurship". London: Cassirer.

GOMBRICH E.H. (1960). "Art and Illusion". London: Phaidon.

GROHMANN W. (1966). "Willi Baumeister: Life and Work". London: Thames & Hudson.

HAUSER A. (1968). "The Social History of Art". London: Routledge & Kegan Paul.

HUSSEY C. (1929). "Tait-McKenzie". London: Country Life.

JAMES C.L.R. (1963). "Beyond a Boundary". London: Hutchinson.

KEPES G. (Ed.) (1965). "The Nature and Art of Motion". London: Studio Vista.

KOFFKA K. (1935). "Principles of Gestalt Psychology". London: Kegan Paul.

MAGA N.L. (1959). "A. Deineka". Moscow: Soviet Artists.

MARINETTI F.T. (Ed.) (1914). "I Manifesti del Futurismo". Florence: Lacerba.

MASTERSON D. (1971). The image of the sportsman in ancient and modern art: a comparative iconography. Unpublished thesis, University of Salford.

MASTERSON D. (1970). Sport and art. In: A. Schouvaloff (Ed.) "A Place for the Arts". Liverpool: Seel House Press.

MOHOLY-NAGY L. (1947). "Vision in Motion". Chicago: Theobold.

NIKIFOROV B.M. (1937). "A. Deineka". Moscow.

OSTERHOUDT R. (1971). An exposition concerning sport and the fine arts. Unpublished Doctoral Thesis, University of Illinois.

POPPER F. (1968). "Origins and Development of Kinetic Art". London: Studio Vista.

PORTER F. (1959). "Thomas Eakins". New York: Braziller Inc.

RICHTER G.M.A. (1959). "A Handbook of Greek Art". London: Phaidon.
SCHARF A. (1968). "Art and Photography". London: Penguin Press.
SOROKIN P.A. (1962). "Social and Cultural Dynamics. I". New York: Bed-
minster.
SOURIAU E. (1889). "L'esthetique du Mouvement". Paris: F. Alcan.
SOURIAU E. (1956). "Les Categories Esthetiques". Paris: C.D.U.
UMMINGER W. (1958). A stroll along the border. In "Sport and Society".
London: Bowes & Bowes.
VALERY P. (1964). "Aesthetics". London: Routledge & Kegan Paul.

Journal Overviews

ADAMS M. (1969). The concept of physical education. II. *Proc. Phil. Ed. Soc.,*
23-35.
ALLEY R. (1965). William Roberts, A.R.A. *Cat. Arts Coun. Exhib.*
ANTHONY D. (1968). Sport and physical education as a means of aesthetic
education. *Physical Ed.,* **60,** 1-6.
ASHTON D. (1965). The Dante drawings. *Cat. Arts Coun. Exhib.*
CARLISLE R. (1969). The concept of physical education. *Proc. Phil. Ed. Soc.,*
5-22.
GEBELWICZ E. (1965). Aesthetic problems in physical education and sport.
Bulletin F.I.E.P., 3.
GHOSE Z. (1962). Sport and the spectator. *The Listener,* **LXVIII,** 1750, 568-9.
HOHLER V. (1961). Vyznam objevu I.P. Pavlov pro estetiku telesne vychovy V.
Ostace esteticko prozitku. *Telovychovny sbornik.,* **6,** 7-23.
HÖHNE E. (1969a). Coubertin on the place of art in modern olympism. *Bull. Nat.
Olymp. Comm. G.D.R.,* **XIV,** 4; **XV,** 1; **XV,** 2.
HÖHNE E. (1969b). Kunst und sport—eine bemerkenswerte ausstellung. *Korper-
erziehung,* **10,** 506-512.
MASTERSON D. (1968). Le sport et l'art. *Ed. Physiq. Spt.,* **94,** 13-16.
MASTERSON D. (1968). Le sport et la peinture moderne. *Ed. Physiq. Spt.,* **95,**
13-16.
POURET H. (1965). Esthetique des formes dans le sport. *5th Int. Olymp. Acad.
Rep.,* 121-136.
POURET H. (1970). Is sport an art. *10th Int. Olymp. Acad. Rep.,* 129-133.
REID L.A. (1969). Aesthetics and education. In "Conference on Physical
Education and the B.Ed. Degree Report". Liverpool.
ROBERTSON B. (1964). "Robert Rauschenberg". London: Whitechapel Gallery.
RODIONOV A. (1966). Sport as a means of aesthetic education. In: "Conference
on Sport in Physical Education and Recreation Report". London: C.C.P.R.
UMMINGER W. (1969). Encounters with sport and art. *Sport '69,* 2-3.
VIALAR P. (1963). L'esprit du sport. *3rd Int. Olymp. Acad. Rep.*

Specific References

ARNHEIM R. (1956). "Art and Visual Perception". London: Faber & Faber.

BALLÄ G. et al (1910). "Manifesto of Futurist Painting". Comoedia.

BERLIOUX M. (1968). André Dunoyer de Segonzac and sport. *I.O.C. Newsletter,* **15,** 605-9.

ELLIOT R. (1969). Sport and art. Unpublished paper. "Conference on Aesthetic Aspects of Sport, Physical Education and Dance". University of Salford.

FRANCIA P. de (1969). "Peter de Francia on Léger's 'The Great Parade'." London: Cassell.

GARDINER E.N. (1930). "Athletics of the Ancient World". London: Oxford University Press.

GELLERMAN L.W. (1933). Form discrimination in chimpanzees and two-year-old children. *J. Genet. Psychol.,* **42,** 2-27.

GRAY C. (1967). Rediscovery: Jim Davis. *Art in America.,* **55,** 8, 64-70.

HOLME B. (1960). Sport as a challenge to artists. *Design.,* **61,** 125-127.

KELLER H. (1969). Genius and mastery in sport and art. Unpublished paper. "Conference on Aesthetic Aspects of Sport, Physical Education and Dance". University of Salford.

MICHOTTE A. (1946). "La Perception de la Causalite". Paris: Louvain.

MURRAY P. & MURRAY L. (1959). "A Dictionary of Art and Artists". Harmondsworth: Penguin.

OPPENHEIMER E. (1935). Optische versuche über ruhe und bewegung. *Psych. Forschung.,* **20,** 1-46.

RICKEY G. (1965). The morphology of movement. In "The Nature of Art and Motion." London: Studio Vista.

RIDDER A. de & DEONNA W. (1968). "Art in Greece". London: Routledge & Kegan Paul.

RORSCHACH H. (1942). "Psychodiagnostics". New York.

TOYNBEE L. (1961). Some notes on the painting of contemporary games. *Lond. Mag.,* **1,** 8, 57-60.

TOYNBEE L. (1962). Artists and sport. *New Society,* November 8th.

TOYNBEE L. (1963). The artist as fan. *Times Ed. Supp.,* February 15th.

URBSCHEITT L.F. (1968). Football: a subject for art. *Amer. Art.,* **32,** 8, 56-61.

WALLACH H. & O'CONNELL D.N. (1953). The kinetic depth effect. *J. Exp. Psychol.,* **45,** 205-217.

WITT G. (1965). Some Problems Concerning the Relationship Between Sport and Art in the German Democratic Republic. Unpublished Thesis, Karl Marx University.

TOPIC AREAS-DEVELOPED

HUMAN MOVEMENT : A SOCIETAL STUDY

by L.B. HENDRY

While there is not a vast integrated literature on patterns of human movement within society, an attempt will be made here to present a framework around which existing evidence can be examined, together with speculative comments which may lead to future investigations in this area.

In this section, it is intended that movement should be examined from the interactive situations of the individual within the family to the societal and cross-cultural level:

Comparison of Classifications	G4	Cross-cultural Level
Classifications	G3	Societal Level
Typologies and Movement patterns	G2	Sub-Societal Level
'Catching' and learning Movements in interactive situations.	G1	The family
Innate and learned Movements		The individual

G.1. The individual in the Family-Primary socialisation
a. In utero

Movements occur before birth and early traces have been recorded at about 17 weeks after conception (Preyer 1937).

Walters (1965) has indicated a high correlation between pre-natal vigour, in movement terms, and resultant post-natal motor activity.

Interpretations of why movement in utero should lead to precocious development must be approached with caution, though some writers, (e.g. Feldenkrais, 1949; Zubeck, 1954; Zagora, 1959), have drawn attention to the reciprocal influence of myelinisation and movement behaviour. But, this is only one of a number of possible explanations.

Heyns (1963) has proposed that movement precocity can result from freer foetal movement in an oxygen-enriched interuterine environment (using the technique of abdominal decompression). A later study (Liddicoat 1967) using more elaborate controls—particularly the social class factor—failed to confirm these findings.

Fairweather & Illsley (1960) and Stott (1966) suggest that pre, para and post-natal complications (due to oxygen starvation and stress) with resultant movement impairment reveal a greater representation in the lower social classes.

Innate movements create the underpinnings for the development of what Fitts & Posner (1967) have termed 'universal skills'. From these movements bodily control develops sequentially (Gessell & Thompson 1934; Hurlock, 1949; Griffiths, 1954 and Breckenridge & Vincent 1966—but see Pikler 1968).

Environmental supports markedly affect maturation in determining the quality of motor activity. Hence, movement opportunities afforded by the family are important to the individual's movement patterns, interests and aptitudes, and in turn, may be related to intellectual and social development.

b. Social and intellectual development

Social skills depend upon a number of developed abilities in the child. Similarly, control of the musculature is necessary before appropriate gestures—facial and bodily—can be used in social interaction.

Bell (1962) describes socialisation as:-

The process by which the child, whose behaviour and personality are undifferentiated, is moulded into an adult that is acceptable to the society in which he lives.

This implies interaction between mother and child, other members of the family, peers and finally society at large (e.g. Mead 1934).

The effects of emotional deprivation and movement restriction are borne out by studies on hospitalised or institutionalised children (Schaffer, 1965; 1966; Schaffer & Emerson, 1968). Whiting (1967) has suggested that varied, unrestricted movement experiences, aided by parental encouragement, may produce optimum individual development. Stevenson (1965) considers adult approval is the most commonly used means of affecting behaviour. By contrast, Pikler (1968) indicates that in an emotionally sound environment children need no adult reinforcement in achieving 'normal' motor development.

Some evidence for the effects of movement experience on early cognitive development is available. Piaget (1953) in particular has described a sensori-motor period related to the development of intelligence in children. In Piaget's (1953) terms, restricted experience—including restricted movement experience—would affect the development of 'schemata' or in Hebb's (1949) terms the development of 'cell-assemblies' on which later intellectual development depends.

Whilst there exists confusion and disagreement about the precise relationship between motor and mental ability in normal populations (see Ismail & Gruber (1967) for a review), Malpass (1963) concluded that mentally retarded children tend to demonstrate less movement competence and skills than normals of the same age and sex. These findings are supported by the studies of Fish (1961) and Malpass (1959).

Hunt (1961) has proposed that marked motor retardation in children may be due to the homogeneity of input during early years, and in a similar way Fiske & Maddi (1961) suggest that it may be changes in perceptual input brought about by movement rather than movement opportunities per se that are most important in psychological development during the earliest months. Schaffer (1965) showed that children in a hospital in which the staff child ratio was 1 : 6 had lower developmental quotient scores after a period of hospitalisation than children in a baby-home where the staff child ratio was 1 : 2.5. In a follow-up study Schaffer (1966) indicated that children with constitutionally determined low levels of spontaneous movement were more dependent upon the current environment for stimulation and hence were more likely to suffer under conditions of deprivation. Earlier, Goldfarb (1945) had shown how infant isolation and deprivation of maternal affection seemed to have a marked effect on the general personality, often resulting in the child being less active in his movement patterns. Hence, family interactions and movement experiences may form an important chain in moulding the child.

c. Parental influences

Social interactions within the family are mostly reciprocally affective experiences involving rather subtle and unconscious cues—initially between mother and child. Frank (1957) has discussed the reaction of infants to movement signals of the mother.

Fantz (1961) claimed that some responses of the neonate to schematic facial features are unlearned. This finding is supported by Hershenson, et al (1965) though Fantz (1961) agrees subsequent changes reveal effects of age and experience. Spitz & Wolf (1946) reported that smiling is triggered at about the second month, and is extended by the child to others in a meaningful way between the third and sixth month. Spitz (1946) suggested that the movements of face and mouth were the crucial stimuli. Developmentally these important imprinting occurences appear to be sequential. Children begin to smile around three months, show anger between three and four months, and by six months show fear when a stranger approaches (Bridges, 1932). Green & Money (1961) studying effeminacy in boys have speculated that the inappropriate behavioural and movement patterns may be the result of faulty imprinting. If so (and from animal studies imprinting can occur from siblings and not necessarily parental stimuli) then femininity in boys being more prevalent in families where the male is younger than one or more sisters seems a reasonable assumption (Cratty, 1959). This may have implications for institutionalized movement experiences that children undergo in nursery and primary schools, predominantly tutored by women.

Parental influence on movement behaviour of the child is no doubt related to the extent to which the child identifies with one parent. Both sexes initially identify with the mother; then males switch to the father and this masculine identification is usually reinforced by societal demands. This emotional allegiance provides a model for attitudes and behaviour patterns of the identified parent by the child. Largely masculine behaviour is characterised by gait, posture and gesture patterns accorded to males by western society. Cratty (1967) reports remedial movement treatment attempted on boys with inappropriate sex identification. This was manifest amongst other actions by movement patterns, which were generally restricted with regard to the amount of space used, especially in throwing, running, walking movements, and subtle 'feminine' gesture patterns in social interaction.

Giescke (1935) suggested that hand dominance appears as early as three or four months of age, while Jenkins (1930) pointed out that, in general, the dominant hand influences early skill acquisition, the other hand being used to support or stabilize the body. This laterality, reinforced by parental encouragement, no doubt forms the basis of many regularized movement patterns within society. Whilst children who are not handicapped will evidence the same basic motor patterns that underlie the fundamental movements being discussed here regardless of their upbringing, these basic patterns will be affected by observation of parents and other sub-cultural influences (Sears 1950).

Fathers seem to figure significantly in achievement training, and parental encouragement and attitudes to achievement in movement tasks has been investigated by Rosen & D'Andrade (1959). Cratty (1967) wrote that Anderson (1937) in the United States has shown that minority groups (e.g. negroes) have available about the same amount of play facilities as white labouring classes, but they reveal greater social interaction at play. Studies by Knobloch & Pasamanick (1953) and Williams & Scott (1953) indicate greater precocity in gross motor behaviour of negro children when compared with whites.

If parental attitudes are communicated to children, then it is reasonable to suggest that attitudes towards movement participation and evaluation of specific activities take place within the family. Cratty (1959; 1960) provides some father-son correlational evidence and suggests a 'familial similarity', combining parental attitude, opportunity and inherent factors. The 'catching' of movements from parent to child has also been demonstrated (Birdwhistell, 1960; 1962). Thus in a general sense the influence of the family will range from the function of movement in the process of socialization into physical activity to the more specific induction into particular activities by interested family members. Dennis (1957) has shown how parental reinforcement of children's participation in games revealed certain cultural differences.

The development of human movement patterns, which have genetic under-pinnings, are also related to the cultural background. These have great relevance to the dynamics of social organisation and education. Imitative play is the way in which children in primitive societies acquire proficiency in motor skills, many of which relate to the needs of the particular society. Whilst adult approval is a known social reinforcer Baer et al (1963) show that as the child grows older social reinforcers become less important in guiding behaviour, and it is more likely that direct situational information is used. The child having first learned by imitation, then develops the social skills himself in the process of becoming socialized.

d. Status

Competence in movement skills (e.g. skipping, catching, riding a bicycle) carries a 'status significance' a particular stages of childhood, and success in these activities may affect a child's self-esteem. Northway (1966) has speculated that precocity of social skills in nursery aged children is related to a higher degree of sensory-motor skill development—the more skilled child, in movement terms, can be more involved in rewarding social interactions with others, while the less skilled is delayed and handicapped.

Hewett et al (1970) point out there is loss of status and family difficulties associated with the physically clumsy child because his movement capabilities render him outside 'normal' family rules of behaviour and punishment. Biesheuvel (1963) suggests that failure to provide the right movement experience at particular maturational stages will prevent full realisation of potential ability. While the maturational concept might be questioned, he raises the interesting point as to whether limited opportunities for learning certain basic movement habits in a particular social evironment (in this case tribal African society) may be responsible for subsequent manual skill deficiencies for both social and employment situations.

e. Social development

Those researchers who have attempted to assess the social maturity of children accord great importance to movement development as assessors of these skills (Doll, 1937). Developmentally, Buhler (1935) has given some indication of how movement, both individual and group, can reveal various social stages (e.g. competitive, co-operative, etc.,) the child passes through in the assimilation of cultural norms. Piaget (1932) in relation to understanding rules to a conceptual level has postulated stages which are passed through by the child. These include movement experiences within a games setting, which he has also related to moral development. Mead (1934) has indicated the value of movement experiences within group settings for role taking and role learning.

f. Bizarre movements

Bizarre movements of an individual within the family can range from temper tantrums to autistic 'rocking' movements. Kulka et al (1960) have proposed that rocking and head banging and other such rhythmic movements which, as seen in infants with prolonged deprivation, are attempts to gratify kinaesthetic needs. Vaizy & Hutt (1966) in an experiment in which playground density was varied, showed that while normal and brain-damaged children of ten became aggressive this was not the case with autistic children.

A syndrome of touch-hunger has been noted in relation to maladjusted children where the close-contact which normally exists between parent and child has been deficient (Kydd, 1962). Jourard (1967) has pointed out the limited use of body contact in Western societies.

g. Self-image

Freud (1961) postulated a three-stage libidino organisation theory in which the erogenic zones constitute the initial developmental areas. Masturbatory movements in the child are generally considered merely as pleasurable till adolescence (Falkner, 1967). Witkin et al (1962) consider all the experiences which the child encounters during development in relation to his own body and the bodies of others as important in the development of a 'self concept'. (See also Spitz, 1962).

G.2. Sub-cultural influences: class and peer groups—secondary socialisation

As has been pointed out earlier, socialisation is the process by which an individual learns the values, skills and knowledge appropriate to membership of a particular sub-cultural system. Caillois (1961) proposed that particular modes of movement participation reveal:

The character, pattern and values of every society.

a. Social class

Social class has been outlined by such writers as Klein (1964) and Cratty (1967) as having implications for general child-rearing patterns. Possible implications for the movement patterns adopted by various sub-cultural groupings are presented:-

Working Class	Middle Class
Present-time oriented	Future-time oriented
Rough or Respectable	Control of aggression, cultivation of courtesy
Traditional group solidarity	Training for individual responsibility
Inconsistent child rearing	Specific child-rearing values
Restricted language code	Elaborated language code
Early socialization	Ambition, academic achievement valued
Immediate gratification	Deferred gratification

The work of Roberts et al (1959) and Sutton-Smith et al (1959; 1963) is relevant to the function of play and games (i.e. patterned human movement) in the process of socializing individuals into a specific sub-culture and 'life style'. Loy (1969) has attempted to link Merton's (1957) theories on social adaptation to those of Caillois (1961) on games theory. Such studies as Loy's (1969) however, present a theoretical and methodological standpoint, and there are few (if any) investigations into the implications of movement patterns for socialization into British sub-cultures.

The function of human movement in the process of inculcating fundamental human skills and knowledge has been dealt with by Mead (1934) and Helanko (1963). Musgrove (1964) has proposed that children acquire their normative values from their parents, and this will have implications at various class levels for attitudes towards and presumably, participation in, movement. Efron's (1941) study has shown the importance of cultural assimilation, and of class influences on movement patterns. Goffman (1961) indicates class differences in the way people sit, and enter a room (Burns, 1964).

Within higher socio-economic classes there seems to be less need for non-verbal communication and gesture. This may be related to the verbal facility exhibited by the middle class (Bernstein & Henderson 1969).

It is possible to consider the physical attempts to produce speech as movement and Argyle (1969) has written about movements which accompany speech as being acquired within a social context. He considers body postures, gestures, facial expressions and eye-movements as vital elements in what he terms social techniques. Again, this can be related to social class differences via Bernstein's (1961) socio-linguistic code. Ruesch & Kees (1956) are concerned with non-verbal communication patterns such as the sign language of the deaf. These movements serve personal needs of the individual, but also provide statements for those who perceive them.

Sex differences are also apparent in movement patterns. Rosenberg & Sutton-Smith (1959) have shown how children ten to imitate the like parent in physical behaviour. The role of the parent being not only to give formal instruction, but also to influence children as to the worth or otherwise of participation in physical activities, this being a movement model. It was found that boys' activities involve forceful physical contact, dramatization of conflict between males, propulsion of objects in space, and complex team organisation. Girls used games involving relatively static activity, ritualistic non-competitive actions and rhythmic activities.

Sub-cultural differences in movement patterns are to some extent controlled by financial considerations. Emmett (1971) has presented figures to show sex differences in involvement and interest in movement participation in early post-school life. Middle class children are more participant than the working class. Isolates seek activities previously known to them, and those 'with it' have greater peer group allegience and are less participant. The 'squares' being more conformist will play traditional activities.

The Keele University study (1967) also shows class differences in interests and participation patterns of movement activities within the leisure-sphere. These studies indicate the powerful influence of earlier socialisation processes within the family, and of how the class characteristics are apparent in all life sectors.

b. Peer groups

In certain areas peer group influence may be important in creating a value system outside both home and school. Those physical skills acquired are supported by the norms of the peer group (although there will be good reasons why particular members do not acquire them). This has been investigated to some extent by studies demonstrating the relatively high status of the skilled movement performer in his group. In childhood and adolescence, particularly among males, social success may to a large extent be based upon skilled performance in certain sporting abilities and movements (Coleman, 1961).

After the age of three, children prefer social interaction to accompany their movement activities and with age, size, sex and complexity of the play-group increases. Bandura & Walters (1964) have demonstrated the way a 'peer model' can affect the behaviour of another child involved in a physical activity. Children also appear to strive harder at physical tasks to show their superiority in the presence of a disliked peer (Hartup 1964).

Amongst others, Kane (1965) and Coleman (1961) have shown how movement ability in certain sub-culturally valued physical skills enhance the individual's social status. Coleman (1961) suggested that the peer group tends to be a feature of the working class, while Helanko (1963) contended that sporting activities are born in the gang and consolidate the group. The peer group functions as a protection against relationships with the other sex (Whyte 1943; Helanko 1963) until the aggregation stage. In the gang there is support for independence against adult values, and the opportunity to learn adult roles in a 'non-realistic' setting by physically enacting many roles. Hence Loy's (1969) nomenclature of 'pseudo rebellion'. Whyte (1943) has discussed the implications of movement skills in relation to status within a gang structure. Or conversely, that status via expectations (Rosenthal, 1964) produce differential standards of competitive movement performance in different situations.

c. Adolescence

The function of human movement for the adolescent is shown to diversify according to age, sex and social class. There are no adolescent signposts in our society which clearly point towards adulthood, whereas in certain primitive societies it is a continuous process, often with severe trials of initiation, then different roles are expected from the individual. Differing child rearing practices will have produced role and gender differences, and adolescents will relate to available activities in terms of role perceptions (e.g. Do they want to emulate *this* particular role?) Boys appear to adhere to interest in traditional activities (Allot 1966), whereas there is rejection and apathy from girls (Smallridge 1967). In the United States, Jones (1946) has shown that for boys skill in competitive movement is important even at 19 years of age. For women athletic prowess had little prestige by itself, but skill in dance, swimming and tennis carried some prestige, these kinds of activities having social implications.

Class difference in patterns of sexual behaviour now become apparent. Middle class boys are more likely to masturbate while working class will seek actual coitus (Kinsey 1948). The particular movement behaviours can be related back to the social class typology where working class boys are socialised to seek immediate gratification. Sex within the peer group, enables experimentation without adult approval, but where there is shared confidence. Initially the gang is single-sexed, but this leads to smaller groups in later adolescence, where sex activities may again reveal class-based values (Nuttall 1968). Nielsen (1970) has written about the important use of body language in what he calls the 'courtship dance' of the adolescent.

d. Maladjustment

Whiting et al (1969) indicate that some evidence is available to suggest that motor retardates' (i.e. those whose movement qualities are restricted) movement patterns, can be categorised as 'slow and careless'. Approved school boys show a greater incidence of movement impairment (Whiting et al, 1970). This can be linked with Stott's (1959; 1962; 1964) postulations of a syndrome of movement deficiency, maladjustment and delinquency. Bamber (1966) confirms such a syndrome.

e. Schools

Little systematic evidence is available on the influence of secondary groups (as in school or the youth club) on movement patterns. Some schools use a 'play method' of teaching to evoke the *active* participation and interest of the children. There is learning through active movement experiences with the teacher guiding the pupils' experience, controlling and manipulating the environment. Humphrey (1965) compared the use of active games and language workbook exercises as learning media in the development of language and understanding with children. While children learned through both media, the active games group were superior. The use of active games in the learning of number concepts by children has also been investigated (Humphrey 1966). The important role played in the learning process by kinaesthetic feedback produced by body movement is pointed out. The skill of the teacher in this setting would be to provide 'play with a purpose'. Stephenson & Robertson (1965) (on retardates) point out that children must be assisted in getting all the movement experience they can, and being aware of all their own movements. Within schools certain institutionalized movements such as educational gymnastics and dance might be considered as 'movement outlets' for working class children, though there may well be other explanations.

Hewett et al (1970) have said that apart from the mentally handicapped (for whom allowances can be made) children least favoured in play situations were those who lacked sufficient movement competence.

It has also been shown with older children (Sutton-Smith & Roberts 1964) that those exhibiting particular modes of movement behaviour (physically skilled 'potentists') were classified as 'playground successes'. Sutton-Smith & Roberts (1963) and Jones & Bayley (1950) point out that children who are physically precocious do not need to strive for status within the group.

G.3. A societal classification

The movement classification given below is neither original nor mutually exclusive, but will serve to enable certain aspects of movement (at a societal level) to be discussed under separate headings. It will quickly become obvious to the reader that several of these areas are in fact inter-related, but the unidimensional form utilised may give some clarity to description and explanation:-

a) GOAL-DIRECTED   f) GROUP-STATUS
b) EXCITATORY   g) RITUALISTIC
c) BIZARRE   h) PROTEST
d) AESTHETIC   i) STILLNESS
e) COMMUNICATIVE

For example, the Aldermaston March classified as Protest (h), in the years since these marches began may now contain certain Ritualistic (g) components. Nevertheless, with this sort of limitation in mind these headings will be discussed as separate sections for the purposes of this analysis. The important influence of culture on movement is vital. A society's culture consists of relationships and social arrangements passed on and institutionalized to handle routinely the characteristic problems of that society (Hollander 1967). By providing social reality the basic psychological effect of culture is to influence individuals towards distinctive ways of thinking and acting. It is important then to understand the effects of cultural norms on movement participation.

(a) Goal-directed

Within the context of this section 'goal-directed' movements include everyday functional actions involved in carrying out objective tasks in various aspects of societal behaviour. Movement patterns concerned with eating, which whilst following broad similarities, entail quite different actions dependent upon sub-cultural norms. Manuals on gracious manners substantiate this (Fast 1970). Morris (1967) has written about eye movements, innately determined he suggests, involved during the movement patterns in eating. Jourard (1967) recounted the social touching patterns involving in meal-taking in various countries. Walking, running, sitting movements are completely functional, yet these have been influenced by types of clothes worn. Fashion, according to Broby-Johanson (1968) is an attempt to unify all the expressive capacities of language, gesture, physiognomy in a given society—a display art. Clothing reflects attitudes and values, but also influences day-to-day movements within society. There are clothes to suit the climate, to attract and emphasise sex differences, to reveal dominance, to restrict or facilitate movement.

Certain stereotype movement patterns can be noted. There are male, female differences; which are partly anatomical, partly cultural expectations and imprinting. Cratty (1967) has discussed reasons for males with a 'homosexual' gait.

Class differences are hard to substantiate though Melly (1965) has indicated social class differentials in posture and walking in London. Physique apparently is influential in movement participation in various activities (Parnell 1958; Tanner 1964).

The relation of man to his environment also requires the attention of research. Both the housewife and industry wish to utilise man-movement in the most efficient way for economy. There have been time and motion studies (Mundel, 1950; Etzioni, 1970; Seymor, 1968; Singleton, 1970) and analyses of efficient movement (Laban & Lawrence, 1947; Broer, 1966; 1968; McClurg Anderson 1951).

Efficient movement in sporting endeavours has also been investigated (e.g. Bunn, 1955) and patterns of movement within team games studied (Elias & Dunning, 1966). Effective patterns emergent from tactical moves are discussed in various coaching manuals.

Cross-cultural and sub-cultural differences in anthropometric measurements, together with direction-of-movement stereotypes, and life-style differences raise important questions about the universality of human factors principles (Ergonomic News 1971). Glazer & Moynihan (1963) discuss ethnic groupings in cities, and the laws to control movement, including human movement, in cities have been considered (Mumford 1961). The effects of overcrowding have been discussed (Chombart de Lauwe, 1959) and this has obvious implications for the movement patterns of various sub-cultural groupings. The new skyscraper flats may also influence movement opportunities from childhood to old age. Hall (1959) has pointed out the kinds of sensory deprivation which result from living in skyscrapers. Modern architecture insulates man from environmental and human contact (Hall, 1959), and the reduced 'flight distance' in modern man has led to movement in cities being designed for the avoidance of touch (Morris, 1967; 1969) despite the crowds of people who move around daily.

Man has a need to experiment and explore physically and mentally. The rhythmic physical play movements of children become elaborated in many complex forms of exercises and sports in adulthood as a means of expanding specialized forms of exploration and experimentation even in urban societies (Morris, 1967).

(b) Excitatory

In a society so geared to sex and 'status sex' (Morris, 1969) excitatory movements will be in evidence. Sexual movements are obviously not easy to document. However, Kinsey et al (1948; 1953), Ford & Beach (1952) and Masters & Johnson (1966) report certain class differences that could be related to values of immediate or deferred gratification patterns. Again, the availability of literature would indicate an advantage to the literates in experimentation and variation in sexual movements. There is a trend towards pictorial presentations which could be explained as a way of catering for the less literate.

Kaelin (1964; 1968) examined the aesthetic aspects of coitus. Pre-copulatory movement entail complex facial expressions, bodily postures, vocalization (Morris, 1967). Despite the evidence of excitatory movement in society, there is retention of the 'play boy' philosophy of 'look, admire, but do not touch'—strip-tease, go-go discotheques, etc. These are ritualized movements of the earlier stages of sexual sequence. Postures and movements are exaggerated to compensate for incompleteness (Morris, 1969). Morris (1969) and Nuttall (1968) have suggested that even the guitar has changed its shape to become a phallic symbol and used in conjunction with the performers' movements at pelvic level fulfils its exitatory potential—symbolic masturbation. Goffman (1961) speculated that individuals are so afraid of revealing themselves fully by body language in love-making and thus only 'unmask' in complete darkness—though sometimes even this is not sufficient. In terms of sexual practice, Kinsey et al (1953) has shown that if anything, the middle class tends to be more experimental than the working class, and less apt to shield its emotions.

Fast (1970) suggests that before the child is inhibited by social mores he loves to touch a variety of objects including his own genitals. Body movements both conscious and unconscious reveal sexual availability in women. This is characterised by movie sex-symbols. Scheflen (1968) talks about 'quasi-court-ship'—the use of flirting movements to achieve non-sexual goals. In sexual behaviour eye movements and contacts play an important part though there is little systematised evidence (e.g. Fast 1970). There are taboos on mutual looking which are widespread as being associated with the sexually arousing properties of eye-contact. Hess (1960) suggested that enlarged pupil size in a woman is sexually arousing to a man, whereas narrow pupils indicate an aggressive look. Argyll (1967; 1969) has commented on the implications of bodily contact. Cultural differences, sex differences and physical proximity have been documented (Kinsey et al, 1948; 1953). Jourard (1966) has discussed access to body zones. Not all exitatory movements are directly sexual. 'Pop' allows an outlet of aggressive release (Guppy 1970). There is plenty of movement, but again lack of touch. Movement associated with colour and disguise can excite (Frankenburg 1957). Dennis et al (1956) and Williams (1956) felt that this in part was the attraction of football matches for the working class. Dress, posture and colour are seen to be related to certain cultural values (Bantock 1963; Musgrove 1964). Models in beauty contests use movement-aids to enhance sexuality, e.g. high heels to exaggerate buttock movements (Morris 1967). Play movements can also be considered to excite as Huizinga (1938) pointed out; dress for games is a disguise and the player is intensely and utterly absorbed. Caillois (1961) saw play as spectacular and ostentatious movement. The costume for the dance no doubt enhances the dancer by movement to reveal through emotion the abandonment of self to the actions, moods of the performance as exemplified by Isadora Duncan. From the beginning, modern dance utilised costume in various ways—some dancers reduced clothing to the minimum (even dancing in the nude in certain productions) to avoid hampering body movement. In dance styles other than modern dance, costume can be used to decorate (as in most ethnic dances), to deceive (e.g. some primitive dance) or to distort (dance performance inside net tubes). In classical ballet costume has been traditionally used to add height and give added fluency of movement.

(c) Bizarre movements

Movements under the influence of drugs, in modern parlance 'freaked out' are relevant here, as are the withdrawal symptons of the drug addict or alcoholic or the passivity of schizophrenics. Movements prior to death might also be categories in this section.

Agitated movements under fear stress as with the fictitious martian invasion inspired by Orson Wells on American radio can be communicated within a crowd, and reverberate wider and wider through the community (Cantril, 1940). Szasz

(1961) has developed the view that hysterical bodily symptoms are a kind of non-verbal communication resorted to when words have failed. It has been shown that as personality disintegrates movements become more stereotyped and restricted (Wolff 1943). Argyle (1969) has reported on the jerky, poorly controlled, overdramatic movements of maniacs and neurotics. Dancing mania of the Middle Ages—where the individual moved till unconsciousness or even death—has been confused with the concept of the 'danse Macabre'. In fact, so-called dancing manias were considered by Backman (1952) to be a form of ergot poisoning. Similarly, it should be remembered that any form of gross, bizarre movement was termed 'dance' in medieval times (e.g. St. Vitus dance). Shipman (1969) noted that such movements of a maniac nature historically and cross-culturally occurred at times of severe stress—famine, war, epidemic—the actions were a way of dispelling fear.

Cross-laterality may be looked upon as bizarre movement of a sort in our right-handed society (Barsley 1970). Autism manifests itself as a preference for taste and touch, bizarre movement patterns, ritualised gesture (Savage, 1968). Skinner's (1953) techniques of operant conditioning in many cases treat the overt responses instead of seeking a root cause. Hunt's (1962) suggested treatment of brain-damaged children is for continued body contact in play situations to improve social behaviour. Psychodrama is a recognised technique in aiding the maladjusted by 'playing out' through physical movements their fears, anxieties and phobias (Morgan & King 1967). Relearning the language of the body has been suggested as a method of sustaining mental health, and is concerned with touching procedures in a group setting (Schutz 1968).

(d) Aesthetic movements

Aesthetic movements have been dealt with elsewhere. This writer would only point out that as Kaelin (1968) states, in sporting activities the outcome is aesthetically irrelevant though competitive sport reaches into aesthetics when the winner narrowly surpasses an opponent. Lamb (1965) has drawn an important distinction here between performer and observer. There is an additional sex variable as when a male judges notions of the beauty of the female in motion. The spectator may be involved with sporting interest disregarding the aesthetic, and the activity is thus a process of identification and catharsis (Elias & Dunning 1967) though with more detachment the movements can be viewed aesthetically. Carlisle (1969) has suggested that the competitive element is part of the aesthetic medium but this seems most likely post-activity. Lamb (1965) however contends that there is beauty in all physical movements—the labourer, walker, athlete, dancer, etc., while Henry (1965) distinguishes between everyday movements and other activities yielding aesthetic content. Grossner (1951) has pointed out that Greek architecture was hundreds of years ahead of painting because it was primarily a tactile and kinaesthetic art, and Hall (1959) considers that one mistake made by museums is that they do not allow people to touch sculpture. It is also suggested by

Hall (1959) that certain painters force the viewer to find an appropriate distance, focus on one central aspect of the canvas, or like Hobbema's painting, the viewer is made to turn head, bend neck, and move to appreciate the total aesthetic scene. The movements are vital to a complete appreciation of the particular art form. Hinks et al (1971) have outlined the central role dance plays in aesthetic aspects of movement and Reid (1969; 1970) has indicated the range of movements including borderline activities in the aesthetic realm.

(e) Communicative movement

A number of studies have examined the implications of non-verbal communication in social interaction and interpersoi.al perceptions. (See Section D). Argyle (1969) suggested interpersonal attitudes are revealed by such cues as spatial proximity and posture. Ekman & Friesen (1967) have indicated the non-verbal gestural patterns that accompany speech. Non-verbal cues have greater impact than the verbal, and are of use in signifying interpersonal relationships as in bodily contacts (Frank, 1957). Kendon (1967) and Birdwhistell (1968) have devised approaches to analyse the basic movements of face and body, and Krout (1954) has related different emotional and motivational states to particular hand and body movements. Sarbin & Hardyck (1953) have outlined a number of emotional states communicated to others by posture. Weisbrod (1965) has shown how dominance seems to be established by staring others down, while Scheflen (1965) and Goffman (1961) have stated that posture can reflect reactions to, and status with, the group.

Secord (1958) suggested inferences from bodily appearance and movements to personality are made by analogy or metaphor—e.g. a person with coarse skin is coarse, full lips and swaying hips *is* highly sexed, etc.

Bodily proximity is another relevant way of considering social reactions, and Sommer (1965) and Cook (1968) have shown how extraverts tend to choose 'competitive' sitting positions vis-a-vis others. Williams (1970) determined that introverts tend to keep people at a greater conversational distance than extraverts.

It is known subjectively in relation to the body-taboo that contact, with differing connotations, can be established by 'light' or 'heavy' hands. Some people are compulsive 'touchers' even in the wrong social context (Fast 1970). Encounter groups (Schutz, 1968) have been established to encourage 'touch' for those starved of tactile communication in our society. Jourard's (1966; 1967) findings suggest links with movement within a sporting context—physical contact games, non-contact sports, etc.—a continuum of 'proximity-distance' could be applied. As well as for other well-documented reasons (e.g. physique, temperament) 'touch hungry' individuals could well be attracted towards certain types of movement situations as one sanctioned acceptable outlet for bodily contact in present-day western society, others could well be repelled by such contacts. Stokes (1968) has suggested a sexual element in games, and again one can see connections with

Jourard's (1966) postulations. Efron's (1941) study has revealed sub-cultural (immigrant) differences in movement patterns in New York, and also the gradual assimilation culturally of their movements, which produced some 'bi-gestural' people. Movement patterns are often reflective of cultural attitudes. Threatening and defensive gestures can also be communicated (Brannigan & Humphries, 1969). Opera relies to an extent on communication and interest in the movement and colour, as the story is known though often sung in a language foreign to the land of its performance. Ballet also is reliant upon recognisable movement forms and specific actions (for example, the kiss). Communication at a non-verbal level can utilise movement cues and physical distance. The need for space and movement is so strong that even in a crowd each individual will demand a given amount of space (Fast, 1970). As a crowd gets larger and tighter and compact it may also become uglier. A loose knit crowd may be easier to handle. Fast (1970) outlines the use of space in psycho-analysis and police interrogation—movement into and out of 'personal territory'. Sommer (1969) talks about 'personal buffer zones' and 'recognition ceremonies' related to body movements utilised in social interactions in cafeterias etc. These actions intimate willingness or reluctance to share seats, tables and so forth.

(f) Group-Status

The number of people involved apparently has some effect on movement efficiency within the group. Weyner & Zeaman (1956) report that in an accuracy task using up to four subjects per team that better performance was produced by the larger teams. Cratty (1967) also recorded size as an important group variable in efficiency of movement—in this case a carrying task. Comrey & Deskin (1954) have ascertained that movement efficiency within a group setting is relatively independent of how the individuals perform separately. Sherif & Sherif (1953) have shown how shared movement activities enhance friendship, and can specifically split previously formed alliances (i.e. propinquity and friendship.)

Mutual attraction can be developed by external group threat (Pepitone & Kleiner 1957), but whereas success in competitive movement situations heightens interpersonal liking, failure reduces friendship within the group (Wilson & Miller 1961).

Fieldler (1967) in basketball studies has reported that more successful teams are those who subordinate mutual attractiveness to the desire to win by efficiency of movement; hostility generated for the less skilled. Cratty & Sage (1964) found sociability can adversely affect motor skill performances. In sum, the affiliation individuals have in groups formed for the purpose of engaging in human movement have the potential for affecting both the individual's actions and psychological state.

The emergence of leaders has implications for movement study. Weber (1950) has talked of charismatic leaders, and Lamb (1965) discussed 'charismatic movement' leaders. 'Kinesic leaders' who are not 'meta-incongruent' (i.e. whose movements are not at variance with the verbal statements they accompany) are explained by Birdwhistell (1960). He also discusses the importance of movement dominance despite having a low percentage of verbalisation in the group setting. Grusky (1963) noted managers (previously baseball players) tended to be infield players—close physical proximity with others. Kiker & Miller (1967) have shown that the charismatic leader is recognised by physique, facial expressions and posture. Carter et al (1951) have stated that leaders in physical performance situations are those who act and make decisive decisions. Lippitt et al (1952) stated that children's size, height and age, tend to be important factors in dominance. In maintaining leadership status, perhaps gained in another situation individuals take physical risks (Cratty & Sage, 1964). Hardy (1937) has shown that group leaders in the young are distinguished in some form of valued movement activity—running, throwing, etc. In adult leadership role, stature seems more important than direct physical action—staring down, imposing a physical presence. In dominance-submission situations Morris (1967) stated that individuals 'lie' more with words than other communication signals. Primate studies indicate that dominance is established by staring down, facial expressions and colour-ritualised intention movements. In human situations there is the elevation of orators, appropriate seating arrangements etc. Use of distance and space zones by circus trainers utilises knowledge of dominance-submission movements. 'Feminine' actions emerge in submission, e.g. being caned on the bottom by teachers (Morris 1969). Displacement activities in aggressive encounters are not restricted to a few species-specific displacement patterns (Morris 1967). Gesture has evolved as a bloodless substitute in status struggles (Morris 1969). The importance of even subtle movement cues is emphasised in a study by Eisenberg & Reichlane (1939). Women labelled dominant and submissive (by personality inventory) were successfully differentiated by observers viewing films of their walks.

In social psychological terms norms and roles are in the nature of social expectancies which individuals share in their psychological fields. Roles refer especially to those particular expectations regarding appropriate behaviour for a person occupying a particular position or status:—the movement performance of potential leaders (Cratty & Sage, 1964) group determinants on levels of physical performance in the Norton gang (Whyte, 1943). Movement configurations within the total pattern of the game of 'star' players (Elias & Dunning, 1966) are relevant here. Indeed this consideration can be viewed in relation to royalty. Historically their position in court dance was determined by status. With increased democracy they now engage in sporting endeavour on a class basis against restricted opposition. The reaction of the British press to royalty dancing 'the twist' still reveals their surprise at such 'democratic' movement.

(g) **Ritual**

This section will cover such diverse areas as great sporting occasions and religion.

To Laban (1948) party games, masquerades, dance, games, ceremonies, rituals all form part of the art of movement, but here the concern is with the involvement and understanding and attitude of the spectator in which the participant (eventually) becomes the object of attention.

Older forms of society found their relaxation and emotional outlet in the occasional, traditional festival. Fitzstephen (1772) for instance, in describing aspects of the social history of England in the Middle Ages, referred to a host of practices which offered release from taboo and convention in games, music, dance and convivial buffoonery. Movement appears to be the 'release mechanism' against stringent social control.

Association football is a good example of the particularly appealing powers for ritualistic movement associated with games. Crowd movements reveal a clear pattern—converging on the stadium, expressing emotion and colour at the tides of fortune during the game, spreading out to disperse into the city at the end. These are the ritual movements and displays associated with team support. Ritual either serves to consolidate a group or to divide one group from another. Religious movements and gestures have a symbolic and ritual significance. Symbols that emerged from movements involved in small intimate groups (e.g. laying of hands—Jungmann, 1959) have extended to movements made by the priest to the congregation (Oesterley, 1923; Turner 1969). Dance is essentially ritualistic movement (Shipman 1969) and as the dance once served as part of religious activity, now religious ritual is performed by priest and parishioners playing separate roles (Shipman 1969). Their movement patterns are separate to an extent. Slow march at military funerals is played out as ritualized movement patterns of respect for a departed hero.

(h) Protest

Modern popular music and movement serves an important differentiating function at a time when rapid social change has created the generation gap (Shipman, 1969). Movements, dress, music, language, emphasise this gap. Patterson (1966) suggested that these movement forms appeal to the young, as one of the few socially acceptable outlets for aggression in western society. Excitatory 'pop' heroes gyrate and move upon the stage to evoke imitation and involvement in a protest movement.

Protest marches have their own special rituals. They contain individuals from all classes at times, or from a particular group making a special protest. There is a chain of marchers who disrupt traffic, chanting slogans, waving arms, 'sitting down' or making a symbolic presentation of a document to an authority figure. Often there is a clash with police that, whilst violent, as with the Japanese students, again provides an outlet in society with strict control. Involvement theatre like 'Hair' requesting the audience to participate physically, and 'happenings' to cause real physical movement participation instead of a passive audience, are all part of this new trend. Pop festivals with Hell's Angels to contain the surplus violence: there is noise, sex, music and movement of great crowds of people.

(i) Stillness—The containment of movement

Morris (1969) uses the term 'chemical dreaming' to describe the state where movement is suspended under the influence of L.S.D. or the cut-off trance-like conditions of voodoo, yoga and hypnotism.

An important aspect of human movement (or its antithesis) in a busy, noisy, hurrying modern society are institutionalised areas where movement is restricted or reverently hushed. Throughout society there are restrictions and rules related to stringency and sanctions in different areas of that society. There may well be cross-cultural differences in movement restrictions. Areas set aside for particularly restricted movements or suspension of movement such as:

i) Churches—for spatial communication (Greek temples are small and intimate—Hall 1959). They are open for quiet meditation.

ii) Libraries—hushed, stilled for concentration and study.

iii) Underground trains—restricted, people being transported passively.

iv) Prisons—movement contained 'for the good of society'. Space is limited.

v) Parks—show the leisurely movements of the old in a 'safe' environment and the exhuberant movements of childhood.

vi) 'Captive populations'—on the parade ground or in the classroom.

vii) Meditation and other methods that lead to spiritual experiences—fasting, breathing and other exercises of yoga (Argyle, 1958).

viii) Sky-scraper flats—restricted sensory input and movement.

ix) As an artistic concept (Clark, 1956; Aldrich, 1963)

G.4. Some Cross-Cultural Comparisons of Classifications

This section is presented merely to give some examples of the previous section in societies other than our own, and to consider the movement classification presented so that cultural variations or universalities in movement patterns can be noted.

a) Goal-directed

Caillois' (1961) classification has been used for cross-cultural studies, and Sutton-Smith & Roberts (1963) have considered cultures without games, and cultures where particular kinds of movement activities predominate. They hypothesised that these movement activities were expressive models of culturally held values in a society, produced by anxieties and conflicts stemming from particular child-rearing practices. The vital role of games, sports and other movement activities in three Pacific cultures in the total cultural pattern of these societies has been analysed (Stumpf & Cozens, 1947; Dunlap, 1951).

How long man can sustain his socio-cultural advance without biological change or alterations to his movement patterns in Western society is questionable. He is over-fed and/or under-active, and makes great use of the automobile for transportation. A need for movement may exist to offset hypokinetic diseases (Kraus & Raab 1961). Increased research support indicates the importance of movement in the prevention of somatic and mental derangement. It might be suggested that modern man is heading towards a kind of cultural motor impairment . . . . . unless he becomes involved again in movement activities of a relatively strenuous type. Preliminary evidence from America indicates the physical and emotional values of vigorous movement (Greenwood 1967).

It has already been stated that motor skills are related not only to the child's developmental status, but also to his cultural background. New Guinea coastal people have a 'boat-dependent' culture and children learn to balance at an early age standing upright in long narrow canoes (Sorenson & Gajdusek, 1966). Cross cultural movement studies are scarce. However, it is possible to suggest that cultural differences in stature may produce some ethnic movement differences. For example, pygmies' movement patterns tend to be play activities—an imitation of hunting and gathering activities (Turnbull, 1965). The masai, on the other hand, are tall, used to walking and have exceptional cardio-vascular fitness. Obviously, cultural differences in movement patterns obtain here. Cultural universality of various forms of patterned human movement—play, games, sports, have been recognised (Murdock, 1945). Yet there seem to be few universals in communicative movements (Fast 1970). Dance, ritualistic and dramatic, takes on individual cultural patterns to fulfil needs of the people (Nida, 1954). Movement patterns which interest the populace reveal the cultural continuity which is reflected in games. (See for example Read, 1959; Maccobyn et al, 1964; Zurcher & Meadow, 1967; Luschen 1967; Riesman & Denny 1963). Climatic conditions may also affect movements if only because man, as with other mammals, has been biologically designed to withstand stress. Indeed, there will also be variations in the movements involved in modes of transportation, e.g. Eskimo's snow-shoes.

Racial differences, whilst to date unsubstantiated by research, may be linked to climatic states. Certainly the vigour and striving of the European stereotype may be contrasted with the slower movement patterns of the stereotype of sunnier shores! Ethnic postures have been alluded to earlier and this can be exemplified by subjective reports that to cater for the particular squatting posture of the Japanese, 'one arm bandits' are placed at ground level in the most unusual sites! Japanese geishas bind their middles tightly because protruding bottoms are considered unsightly. They also walk with feet turned in. This produces a gliding movement of the body in motion, which the kimono emphasises. Their squatting posture from childhood gives beautifully developed shoulders and strong ankles. The European slender ankle is a product of high heels giving alteration to the leg muscles (Broby-Johansen, 1968).

Differences between Japanese, Germans and Americans in social proximity within domestic settings have also been reported (Fast, 1970).

The relationship between fashion, movements, and social attitudes has been dealt with by Broby-Johansen (1968). In relation to movement, the mini-skirt gives complete freedom. Broby-Johansen (1968) makes reference to this by suggesting that modern apparel not only shows shapely legs, but actually encourages these curves by laying down an extra layer of fat. Morris (1969) suggests short skirts are used to emphasise the length of leg and movements. Fashion then can influence movements. One has only to think of the movement of nuns, or the newspaper reports of the dangers of movements while wearing a maxi, such as driving, crossing busy streets quickly or coming down stairs.

In India the sari and dhoti are worn for purposes of ventilation, but binding the waist also emphasises exotic movements in dance. Historically foot-binding in Japan was carried out to provide visual evidence of women's inability to work, man was the provider. But a hidden consequence was to guarantee fidelity by restricted perambulations. Eating patterns within broad general movement categories again reveal great cultural variations. Hewes (1957) has reported that there are a vast number of movements patterns characteristic of ethnic groups which are apparently learned, since they seem to be unique to specific groups, (c.f. Efron, 1941). For example, there are a number of specific and non-specific movement customs which govern the action of individuals eating, playing, etc. unique to Americans. While the influence of fashion on movement has been emphasised here, it is obvious that, to some extent, this influence is reciprocal.

b) Excitatory movements

Within basic patterns excitatory movements have specific cultural implications. Compare, for example, Egyptian belly-dancers, western strippers, hindu temple dancers in paintings. Women from Spanish-speaking countries can flaunt their sexuality in their walk because of the strict code of social behaviour. In western societies these movements would be considered exaggerated, graceless and in some sub-cultures, unnatural (Fast, 1970).

c) Bizarre movements

There is such contrast as the trance-like state of South American Indians under the influence of mescaline or cocaine, the violent movements of voodoo and limbo dancers, the colour, riot, excitement and movement of South American carnivals where all social norms are relaxed.

To some extent the movements of the left-hander at work or play in this 'alien' right-handed world may also be categorized as bizarre (Barsley, 1970).

d) Aesthetic Movement

Movements may reveal much of social significance. Kealiinowomoku (1970) has considered ballet as ethnic dance, and modern dance in Western Europe may exemplify the high valuation of individualism, love of freedom, democratization and self-expression (Veerhault, 1969).

e) Communicative Movement

Hall (1963) has outlined cross-cultural differences in body proximity involving social interaction, where western Europeans dislike intrusion by others into their personal 'bubble', whereas Arabs and Latin American stand closer, Campbell et al (1966) found subjects moved closer to preferred racial groups. Differences in proximity within cultures and in different social settings have also been investigated (Sommer 1959). There are also reported variations in permitted social touching as a means of communication and contact (Jourard, 1967).

Frank (1957) has stated that physical contact is used far more in some cultures than in others. Mediterranean people for instance use touch considerably, whereas it is used relatively little in north-east Europe. Watson & Greaves (1966) discovered that Arabs in conversation faced directly, touched more often, looked oftener than Americans.

Within different cultures emphasis on particular modes of information processing may differ. Wober (1966) coined the term 'sensotype' to represent the pattern of relative importance of the different senses by which a child learns to perceive the world. He found in African cultures an emphasis on proprioceptive and auditory information rather than visual and suggested that such a 'sensotype' would be different from that of an individual skilled in the visual world. In this culture at least movement experiences would seem to predominate.

Innumerable cultural variations in gestures which communicate can be cited (Kohler, 1927; Fast, 1970). Labarre (1964) reported cross-cultural differences in the movements involved in greeting and parting from people. Laban (1960) also mentioned cross-cultural gesture interpretation. Dusenbery & Knower (1938) found that when people of one culture attempted to judge the communicative movements of another little accuracy was possible.

International deaf language, however, exists. Ability of the deaf to 'read' body language conveying emotions has been reported (Fast, 1970). Yet generally the physical activity of man is dependent upon, conditioned by, and adapted to, cultural influences.

Technology of a complex nature may have implication on the movement patterns of particular societies, and increased technological change throughout the globe may reduce the indiginous differences in movement patterns.

f) Group-Status

The leader's role in Chiricahan society is related, not to age and wisdom, but to his movement abilities, particularly in connection with hunting (Opler, 1946).

Certain aborigines, to be polite, must talk to each other without looking (Fast, 1970).

g) Ritual Movement

Ritual movements in societies other than ours reveal certain differences. For example, there is colour, music, noise and movements at Indian funerals. Dancing at funerals reunites a group after a loss. Emotions are released and group unity is symbolised in the movements of the dance. Amongst West Indians, for example, movement and emotional expression promotes a rapid return to normal life (Shipman, 1969). In certain societies elements of consolidating movement rituals survive (e.g. the kibbutz) (Shipman, 1969). The ritual patterns of team support of South American football teams—noise, colour, agitation, gesticulation—is well documented. To some extent the turning movements of the Dancing Dervishes in their attempts to achieve religious ecstacy contains a ritualistic element.

h) Protest

Encounters between students of various societies (e.g. Japanese, German) and police are a recognised (if not legitimate) outlet where strict social controls exist. Perhaps protest movements can be considered as a universal behaviour pattern, though in certain cultures there may be greater violence and physical contact in the clashes between authority and the protesters.

i) Stillness

The stillness of meditative states in eastern communities differs from our relatively active approach to worship. This control of movement renders certain individuals capable of bodily responses not normally under conscious control.

ALDRICH V.C. (1963). "The Philosophy of Art". New Jersey: Prentice-Hall.

ALLOTT R. (1966). An investigation into immediate post-school physical recreation. *Res. Phys. Educ., 2,* 29-40.

ANDERSON H.H. (1937). Domination and integration in the social behaviour of young children. *Genet. Psychol. Monog.* **19,** 343–408.

ARGYLE M. (1958). "Religious Behaviour." London: Routledge & Kegan Paul.

ARGYLE M. (1967). "The Psychology of Interpersonal Behaviour". Harmondsworth: Penguin.

ARGYLE M. (1969). "Social Interaction". London: Methuen.

BACKMAN E.L. (1952). "Religious Dances in the Christian Church and in Popular Medicine". London: Allen and Unwin.

BAER D.M., HARRIS R.F. & WOLF M.M. (1963). Control of nursery school childrens behaviour by programming social reinforcers from their teachers. Unpublished manuscript, Univ. Washington.

BAMBER J. (1966). Motor impairment. M.A. Thesis, Univ. Glasgow.

BANDURA A. & WALTERS R.H. (1964). "Social Learning and Personality". New York: Holt, Rinehart & Winston.

BANTOCK G.H. (1963). "Education in an Industrial Society". London: Faber and Faber.

BARSLEY M. (1970). "Left-handed Man in a Right-handed World". London: Pitman.

BERNSTEIN B. (1961). Social class and linguistic development: a theory of social learning. In A.H. Halsey, J. Flood and C.A. Anderson (Eds.) "Education, Economy and Society". Glencoe: Free Press.

BERNSTEIN B. & HENDERSON D. (1969). Social class differences in the relevance of language to socialisation. *Sociol., 3,* 1-20.

BIESHEUVEL S. (1963). The growth of abilities and character. *S. Afr. J. Sci.,* **59,** 375-385.

BIRDWHISTELL R.L. (1960). "Kinesics and Communication: explorations in communication". Boston: Beacon Press.

BIRDWHISTELL R.L. (1962). "Introduction to Kinesics". Louisville:

BIRDWHISTELL E.L. (1968). Kinesics. *Int. Encyclo. Soc. Sci.,* **8,** 379-385.

BRANNIGAN C. & HUMPHRIES D. (1969). I see what you mean. *New Scientist,* 22nd May, 406-408.

BRECKENRIDGE M.E. & VINCENT E.L. (1966). "Child Development". London: Saunders.

BRIDGES K.M.B. (1932). Emotional development in early infancy. *Child Dev., 3,* 324-341.

BROBY-JOHANSON R. (1968). "Body and Clothes". London: Faber and Faber.

BROER M.R. (1966). "Efficiency in Human Movement". Philadelphia: Saunders.

BROER M.R. (1968). "An Introduction to Kinesiology". London: Prentice-Hall.

BUHLER C. (1935). "From Birth to Maturity". London: Kegan Paul.

BUNN J.W. (1955). "Scientific Principles of Coaching". New York: Prentice-Hall.

BURNS T. (1964). Non-verbal communication. *Discovery,* **25,** 10, 30-37.

CAILLOIS R. (1961). "Man, Play and Games". Glencoe: Free Press.

CAMPBELL D.T., KRUSKAL W.H. & WALLACE W. (1966). Seating aggregation as an index of attitude. *Sociometry,* **29,** 1-15.

CANTRIL H. (1940). "Invasion from Mars." Princeton: University Press.

CARLISLE R. (1969). The concept of physical education. Conference proceedings, Philosophy of Education Society of Great Britain.

CARTER L.F., HAYTHORN W., SHRIVER B. & LANZELLA J. (1951). The behaviour of leaders and other group members. *J. Abnorm. Soc. Psychol.,* **46,** 589-595.

CHOMBART DE LAUWE P. (1959). "Famille et Habitation". Paris: Editions du Centre National de la Recherche Scientific.

CLARK K. (1956). "The Nude". Harmondsworth: Penguin.

COLEMAN J.S. (1961). "The Adolescent Society". Glencoe: Free Press.

COMREY A.L. & DESKIN G. (1954). Further results on group manual dexterity in men. *J. Appl. Psychol.,* **116.**

COOK M. (1968). Studies of orientation and proximity. Unpublished paper, Oxford Institute of Experimental Psychology.

CRATTY B.J. (1959). Athletic and physical experiences of fathers and sons who participated in physical fitness testing at Pomona College. *Calif. J. Educ. Res.,* 207-211.

CRATTY B.J. (1960). A comparison of fathers and sons in physical ability. *Res. Quart.,* **31,** 12-15.

CRATTY B.J. (1967). "Social Dimensions of Physical Ectivity". New Jersey: Prentice-Hall.

CRATTY B.J. & SAGE J.N. (1964). The effects of primary and secondary group interaction upon improvement in a complex movement task. *Res. Quart.,* **35,** 265-274.

DENNIS N., HENRIQUES F. & SLOUGHTER C. (1956). "Coal is Our Life". London: Eyre and Spottiswood.

DENNIS W. (1957). A cross cultural study of the reinforcement of child behaviour. *Child Dev.,* **28.**

DOLL E.A. (1937). The inheritance of social competence. *J. Heredity,* **128.**

DUNLAP H.L. (1951). Games, sports, dancing, and other vigorous recreational activities and their function in Samoan culture. *Res. Quart.,* **22,** 298-311.

DUSENBERY D. & KNOWER F.H. (1938). Experimental studies of the symbolism of action and voice. *Quart. J. Speech,* **24,** 424-435.

EFRON D. (1941). "Gesture and Environment". New York: King's Crown Press.

EISENBERG P. & REICHLINE P.B. (1939). Judging expressive movements; II Judgment of dominance feeling from motion pictures of gait. *J. Soc. Psychol.,* **10,** 345-357.

EKMAN P. & FRIESEN W.V. (1967). Head and body cues in the judgment of emotion: a reformulation. *Percept. Motor Skills,* **24,** 179-215.

ELIAS N. & DUNNING E. (1966). Dynamics of group sports with special reference to football. *Brit. J. Sociol.,* **18,** 4.

ELIAS N. & DUNNING E. (1967). The quest for excitement in unexciting societies. Paper to B.S.A. Conference, London, April, 1967.

EMMETT I. (1971). "Youth and Leisure in an Urban Sprawl." Manchester: University Press.

E.R.S. NEWS (1971). Preliminary announcement on: NATO Symposium on national and cultural variables in human factors engineering. Newsletter of the Ergonomics Research Society, Number 19.

ETZIONI A. (1970). "A Sociological Reader on Complex Organisations". London: Holt, Rinehart & Winston.

FAIRWEATHER O.V. & ILLSLEY R. (1960). Obstetric and social origins of mentally handicapped children. *Brit. J. Prev. Soc. Med.,* **14**, 149-159.

FALKNER F. (1967). "Human Development". London: Saunders.

FANTZ R.L. (1961). The origin of form perception. *Sci. Amer.,* **204**, 459-475.

FAST J. (1970). "Body Language". New York: Evans.

FELDENKRAIS M. (1949). "Body and Mature Behaviour". London: Routledge & Kegan Paul.

FIELDLER F.E. (1967). "A Theory of Leadership Effectiveness". New York: McGraw-Hill.

FISH B. (1961). The study of motor development in infancy and its relation to psychological functioning. *Amer. J. Psychiat.,* **17**, 113-118.

FISKE D.W. & MADDI S.R. (1961). "Functions of Varied Experience". Illinois: Dorsey.

FITTS P.M. & POSNER M.I. (1967). "Human Performance". California: Brooks-Cole.

FITZSTEPHEN W. (1772). "Description of the City of London". London: Printed for B. White.

FORD C.S. & BEACH F.A. (1952). "Patterns of Sexual Behaviour". London: Eyre & Spottiswoode.

FRANK L.K. (1957). Tactile communication. *Genet. Psychol. Monogr.* **56**, 209-225.

FRANKENBURG R. (1957). "Village on the Border". London: Cohen and West.

FREUD S. (1961). "The ego and the id". In Collected Works, Vol. 19. London: Hogarth.

GESELL A. & THOMPSON H. (1934). "Infant Behaviour, its Genesis and Growth". New York: McGraw-Hill.

GIESCKE M. (1935). The genesis of hand preference. *Memo. Soc. Res. Child Dev.,* **5**. National Research Council, Washington, D.C.

GLAZER N. & MOYNIHAN D.P. (1963). "Beyond the Melting Pot". Cambridge: MIT Press.

GOFFMAN E. (1961). "Asylums". New York: Anchor Books.

GOLDFARB W. (1945). Effects of psychological deprivation in infancy and subsequent stimulation. *Amer. J. Psychiat.,* **102**, 18-22.

GREEN R. & MONEY J. (1961). Effeminacy in prepubertal boys, summary of eleven cases and recommendations for case management. *Pediatrics,* **27**, 2.

GREENWOOD E. (1967). Emotional well-being through sports. *J.O.H.P.E.R.,* **38**, 9.

GRIFFITHS R. (1954). "The Abilities of Babies". New York: McGraw-Hill.

GROSSNER M. (1951). "The Painter's Eye". New York: Holt, Rinehart & Winston.

GRUSKY O. (1963). The effects of formal structure on managerial recruitment: a study of baseball organization. *Sociometry*, **3**, 147-152.

GUPPY N. (1970). "Wai-Wai". London: Penguin.

HALL E.T. (1953). "The Silent Language". New York: Premier Books.

HALL E.T. (1959). "The Hidden Dimension". New York: Doubleday.

HALL E.T. (1963). A system for the notation of proxemic behaviour. *Amer. Auth.*, **65**, 1003-1026.

HARDY M.C. (1937). Social recognition at the elementary school age. *J. Soc. Psychol.*, **8**, 365-386.

HARTUP W.W. (1964). Friendship, status and the effectiveness of peers as reinforcing agents. *J. Exp. Child Psychol.*, **1**, 154-162.

HEBB D.O. (1949). "Organisation of Behaviour". New York: Wiley.

HELANKO R. (1963). Sports and socialization. *Acta Sociologica* 2. Reprinted in N.J. Smelser and W.T. Smelser (Eds.) "Personality and Social Systems". New York: Wiley.

HENRY F.M. (1965). Physical education: an academic discipline. *The Leaflet*, Jan-Feb.

HERSHENSON M., MUSINGER H. & HESSEN W. (1965). Preferences for shapes of intermediate variability in the newborn human. *Science*, **144**, 315-317.

HESS E.H. (1960). Pupil size as related to interest value of visual stimuli. *Science*, **132**, 349-50.

HEWES G.W. (1957). The anthropology of posture. *Sci. American*, **196**, 123-132.

HEWETT S., NEWSON J. & NEWSON E. (1970). "The Family and the Handicapped Child". London: Allen & Unwin.

HEYNS O.S. (1963). "Abdominal Decompression". Johannesburg: Witwatersrand University.

HINKS E., ARCHBUTT S. & CURL G. (1971). Movement studies—a new standing committee *Bull. Univ. London, Institute of Education*. **23**, 4-10.

HOLLANDER E.P. (1967). "Principles and Methods of Social Psychology". New York: Oxford Univ. Press.

HUIZINGA J. (1938). "Homo Ludens". London: Routledge & Kegan Paul.

HUMPHREY J.H. (1965). Comparison of the use of active games and language work book exercises. *Percept. Motor Skills*, **21**, 23-26.

HUMPHREY J.H. (1966). An exploratory study of active games in learning of cumber concepts. *Percept. Motor Skills*, **23**, 341-342.

HUNT J.McV. (1961). "Intelligence and Experience". New York: Ronald Press.

HUNT V (1962). Cerebral palsied youngsters and body-image problems. Unpublished Report, Univ. California.

HURLOCK E.B. (1949). "Developmental Psychology". New York: McGraw-Hill.

ISMAIL A.H. & GRUBER J.J. (1967). "Motor Aptitude and Intellectual Performance". Ohio: Merrill.

JENKINS L.M. (1930). Comparative study of motor achievements of children five, six and seven years of age. New York: Teachers College, Univ. Columbia. Contributions of Education, 414.

JONES H.E. (1946). Physical ability as a factor in social adjustment in adolescence. *J. Educ. Res.*, **40**, 287-301.

JONES M.C. & BAYLEY N. (1950). Physical maturing among boys as related to behaviour. *J. Educ. Psychol.,* **41,** 129-148.

JOURARD S. (1966). An exploratory study of body accessibility. *Brit. J. Soc. Clin. Psychol.,* **5,** 221-231.

JOURARD S. (1967). Out of touch: the body taboo. *New Society,* 9th Nov.

JUNGMANN J. (1959). "Handing on Faith". London: Burns, Oates.

KAELIN E.F. (1964). "Being in the Body". N.A.P.E.C.W. Report. Washington: A.A.H.P.E.R.

KAELIN E.F. (1968). The well-played game. Notes towards an aesthetics of sport. *Quest Monograph,* **10.**

KANE J.E. (1965). Physique and physical abilities of 14 year old boys in relation to their personality and social adjustment. M. Ed. Thesis, Univ. of Manchester.

KEALIINCWOMOKU L. (1970). An anthropologist looks at ballet as a form of ethnic dance. *Impulse:* Extension of Dance, 24-33.

KEELE UNIVERSITY (1967). National recreation survey, Report No. 1, Univ. Keele/British Travel Assoc.

KIKER V.L. & MILLER A.R. (1967). Perceptual judgement of physiques as a factor in social image. *Percept. Motor Skills,* **24,** 1013-1014.

KINSEY A.C., POMEROY W.B. & MARTIN C.E. (1948). "Sexual Behaviour in the Human Male". Philadelphia: Saunders.

KINSEY A.C., POMEROY W.B., MARTIN C.E. & GEBHARD P.H. (1953). "Sexual Behaviour in the Human Female". Philadelphia: Saunders.

KLEIN J. (1964). "Samples from English Cultures". London: Routledge & Kegan Paul.

KNOBLOCH H. & PASAMANICK B. (1953). Further observations on the behavioural development of Negro children. *J. Genet. Psychol.,* **83,** 137-157.

KOHLER O. (1927). Ober den Gruppenwirkugsgrad der nemschlichen Kooperarbeit und die Bendingung optimaler Kollektivkraftreaktion. *Indust. Psychotech.,* **4,** 209-226.

KRAUS H. & RAAB W. (1961). "Hypokinetic Disease". Springfield: Thomas.

KROUT M.H. (1954). An experimental attempt to determine the significance of unconscious manual symbolic movements. *J. Genet. Psychol.,* **51,** 93-120.

KULKA A., FRY C. & GOLDSTEIN F.J. (1960). Kinaesthetic needs in infancy. *Amer. J. Ortho. Psychiat.,* **30,** 306-314.

KYDD R. (1962). Touch hunger. In R. Johnson (Ed.) "An ABC of Behaviour Problems, Vol. 10". Magazine of the Residential Child Care Association.

LABAN R. & LAWRENCE (1947). "Effort". London: MacDonald & Evans.

LABAN R. (1948). "Modern Educational Dance". London: MacDonald & Evans.

LABAN R. (1960). "The Mastery of Movement." London: MacDonald & Evans.

LABARRE W. (1964). Para-linguistics, kinesics and cultural anthropology. In T.A. Sebesk (Ed.) "Approaches to Semiotics". The Hague: Mouton.

LAMB W. (1965). "Posture and Gesture: An Introduction to the Study of Physical Behaviour". London: Duckworth.

LIDDICOAT R. (1967). The effects of maternal antenatal decompression treatment on infant mental development. *S. Afr. Medic. J.,* March, 203-212.

LIPPITT R., POLANSKY N., REDL F. & ROSEN S. (1952). The dynamics of power. *Human Rel.,* **5,** 37-64.

LOY J.W. (1969). Game forms, social structure and anomie. In R.C. Brown, & R.J. Cratty (Eds.) "New Perspectives of Man in Action". London: Prentice-Hall.

LÜSCHEN G. (1967). The interdependence of sport and culture. *Int. Rev. Sports Sociol.,* **2,** 127-141.

McCLURG-ANDERSON T. (1951). "Human Kinetics and Analysing Body Movements". London: Heineman.

MACCOBYN M., MODIANO H. & LANDER P. (1964). Games and social character in a Mexican village. *Psychiat.,* **27,** 150-162.

MALPASS L.F. (1959). "Responses of Retarded and Normal Children to Selected Clinical Measures". South Illinois: Univ. Press.

MALPASS L.F. (1963). Motor skills in mental deficiency. In N.R. Ellis' (Ed.) "Handbook of Mental Deficiency". New York: McGraw-Hill.

MASTERS W.H. & JOHNSON V.E. (1966). "Human Sexual Response". London: Churchill.

MEAD G.H. (1934). "Mind, Self and Society". Chicago: University Press.

MELLY G. (1965). Gesture goes classless. *New Society,* 17th June, 26-27.

MORGAN C.T. & KING R.A. (1967). "Introduction to Psychology". New York: McGraw-Hill.

MORRIS D. (1967). "The Naked Ape". London: Cape.

MORRIS D. (1969). "The Human Zoo". London: Cape.

MUMFORD L. (1961). "The City in History". London: Secker & Warburg.

MURDOCK G.P. (1945). The common denominator of cultures. In R. Linton (Ed.) "The Science of Man in the World Crisis". New York: Columbia Univ. Press.

MUNDEL M.E. (1950). "Motion and Time Study". London: Prentice-Hall.

MUSGROVE F. (1964). "Youth and the Social Order". London: Routledge & Kegan Paul.

NIDA E.A. (1954). "Customs, Culture and Christianity". London: Tyndale.

NIELSEN G. (1970). "Studies in Self Confrontation". Copenhagen: Munksgaard.

NORTHWAY M.L. (1966). Unpublished lecture, London University, Institute of Education.

NUTTALL J. (1968). "Bomb Culture". London: MacGibbon & Kee.

OESTERLEY W.O.E. (1923). "The Sacred Dance". New York: Dance Horizons.

OPLER M.E. (1946). Themes as dynamic forces in culture. *Amer. J. Sociol.,* **51,** 198-206.

PARNELL R.W. (1958). "Behaviour and Physique". London: Arnold.

PATTERSON O. (1966). The dance invasion. *New Society,* 401-403.

PEPITONE A. & KLEINER R. (1957). The effects of threat and frustration on group cohesiveness. *J. Abnorm. Soc. Psychol.,* **54,** 192-199.

PIAGET J. (1932). "The Moral Judgement of the Child". New York: Harcourt Brace.

PIAGET J. (1953). "The Origins of Intelligence in Children". New York: International Universities Press.

PIAGET J. (1965). "The Moral Judgement of the Child". London: Routledge & Kegan Paul.

PIKLER E. (1968). Some contributions to the study of the gross motor development of children. *J. Genet. Psychol.* **113**, 27-39.

PREYER W. (1937). Embryonic motility and sensitivity. *Mono. Soc. Res. Child Dev.,* **2,** 6.

READ K.E. (1959). Leadership and consensus in New Guinea society. *Am. Anthrop.,* **61,** 425-436.

REID L.A. (1969). "Meaning in the Arts". London: Allen & Unwin.

REID L.A. (1970). Movement and meaning. *L.A.M.G. Magazine* **45,** 5-31.

RIESMAN D & DENNY R. (1963). Football in America. In D. Riesman (Ed.) "Individualism Reconsidered". Glencoe: Free Press.

ROBERTS J.M., ARTH J. & BUSH R.R. (1959). Games in culture. *Am. Anthrop.* **61,** 597-605.

ROSEN B.C. & D'ANDRADE R. (1959). The psychosocial origins of achievement motivation. *Sociometry,* **22,** 185-218.

ROSENBERG B.G. & SUTTON-SMITH B. (1959). The measurement of masculinity and femininity in children. *Child Dev.,* **30,** 373-380.

ROSENTHAL R. (1964). Experimenter outcome orientation and the result of the psychological experiment. *Psychol. Bull.,* **61,** 405-412.

RUESCH J. & KEES W. (1956). "Non Verbal Communication". London: Cambridge University Press.

SARBIN T.R. & HARDYCK C.D. (1953). Contributions to role-taking theory: role-perception on the basis of postural cues. Cited by T.R. Sarbin: Role Theory, in G. Lindzey (Ed.) "Handbook of Social Psychology". Cambridge, Mass: Addison-Wesley.

SAVAGE D.R. (1968). "Psychometric Assessment of the Individual Child". Harmondsworth: Penguin.

SCHAFFER H.R. (1965). Changes in the developmental quotient under two conditions of maternal separation. *Brit. J. Soc. Clin. Psychol.,* **4,** 39-46.

SCHAFFER H.R. (1966). Activity level as a constitutional determinant of infantile reaction to deprivation. *Child Dev.,* **37,** 592-602.

SCHAFFER H.R. & EMERSON P.E. (1968). The effects of experimentally administered stimulation on developmental quotients of infants. *Brit. J. Soc. Clin. Psychol.,* **7,** 61-67.

SCHEFLEN A.E. (1965). "Stream and Structure of Communicational Behaviour". Pennsylvania: Eastern Pennsylvania Psychiatric Institute.

SCHEFLEN A.E. (1968). Human communication. *Beh. Sci.,* **13.**

SCHUTZ J. (1968). "Joy". New York: Grove Press.

SEARS R.R. (1950). Ordinal position in the family as a psychological variable. *Amer. Sociol. Rev.* **15,** 397-401.

SECORD P.F. (1958). The role of facial features in interpersonal perception. In R. Tagiuri & L. Petrullo (Eds.) "Person Perception and Interpersonal Behaviour". Stanford: University Press.

SEYMOUR W.D. (1968). "Skills Analysis Training". London: Pitman.

SHERIF M. & SHERIF C. (1953). "Groups in Harmony and Tension". New York: Harper.

SHIPMAN M.D. (1969). A Sociological Perspective of Dance. A.T.C.D.E.: Dance Section Report.

SINGLETON W.T. (1970). "Measurement of Man at Work". London: Taylor & Francis.

SKINNER B.F. (1953). "Science and Human Behaviour". London: MacMillan.

SMALLRIDGE K.C. (1967). The pattern of participation in physical activities of 16 and 17 year olds as an aspect of leisure. *Res. Papers P.E. Carnegie Coll.*, **4**, 36-47.

SOMMER R. (1965). Further studies of small group ecology. *Sociometry, 28,* 337-348.

SOMMER R. (1969). "Personal Space". New York: Prentice Hall.

SORENSON R.E. & GAJDUSEK C.D. (1966). The study of child behaviour and development in primitive cultures: a research archive for ethnopaediatric film investigations of styles in the patterning of the nervous system. *Pediatrics, 37,* 149-243.

SPITZ R.A. (1962). Auto-eroticism re-examined. The role of early sexual behaviour patterns in personality formation. "The Psychoanalytic Study of the Child", New York: I.U.P.

SPITZ R.A. & WOLF K.M. (1946). The smiling response: a contribution to the ontogenesis of social relations. *J. Genet. Psychol. Mono., 34,* 57-156.

STEPHENSON E. & ROBERTSON J. (1965). Normal child development and handicapped children. In J.G. Howells (Ed.) "Modern Perspectives in Child Psychiatry". Edinburgh: Oliver & Boyd.

STEVENSON H.W. (1965). Social reinforcement in children's behaviour. In R.H. Walters and R.D. Parkes (Eds.) "Advances in Child Development and Behaviour", New York: Academic Press.

STOKES A. (1968). Psycho-analytic reflections on the development of ball games. In A. Natan (Ed.) "Sport and Society". London: Bowes & Bowes.

STOTT D.H. (1959). "Unsettled Children and Their Families". London: University Press.

STOTT D.H. (1962). Evidence for a congenital factor in maladjustment and delinquency. *Am. J. Psychol.,* **118,** 781-794.

STOTT D.H. (1964). Why maladjustment? *New Society,* Dec. 10th

STOTT D.H. (1966). A general test of motor impairment for children. *Dev. Med. Child. Neur., 8.*

STUMPF F. & COZENS F. (1947). Some aspects of the role of games, sports and recreational activities in the culture of modern primitive peoples. *Res. Quart.,* **18,** 198-218.

SUTTON-SMITH B. (1959). "Games of New Zealand Children". Berkeley: University California Press.

SUTTON-SMITH B. & ROBERTS J.M. (1963). Game involvement in adults. *J. Soc. Psychol.* **34,** 119-126.

SUTTON-SMITH B. & ROBERTS J.M. (1964). Rubrics of Competetive behaviour. *J. Genet. Psychol.,* **105.**

SZASZ T.S. (1961). "The Myth of Mental Illness". London: Secker & Warburg.

TANNER J.M. (1964). "The Physique of the Olympic Athlete". London: Allen & Unwin.

TURNBULL C.M. (1965). The Mbuti pygmies of the Congo. In J.L. Gibbs, Jr. (Ed.) "People of Africa". New York: Holt, Rinehart & Winston.

TURNER V.W. (1969). "The Ritual Process". London: Routledge & Kegan Paul.

ULRICH L. & TRUMBO D. (1965). The selection interview since 1949. *Psychol. Bull.,* **63,** 100-116.

VAIZY M.J. & HUTT C. (1966). Differential effects of group density on social behaviour. *Nature,* **209,** 1371-1372.

VEERHAULT P. (1969). A sociology of dance. Assoc. of Principals of Women's Colls. of Phys. Educ. Conference Report

WALTERS C.E. (1965). Prediction of post-natal development from foetal activity. *Child Dev.,* **33,** 801-808.

WATSON O.M. & GREAVES T.D. (1966). Quantitative research in proxemic behaviour. *Am. Auth.,* **68.**

WEBER M. (1950). From: Gerth and Mills (Eds.) "Max Weber". New York: Harper.

WEISBROD R.M. (1965). Looking behaviour in a discussion group. Quoted in M. Argyle & A. Kendon (1967). The experimental analysis of social performance. *Adv. Exp. Soc. Psychol.,* **3,** 55-98.

WEYNER A. & ZEAMAN D. (1956). Team and individual performances on a motor learning task. *J. Genet. Psychol.,* **55,** 127-142.

WHITING H.T.A. (1967). Significance of movement for early development. *Remed. Gym. Rec. Therapy,* **45,** 14-19.

WHITING H.T.A., JOHNSON G.F. & PAGE M. (1969). Factor analytic study of motor impairment at the ten-year age level in normal and E.S.N. populations. Unpublished paper, P.E. Dept., Univ. Leeds.

WHITING H.T.A., MORRIS P.R., DAVIES J.G., GIBSON J.M., LUMLEY R. & SUTCLIFFE R.S.E. (1970). Motor impairment in an approved school population. Unpublished paper, P.E. Dept., Univ. Leeds.

WHYTE W.F. (1943). "Street Corner Society". Chicago: University Press.

WILLIAMS J.L. (1970). Personal space and its relation to extraversion-introversion. M.A. Thesis. University Alberta.

WILLIAMS J.R. & SCOTT R.B. (1953). Growth and development of Negro infant IV: Motor development and its relationship to child rearing practice in two groups of Negro infants. *Child Dev.,* **24,** 103-121.

WILLIAMS W.M. (1956). "The Sociology of an English Village". London: Routledge & Kegan Paul.

WILSON W. & MILLER N. (1961). Shifts in evaluations of participants following inter-group competition. *J. Abnorm. Soc. Psychol.,* **63,** 428-431.

WITKIN H.A., DYK R.B., FATERSON D.R. & KARP S.A. (1962). "Psychological Differentiation". New York: Wiley.

WOBER M. (1966). Sensotypes. *J. Soc. Psychol.,* **70,** 181-189.

WOLFF W. (1943). "The Expression of Personality". New York: Harper.

ZUBECK J.P. (1954). "Human Development". London: McGraw-Hill.

ZAGORA E. (1959). Observations on the evolution and neurophysiology of eye-limbs co-ordination. *Opthalmologica,* **138,** 241-254.
ZURCHER L.A. & MEADOW R. (1967). On bullfights and baseball: an example of interaction between social institutions. *Int. J. Compar. Sociol.,* **8.**

Stage

TOPIC AREAS - DEVELOPED

COMPARATIVE STUDIES

by JEAN CARROLL

General

Movement behaviour is a result of both biological and social processes and these dimensions must be taken into account in comparative studies. Darwin (1872) recognised this when he pointed out that some patterns of expression, like weeping, baring of teeth, blushing and pupil expansion seemed innate whilst other human signals, like head-nodding and shaking seemed to be culturally influenced. Comparative studies in movement behaviour can be developed against a variety of frameworks each with its particular insights. There is no general text on comparative studies in movement behaviour, but useful texts, from different disciplines, utilizing comparative methodology are:-

Andreski (1964); Bereday (1967); Hinde (1966); Waters, Rethlingschafer and Caldwell (1960).

Comparative studies focus on similarities and differences. The method is an application of the general rule of logic in science, but the cases are ready-made rather than experimentally constructed (Mitchell 1968). Comparative analyses can be carried out using both wide range and narrow range data and can utilize both field and laboratory techniques.

This survey will indicate some of the work from which comparative studies in movement behaviour could be developed. Movement behaviour within ethological studies and against an evolutionary framework is examined first. [The work on human primate movement behaviour in historical time is a logical development of this and the subject of another section of the book.] Movement behaviour across cultures is then examined at two levels—that of everyday behaviour and that of institutionalised activities. Finally, movement patterns that are labelled abnormal in our particular culture are compared with those labelled normal.

H.1. Ethological Studies investigate the behaviour of animals in their natural habitats: they are typically field studies carried out under the procedures of naturalistic observation but also employing laboratory experiments closely related to the normal life of the animal. The behavioural phenomena studied are usually

167

the gross behavioural adjustments the organism makes to its natural environmental conditions and are, therefore, of interest to the student of non-verbal behaviour as they are, in fact, movement sequences.

The study of the inter-relationship of structure, behaviour and mode of life in animals was first named 'ethology' by Geoffrey Saint-Hilaire in 1854 (Ewer 1968); Darwin's (1872) emphasis on the biological determinants of movement patterns and Darwin's (1859, 1871) evolutionary hypothesis are the theoretical starting points for ethological studies. Early workers included Spalding, Heinroth, Whitman, and Craig and general introductions to ethology have been provided by Klopfer and Hailman (1967); Scott (1967); Tinbergen (1963). Work relevant to movement behaviour clusters at certain levels of the phylogenetic scale.

(a) Vertebrates other than mammals

Lorenz's (1937) and Tinbergen's (1942) work on 'fixed action patterns' was important in showing that some movement sequences (like the nest weaving movement of the female canary; egg-retrieving of the grey-lag goose; and the soliciting postures of the female chaffinch) unfold systematically once they have been 'released' and seem to be independent of environmental stimuli even though they may be elicited by such stimuli in the first instance. It was suggested that each animal had a repertoire of these 'fixed action patterns' and only a limited ability for developing new ones. Lorenz (1958) and Tinbergen (1960) provided meticulous descriptions of these and Lorenz (1970) surveyed his earlier work in this area.

These classic ethological studies have been rich in concepts and hypotheses which have been used at other levels of the phylogenetic scale, including that of human primates. Foremost among these is the concept of *imprinting* which has been further explored among avian species, where it originated, and in mammals, including human infants (Gray 1958). Some studies have been directed towards discovering which patterns of response were released by particular stimuli; and this can be linked with work on the *seeking of* sensory stimulation (Butler 1954) and the studies of *sensory deprivation* in humans (Vernon 1966). Other studies have investigated the effects of *overall deficiency of stimulation;* Harlow's (1961, 1963) work with monkeys is well known here and has been used in conjunction with work on *maternal deprivation* (Bowlby, 1951, W.H.O. 1962). The concept of the *critical period,* which originally referred to the time at which the following response in nidifugous birds was established, has been applied to the sucking response (Gunther 1961) and the smiling response (Ambrose 1961, 1963; Wolff 1963) in human babies. All this work has been the subject of much controversy (8th Symp, Zool. London 1962; Hinde 1963; Sluckin 1964, 1966; Klopfer 1967 and articles in Endler, Boulter & Osser 1968).

Similarly, *redirected* and *displacement activities,* i.e. seemingly irrelevant patterns of behaviour like the plucking and flicking of grass, were observed by Tinbergen (1960) in gulls. These were examined also by Bastock, Morris & Moynihan (1953) and shown to be important in the communication of social signals. Observers of primate behaviour have used these concepts. Russell & Russell (1968) Ingram (1960) and Morris (1967, 1969) have applied them to human primate behaviour. *Ritualized behaviour* was recognised by ethologists as consisting of displays derived from other behaviours and having developed conspicuous structures, like the 'head-up' threat position of some birds with brightly coloured necks and breasts; the zig-zag dance of sticklebacks; and the sexual and agonistic displays of cichlid fish. Morris (1956) examined the sequence of events in the stickleback's courtship dance and suggested some of the ways in which the basic patterns were ritualized to form signals. A conference of the Royal Society discussed ritualization in animals and men (Huxley 1966); Loizos (1967) related ritualized behaviours to *play behaviours* in a very thorough survey of the area. Millar (1968) drew upon a wide range of work from animal studies in her analysis of play.

(b) Mammals, other than primates

Ewer (1968) discussed some of the basic concepts of ethology and summarised the work on mammals in sections dealing with food finding and storing; social organisation and territory; scent marking; fighting, threat and appeasement behaviour; and amicable and courtship behaviour. Particularly relevent to movement studies are the chapters on expression and communication, and on play. The infant, parental, and social behaviour patterns of mammals were examined by Dimond (1970) who referred to Beach's extensive work on the sexual behaviour of rats. A collection of papers on the social organisation of animal communities was produced by the 14th Symposium of the Zoological Society in London 1965. Scott (1968) summarised the work on early experience and the organisation of behaviour across species, and included many references to mammalian studies.
This work tends to be clustered around topics, such as parental behaviour and play, and so a search has to be made for specific references to movement behaviour. A simple, mechanical movement classification was suggested by Gray (1953).

(c) Primates, other than humans

As primates, among mammals, contain both human and non-human varieties they provide, within the evolutionary framework, essential comparisons of movement behaviour.
Early field studies of howler monkeys (Carpenter, Altmann), rhesus monkeys (Carpenter, Southwick, Altmann, Koford); Zuckerman's study of hamadryas baboons in Regents Park; and some Japanese studies using provisionisation techniques were summarised in Southwick (1963). Later studies of rhesus monkeys in Regents Park were made by Chance, and in Whipsnade by Reynolds; urban and forest languars were studied by Jay. Reynolds in the Budongo Forrest and Goodall in the Gombe Stream reserve and Kortlandt in the Congo observed

chimpanzees in the wild whilst Loizos and Manley studied them in Regents Park. Similarly, hamadryas baboons were studied in Zurich Zoo by Kummer, in Bloemfontein by Hall, and in the mountains of Ethiopia by Kummer and by Crook. Useful surveys of these studies were provided by de Vore (1965) who also included a comparative section on reproductive behaviour, social development, and communication. Morris (1967) and Russell & Russell (1968) also gathered together and discussed important studies. Crook (1970) focused on social behaviour and stressed the importance of ecological factors.

Comparisons between the behaviour of free and captive night monkeys on Barro Colorado were made by Moynihan; of patas in captivity and in Uganda by Hall; and of wild howlers and urban rhesus monkeys by Southwick. Rowell (1967) discussed various studies with regard to social organisation and the problems of comparative primate work; Russell & Russell (1968) focussed on comparisons between free ranging and wild monkeys with regard to violent behaviour. Moynihan (1967) looked at comparative aspects of communication in new world primates.

The movement behaviour of non-human primates is more varied and flexible than that of animals studied by the classical ethologists. Hall (1962) described the variety of *bodily contacts* like nuzzling, clinging, and grooming that differ with social relations in baboons. There has been extensive work on *facial expressions* (Darwin 1872; Van Hooff 1962, 1967; Andrew 1963, 1965). Vine (1970) surveyed communication by facial-visual signals in man and in infra-human species and suggested a model for analysis. Chance (1971) and Chance and Jolly (1970) brought together the information on bodily contact and the spatial properties of social organization in relation to group coherence. Such work has been central to studies of *communication processes* (Marler 1965, Altmann 1967, Sebeok 1967, 1968).

Of particular interest to students of movement are the *repertoires of behaviour* that have been developed by ethologists to record and analyse the non-verbal behaviour of non-human primates. Chance (1956) produced the first, which included a list of four submissive gestures of rhesus monkeys. Altmann (1962) built up a repertoire of fifty-six units of social behaviour. Hinde & Rowell (1961) used a repertoire of five sitting postures, six attack and threat movements, four fear displays, three units of friendly behaviour, and three miscellaneous movements— chewing/gnashing, pacing, and yawning. Other repertoires have also been constructed and indicate differences and similarities in non-verbal behaviour according to age, sex, social position, social group and ecological setting as well as differences between individuals. The problems arising from these and from the observations on which they are based have been discussed by Rowell (1967) and by Aldrich-Blake (1970). Attempts to found a statistically sound basis of observation have been made by Altmann (1965). Napier (1968) suggested a classification of primate locomotor behaviour.

(d) Human studies

Grant (1963) built up an ethogram showing the relationship of non-verbal elements amongst rats in close proximity and he indicated the possible motivation of these elements in terms of aggression and flight and also drew attention to displacement activities: he then applied this method to an analysis of the behaviour of schizophrenics in a hospital ward and so engaged in human ethology (Grant 1965). Blurton-Jones (1967) also used this method of recording units of non-verbal behaviour in his study of the social behaviour of nursery school children and grouped the units in the contexts of aggregative behaviour, responses to adults, agonistic behaviour, and rough and tumble play. Clark, Wyon, and Richards (1969) have, similarly, looked at the free play of nursery school children and Grant (1970) examined human smiles, frowns, hand-movements, etc., using the ethological method.

Broad generalizations between human and infra-human patterns of aggression have been made by Lorenz (1963) and Ardrey (1967). Morris (1967, 1969) suggested sexual and competitive similarities between species. Warnings on the dangers of such extrapolations have been issued by the contributors to the collection of articles edited by Montague (1968) and by Klopfer (1968). More circumspect use of comparative methodology was made by Andreski (1964); by Russell & Russell (1968); and by Tiger (1969). Hinde (1966) wrote:-

'At some stage it becomes necessary to make statements valid for more than one species: these must be based on comparative study of a number of species. But the comparative method has to be used cautiously: apparently close similarities between species may prove to be merely parallel evolutionary adaptations to a similar environment, and rest on quite different causal bases'. He went on, however—'The differences revealed by comparative study may be as interesting and important as the similarities; both will provide new theoretical insight into problems of causation, function, and evolution'.

Washburn & Jay (1968) discussed several aspects of human evolution. Lockard (1971) provided an interesting analysis of the basis of and changes in comparative psychology and stressed the insights provided about behaviour by the study of ecological factors (Crook 1962, Selander 1965, Orians 1969). He also warned of the dangers of cross-species comparisons.

H.2. Cross-Cultural Studies

Different human societies, over long periods of time, seem to have developed different patterns of movement behaviour whilst sharing the same species—genetic make-up. These differences would seem to be *cultural*. Discussions on the concept of culture can be found in Coser & Rosenberg (1964), Bredemeier & Stephenson

(1962), Mitchell (1968), Theordorson & Theordorson (1970). Argyle (1969) summarised six components of culture as a shared language, a shared way of perceiving and thinking about the world, agreed forms of non-verbal communication and social interaction, rules and conventions, agreed values, and a shared technology and material culture.

Human behaviour in all cultures shares some similarities and some differences. Non-verbal similarities and differences will be considered in both everyday patterns and in institutionalized forms. Problems of cross-cultural research are discussed by Strodbeck (1964); Frijda & Jahoda (1966); Whiting (1968) and in an important survey by De Vos & Hippler (1969).

Studies of everyday movement patterns

Early references to this area include Mead (1928) within the context of Samoan culture; Efron (1941) in his observations of the distinctive gesture patterns of Eastern Jews and Southern Italians in New York City; and Labarre's (1947) survey of information on the cultural basis of gestures. Kurath (1950, 1953, 1954, 1956) in her anthropological studies of American Indians recorded the work patterns and everyday postures of her subjects. For this she used a modified version of Laban's (1956) system of recording movement.

Laban's classification examines movement from four aspects—bodily, rhythmic, spatial and relationship aspects—whilst recognising that human movement is a synthesis of all these aspects. Laban's framework will be used in this survey to provide a means of classifying the material which has resulted from observations and experimental explorations of everyday movement patterns.

(a) Bodily aspects

Allport (1961) summarised the inferences made by Americans from physical cues to personality traits. Argyle (1969) referred to the different attention paid to make-up, hair, etc., by men and women in different cultures. He also commented on the cultural modifications of facial expression, e.g. Chinese girls were instructed not to smile easily and not to show their teeth when they smiled. Vinacke and Fong (1955) compared the judgement of facial expressions by three national racial groups in Hawai. See J.1 for other references to work on facial expression.

Labarre (1964) found that greetings and farewells could involve shaking hands, kissing, or more elaborate processes of striking or stroking within the various cultures. The Copper Eskimo welcomed strangers with a buffet on the head and shoulders with his fist; an Ainu, meeting his sister, grasped her hands in his for a few seconds, suddenly released hold, grasped her by both ears, then they stroked each other down the face and shoulders. Labarre also pointed out that the same non-verbal signals may have quite different meanings in two cultures and so lead to confusion: sticking out the tongue meant an apology in parts of China, the evil eye in parts of India, deference in Tibet, negation with the Marquesans; hissing in Japan showed respect to superiors and expressed disdain in England.

The first substantial work on touch was produced by Frank (1957).

Jourard (1966) counted the frequency of contact of couples in cafes in San Juan, Paris, Florida, and London and recorded 180, 110, 2 and 0 respectively. He also compared the extent to which people from various cultures would let others see and touch their bodies.

Melly (1965) indicated social class variations in posture and walking in London; Goffman (1961) suggested there were status differences in the way people sat in conference; and Burns (1964) showed there were status differences in the way people entered a room.

Davitz (1964) reviewed some cross-cultural research in facial and vocal expressions of emotion. Cüceloglu (1970) used sixty abstract facial expressions in examining perception in three different cultures.

(b) Rhythmic aspects

Hayes (1964) showed that the visible elements of gesture and facial expression needed to be learned if a speaker were to be able to communicate effectively with native speakers of a language. Synchronisation of speech and gesture was found to be important within a culture by Kendon (1963) and may well be an important factor between cultures.

Objective recording of this aspect of human movement has proved difficult and has resulted in very little work in this area.

Brewer (1951) distinguished gestures with symbolic meaning, gestures with pictorial meaning and gestures with emphatic meaning in the Arabs of Bierut and Damascus.

(c) Spatial aspects

Hall's study of 'proxemics' is notable here. When Hall (1959) looked at 'the silent language' one of his concepts was that of the physical distance between two interacting people. He devised a recording system (1963) and found that each culture had its own clearly defined degrees of proximity on his four-point classification (1966) and that different cultures had different rules about proximity, e.g. Arabs and Latin Americans stood considerably closer than Europeans or North Americans. Hall (1968) developed eight dimensions, with scales, in his further 'study of man's perception and use of space'.

Watson & Graves (1966) showed that Arabs in conversation faced each other more directly, sat closer to each other, touched each other more, and looked each other more in the eye than Americans did. Campbell et al (1966) found that subjects moved closer towards members of preferred racial groups.

According to Sommer (1959) there were differences in proximity within cultures and in different social settings. Sommer (1965) also considered orientation and suggested that closer proximity was usually combined with a move from confronting orientation to side-by-side orientation. Sommer's (1967) studies in small group ecology stressed the importance of the environment in which the interaction takes place as a key variable affecting social relations and movement behaviour. Altman and Haythorn (1967) examined the ecology of isolated groups. Lyman & Scott (1967) commented on territoriality in relation to lower-class urban Negro youth.

(d) Relationship aspects

Mead (1937) reported differences in co-operation and competition in various simple societies. Whiting & Child (1953) found that aggression was a dependent variable for the severity of socialisation—especially in the relationship between mother and child.

Patterns of everyday movement behaviour are, of course, tacitly understood and accepted within cultures and in this sense are part of the norms of the society. Scheflen (1964) asserted that human social behaviour was neither universal nor individual and unique; and that non-verbal systems of communication had to be examined in context. He presented findings about the significance and function of posture in American interactors within the cultural frame of reference, and said that the standard units were 'specific constellations of behaviour built into a culture, learned and perceived in communication, as Gestalten'.

Similarly, Birdwhistell's (1960, 1968) studies of 'Kinesics' (which he defined as 'a systematic study of how human beings communicate through body movement and gesture', and as 'the systematic study of the visually sensible aspects of non-verbal interpersonal communication') showed the importance of cultural variables in his triple division of Kinesics into pre-kinesics, microkinesics, and social kinesics. Birdwhistell (1971) provided a selection from his extensive work.

Institutionalized forms of movement

Whereas patterns of everyday movement behaviour are implicitly understood and, in this sense, institutionalized, other activities like forms of dance, games, sports, contests, rituals, and ceremonies are explicitly accepted and organised around the satisfaction of important social needs—they are social and cultural institutions. Theordorson & Theordorson (1970) and Mitchell (1968) distinguished between institutions and institutionalization.

Many definitions of these institutionalized activities have been attempted: *games* by Caillois (1955), Roberts et al (1959), Damm (1960) McIntosh (1963), Glassford (1970); *sport* by Luschen (1967) and Loy (1968); *dance* by Humphrey (1950), Kurath (1966) and Kealiinohomoku (1970); and *physical activity* (Kenyon 1968). The related concepts of *play* (see section K) and *leisure* (Veblen 1925, Dumazadier

1968) and *children's games* (Opies 1969) *rituals and ceremonies* (Mitchell 1968, Theordorson & Theordorson 1970) are important.

(a) Traditional approaches to these forms of movement have been:

(i) *historical examinations,* e.g. dance (Vuillier 1898, Kirstein 1935, Sachs 1937); sports and games (Hackwood 1907, Hole 1949, Wymer 1950). These studies have been compiled both by historians and by practitioners of the activities and provide valuable material for comparative examinations of different periods. They are vital for investigations of social change. See section B.

(ii) *structural-functionalist* descriptions from anthropological data of simple societies. References to the pioneer studies of Tylor, Stevenson, and other recorders of sports and games can be found in the later compilations by Luschen (1967, 1970) and in Loy & Kenyon (1969). References to the work of Rivers, Radcliffe-Brown, Seligman, Mead, Hambly, Gorer, and other recorders of dance can be found in Rust (1969) and Kraus (1969). In these studies, non-verbal activities were perceived as indispensible to the smooth functioning of society and, therefore, highly significant.

(b) Contemporary work has continued with these traditional approaches. McIntosh (1963), Braisford (1969), Howell (1969) looked at historical aspects of sport, Frederickson (1960) at the complexity of the role of sports in various cultures with special reference to spectator behaviour and the ritual function of sport; Dunlap (1951) examined the sports of the Samoans within their cultural setting and in historical perspective Nettleton (1967) commented on sport and social values and Frankenberg (1966) spoke of the way in which an anthropologist could examine sport and Kealiinohomoku (1970) discussed an anthropological examination of ballet. Kraus (1969) produced an historical survey of dance.

Industrialization and increasing social complexity have resulted in the study of formalised movement activities in relation to specific aspects of the socio-cultural system, other institutions within society, and specific social problems. Luschen's (1967) survey and bibliography categorised the studies in this way. Loy and Kenyon's (1969) reader contained sections also on sport in relation to social processes, sport as a micro-social system and sport as a sub-culture. Some studies focus on specific activities, e.g. McGrath on rifle shooting, Klein and Christiansen on basketball, Grusky on baseball, Weinberg and Arond on boxing, and Boroff on skiing (Loy and Kenyon 1969).

Maheu (1962), Jokl (1964) and Daniels (1966) examined sport in relation to aspects of culture. Allardt (1967) classified the approaches to sports sociology as structural-functional, neo-structural, evolutionary-functional, and quantitative. Rust (1969) used Parsons' functional imperatives in her analysis of dance; Veerhault (1969), in a slim paper, suggested approaches to dance through the sociology of the arts; Shipman (1968) looked at the consolidating, differentiating and socialising functions of dance.

Comparisons have been made in both narrow and wide range data. In examining sport or dance phenomena in another age or culture there is the implicit comparison with the observer's own culture. This is particularly true of anthropological studies. Wider ranging comparisons have been made along specific dimensions over two cultures by Pohndorf (1969) and over twelve cultures by Robinson (1970). The use of Human Relations Area Files has been important in wide range work though, of course, these are very often based initially on single culture studies. Seagoe (1962) suggested a method of analysing cross-cultural data on children's play as an index to differential cultural impact: she used Japanese and American data.

Probably the most promising studies currently being produced are those which start from an analysis of games or dance and then structure sociological or psychological hypotheses into these analyses. This approach then ensures that the ensuing works are movement based. Caillois' (1955) categorization has been used in this way and Sutton-Smith et al and Roberts et al have made considerable contributions to work along these lines. In a survey, Sutton-Smith and Roberts (1970) looked at cultures without games, cultures possessing games of physical skill alone, cultures with games of chance, cultures with games of strategy and cultures possessing games of physical skill, chance, and strategy. They correlated individual and psychological motivation for games playing to anxieties and conflicts induced by child-training processes. They proposed that games were models of power—models of ways of succeeding over others—by magical power (as in games of chance), force (physical skill), and cleverness (strategy). Avedon & Sutton-Smith (1971) assembled a comprehensive book of readings on the study of games.

(c) Physical Education within comparative studies has a rather different position from sports, games, and dance. It may include these activities and it is institutionalized, but it is institutionalized within the institution of the school. Here, then, is a deliberate and intentional fostering of values through the teaching of selected forms of movement.

One of the earliest writers on this aspect of education was Spencer (1861) who extrolled the superiority of play to gymnastics. Waller (1932) treated the topic sociologically and examined games and athletic activities within schools as a cohesive force, as a method of social control, and as an elaborate culture pattern.

There have been, recently, attempts to define the concept of Physical Education (Zeigler 1968, Adams 1969, Leeds 1970); there have been historically orientated studies (May, 1967, 1968, Hendry 1969); there have been pleas for recognition of the cultural dimension (Cassidy 1965, Anthony 1966, Johnson 1967). There have also been case studies of Physical Education systems in Wales (Rees 1966), Scotland (Thomson 1966), Italy (Wright 1967), C.S.S.R. (Lopata 1968) U.S.S.R.

(Wohl 1968), Syria (Chombag 1968) Czechoslovakia (Petrak 1966). Dumazedier (1968) commented on the relationship of Physical Education to sports and Hendry (1970) suggested some possible influences of P.E. on the school curriculum in general.

This is an area in which it should be relatively simple to collect information from published documents, time-tables and syllabuses.

H.3. Psychopathological studies seek to explain disorders of behaviour or mental activity (Russell-Davis 1966) and often utilize the observation of movement behaviour in both diagnosis and treatment. These studies have tended to take place *within* rather than *between* cultures, although there has been recognition of the importance of the cultural dimension (Montagu 1955, Wittkower 1969). To justify the inclusion of these studies within a comparative framework it has to be accepted that in any examination of what is labelled 'abnormal' is the implied recognition of the 'normal'.

Freud (1914) wrote of the importance of non-verbal signals—'He that has eyes to see and ears to hear may convince himself that no mortal can keep a secret. If his lips are silent he chatters with his finger tips; betrayal oozes out of him at every pore'. Psychiatrists have continued to utilize non-verbal cues in their analyses of disorders. Feldman (1959) contended that mannerisms of speech and gestures in everyday life reflected the personality and that their analysis enhanced our understanding of the unconscious. Brengelmann (1961) surveyed the work on expressive movements and abnormal behaviour. Hall (1933) suggested that a useful approach was that of the comparative psychologist.

The studies have been classified using Laban's scheme, as in J.2 A.

(a) Bodily aspects

Wolff (1945) working with eighty-eight mental patients constructed a repertoire of thirty-seven psycho-motor reactions like sniffling, giggling, opening and closing the fists, rolling eyes and pulling hair. She showed that the more advanced the disintegration of the personality, the more stereotyped and restricted the gestures became. Krout (1954a, 1954b) explored the symbolic meaning of hand gestures of patients and of autistic gestures. Erikson (1956) referred to non-verbal behaviour in his clinical studies of adolescents and in the postulation of his identity-formation theory: in this, one of the conscious experiences of a person who had achieved ego-identity was 'a feeling of being at home in one's body'. Dittman (1962) observed the movements of the hands, head and legs of patients in interviews and related these to moods. Freedman & Hoffman (1967) put forward a scheme for the analysis of spontaneous bodily movements in psychotherapeutic interviews.

(b) Rhythmic aspects

Argyle (1969) commented that manics made vigorous gestural movements and that the gestural movements of neurotic patients were jerky and poorly controlled and that gestural and facial movements of hysterics were over-dramatic. Bettelheim (1959) found that autistic children engaged in repetitive motor activities in relation to mechanical objects. Hamilton (1967) used excerpts from early writers, e.g. Kraepelin who first observed the rhythmic nature of movements of schizophrenics.

(c) Spatial aspects

Sommer & Osmond (1962) found that schizophrenics either chose very distant seats or sat side by side, but avoided a 90° situation. Schooler and Parkel (1966) found that schizophrenics in mental hospitals spent 45% of time doing nothing and much of this time oriented towards the wall. Lefcourt et al (1967) found partial or complete aversion of gaze among schizophrenics. Argyle & Kendon (1967) found that schizophrenics engaged in very little eye contact, tended to gaze at 90° to the line of regard and used very short glances.

(d) Relationship aspects

Bateson et al (1956) suggested that the origin of schizophrenia was to be found in the discrepancies between the non-verbal signals and the verbal utterances emitted by the parents, especially the mothers, of schizophrenics. This, Bateson called the 'double-bind', a concept which has been used generally in spite of criticism (Schuham 1967; Pease 1970).

Cameron (1959) found the interaction skills of paranoids better than those of schizophrenics, but they seemed unable to profit from feed-back from others and became hostile if criticised. Johannsen (1961) found that schizophrenics performed a task as well as normals did under non-social feedback conditions, but, with social feedback, the normals did better than the schizophrenics. Lerner and Fairweather (1963) found that schizophrenics did more work in unsupervised work groups, but under supervision became passive, hostile and withdrawn. Lefcourt & Steffy (1966) found that process schizophrenics were unco-operative with a female tester and that reactive schizophrenics were unco-operative with a male tester.

Szasz (1961) developed the view that hysterical bodily symptoms were a kind of non-verbal communication resorted to when words failed: in using bodily language, the hysteric was pretending to be ill in order to have some control over other people.

Hutt & Vaizey (1966) found that autistic children were totally unresponsive to other people.

Several writers have pointed out that the longer a patient was institutionalized in a mental hospital the more he became dependent on the institution (Goffman 1961, Wing 1962). Sommer & Osmond (1962) found that schizophrenics, unlike prison inmates or other inmates of closed communities, did not engage in co-operative activities.

Clinicians do not agree upon the aetiology of disorders, especially of schizophrenia, and there is a plethora of hypotheses—genetic, neuro-physiological, biochemical, psychological, and sociological. Psychoanalysts, existentialists, and social psychiatrists have frequently provided meaningful accounts of the schizophrenic's world (Sullivan 1955, Jung 1966, Laing 1960a, 1960b, Lidz et al 1966). But the work of learning theorists and behaviour therapists probably provides the most interesting material for movement specialists. (Beech 1969, Meyer and Chesser 1970). Theoretical and methodological problems in studying behaviour on a normal/abnormal continuum were discussed by Dufrancatel (1968) and Robertson (1969). Comer (1968) wrote a short article on the relationships between sport and mental health.

References

Source Texts

ALLARDT E. (1967). Basic approaches in comparative sociological research and the study of sport. *Int. Rev. Spts. Soc.* **2.**

ALLPORT G.W. (1961). "Pattern and Growth in Personality". New York: Holt, Rinehart, & Winston.

ALTMANN S.A. (1967). The structure of primate social communication. In S.A. Altmann (Ed.) "Social Communication Among Primates". Chicago: University Press.

ANDRESKI S. (1964). "Elements of Comparative Sociology". London: Weidenfeld & Nicolson.

ARGYLE M. (1969) "Social Interaction". London: Methuen.

ARGYLE M. & KENDON A. (1967). The experimental analysis of social performance. In L. Berkowitz (Ed.) "Advances in Experimental Social Psychology III". New York: Academic Press.

AVEDON E.M. & SUTTON-SMITH B. (Eds.) (1971). "The Study of Games". London: Wiley.

BEECH H.R. (1969). "Changing Man's Behaviour". Harmondsworth: Penguin.

BEREDAY G.Z.F. (1967). Reflections on comparative methodolgy in education 1964-1966. *Comp. Ed., ***3,** 3.
BIRDWHISTELL R. (1960). Kinesics and communication. In E. Carpenter & M. McLuhan (Eds.) "Explorations in Communication". Boston: Beacon Press.
BIRDWHISTELL R. (1968). Kinesics. In D. Sills (Ed.) *Int. Encyc. Soc. Sci.* **8.** New York: Macmillan.
BIRDWHISTELL R. (1971). "Kinesics & Context". London: Allen Lane.
BLURTON-JONES N.G. (1967). An ethological study of some aspects of social behaviour of children in nursery school. In D. Morris, (Ed.) "Primate Ethology". London: Weidenfeld & Nicolson.
BOWLBY J. (1951). "Maternal Care and Mental Health". Geneva: W.H.O. Monograph 179.
BRAILSFORD D. (1969). "Sport and Society". London: Routledge & Kegan Paul.
BREDEMEIER H.C. & STEPHENSON R.M. (1962). The analysis of culture. In P.I. Rose (Ed.) "The Study of Society". New York: Random House.
BRENGELMANN J.G. (1961). Expressive movements and abnormal behaviour. In H.J. Eysenck (Ed.) "Handbook of Abnormal Behaviour". London: Pitman.
CAILLOIS R. (1955). The structure and classification of games. In J.W. Loy & G.S. Kenyon (Eds.) (1969). "Sport, Culture and Society". New York: Macmillan.
CHANCE M.R.A. & JOLLY C.J. (1970). "Social Groups of Monkeys, Apes and Men". London: Cape.
COSER L.A. & ROSENBERG B. (1964). (Eds.) "Sociological Theory—A Book of Readings". New York: Macmillan.
CROOK J.H. (1970). "Social Behaviour in Birds and Mammals". London: Academic Press.
DARWIN C. (1859). "On the Origin of Species by Natural Selection". London: Murray.
DARWIN C. (1871). "The Descent of Man". London: Murray.
DARWIN C.R. (1872). "The Expression of the Emotions in Man and Animals". London: Murray.
DeVORE I. (Ed.) (1965). "Primate Behaviour: field studies of monkeys and apes". New York: Holt, Rinehart & Winston.
DeVOS G.A. & HIPPLER A.E. (1969). Cultural psychology: comparative studies of human behaviour. In G. Lindsey & E. Aronson (Eds.). "The Handbook of Social Psychology IV". New York: Addison-Wesley.
DIMOND S.J. (1970). "The Social Behaviour of Animals". London: Batsford.
DUFRANCATEL C. (1968). The sociology of mental illness. (English summary). *Current Soc.* **16,** 67-69.
DUMAZEDIER J. (1968a). Some remarks on sociological problems in relation to physical education and sports. *Int. Rev. Spts. Soc.* **3,** 5-12.
DUMAZEDIER J. (1968b). La sociologie du loisir. (English summary). *Current Sociology* **16,** 1.
EFRON D. (1941). "Gesture and Environment". New York: King's Crown Press.
ENDLER N.S., BOULTER L.R. & OSSER H. (1968). (Eds.) "Contemporary Issues in Developmental Psychology". New York: Holt, Rinehart & Winston.
ERICKSON E.H. (1956). The problems of ego-identity. *Am. J. Psychoanalysis,* **4.**
EWER R.F. (1968). "Ethology of Mammals". London: Logos Press.

FRANK L.J. (1957). Tactile communication. *Genet. Psychol. Mon.,* **56,** 209-225.

FRANKENBERG R. (1966). An anthropologist looks at sport. Sport in Ed. & Rec. International conf.: London.

FREUD S. (1914). "The Interpretation of Dreams". London: Heinmann.

FRIJDA N. & JAHODA G. (1966). On the scope and methods of cross-cultural research. In D.R. Price-Williams (Ed.) (1969) "Cross-cultural Studies". Harmondsworth: Penguin.

GOFFMAN E. (1961). "Asylums". New York: Anchor Books.

GRANT E.C. (1965). An ethological description of some schizophrenic patterns of behaviour. Leeds. Symp. on Beh. Disorders.

GRAY J. (1953). "How Animals Move". Cambridge: University Press.

HACKWOOD F.W. (1907). "Old English Sports". London: Unwin.

HALL C. (1933). A comparative psychologist's approach to problems in abnormal psychology. *J. Abn. Soc. Psychol.,* **28,** 1.

HALL E.T. (1959). "The Silent Language". New York: Premier Books.

HALL E.T. (1966). "The Hidden Dimension". New York: Doubleday.

HALL E.T. (1968). Proxemics. *Current Anthrop.,* Apr/Jun.

HAMILTON M. (1967). (Ed.) "Abnormal Psychology". Harmondsworth: Penguin.

HINDE R.A. (1966). "Animal Behaviour". London: McGraw-Hill.

HOLE C. (1949). "English Sports and Pastimes". London: Batsford.

HOWELL M.I. (1969). Towards a history of sport. *J. Health, P.E. Recreation* March.

HUMPHREY D. (1950). 'Dance' and related entries, "Webster's New International Dictionary", 2nd edition. Springfield, Merriam Co.

HUXLEY J. (1966). (Ed.) A discussion on ritualization of behaviour in animals and man. *Phil. Transact. R.S.,* Series B. vol. 251.

KLOPFER P.H. (1968). From Ardrey to altruism. A discussion on the biological basis of human behaviour. *Beh. Sci.,* **13,** 339.

KLOPFER P.H. & HAILMAN J.P. (1967). "An Introduction to Animal Behaviour: ethology's first century". London: Prentice-Hall.

KURATH G. (1956). Choreology and Anthropology. *Am. Anth.,* **58.**

KURATH G. (1966). 'Dance' and related entries, "Webster's New International Dictionary". 3rd Edition. Springfield: Merriam Co.

LABARRE W. (1947). The cultural basis of emotions and gesture. *J. Pers.,* **16.**

LABARRE W. (1964). Paralinguistics, kinesics and cultural anthropology. In T.A. Sebeok (Ed.) "Approaches to Semiotics". The Hague: Mouton.

LAING R.D. (1960a). "The Self and Others". London: Tavistock.

LAING R.D. (1960b). "The Divided Self". London: Tavistock.

LOCKARD R.B. (1971). Reflections on the fall of comparative psychology: is there a message for us all? *Am. Psychol.,* **26,** 168-179.

LOIZOS C. (1967). Play behaviour in higher primates: a review. In D. Morris, (Ed.) "Primate Ethology". London: Weidenfeld & Nicolson.

LORENZ K. (1963). "On Aggression". London: Methuen.

LORENZ K. (1970). "Studies in Animal Behaviour I". London: Methuen.

LOY J.W. & KENYON G.S. (1969). "Sport, Culture and Society". New York: Macmillan.

LÜSCHEN G. (1967). The interdependence of sport and culture. *Int. Rev. Spts. Soc.*, **2**, 127-141.

LÜSCHEN G. (1970). (Ed.) "The Cross-cultural Analysis of Sports and Games". Illinois: Stipes.

LYMAN S.M. & SCOTT M.D. (1967). Territoriality: a neglected sociological dimension. *Social Problems* **12**, 236-249.

McINTOSH P.C. (1963). "Sport in Society". London: Watts.

MEAD M. (1928). "Coming of Age in Samoa". Harmondsworth: Penguin.

MEAD M. (1937). "Co-operation and Competition Among Primitive Peoples". New York: McGraw-Hill.

MEYER V. & CHESSER E.S. (1970). "Behaviour Therapy in Clinical Psychiatry". Harmondsworth: Penguin.

MILLAR S. (1968). "The Psychology of Play". Harmondsworth: Penguin.

MITCHELL G.D. (1968). "A Dictionary of Sociology". London: Routledge & Kegan Paul.

MONTAGUE M.F.A. (1968). (Ed.) "Man and Aggression". Oxford: University Press.

MORRIS D. (1967). (Ed.) "Primate Ethology". London: Weidenfeld & Nicolson.

OPIE I & OPIE P. (1969). "Children's Games in Street and Playground". Oxford: Clarendon Press.

ROBERTS J.M., ARTH M.J. & BUSH R.R. (1959). Games in culture. *Am. Anthrop.*, **61**, 579-605.

RUSSELL W.M.S. & RUSSELL C. (1968). "Violence, Monkeys and Man". London: Macmillan.

RUST F. (1969). "Dance in Society". London: Routledge & Kegan Paul.

SCHEFLEN A. (1964). The significance of posture in communication systems *Psychiatry*, **27**.

SCOTT J.P. (1967). Comparative psychology and ethology. *Ann. Rev. Psychol.*, 18.

SCOTT J.P. (1968). "Early Experience and the Organisation of Behaviour". California: Brookes-Cole.

SHIPMAN M. (1968). A sociological perspective of dance. *A.T.C.D.E.* Dance section report.

SEBEOK T.A. (Ed.) (1968). "Animal Communication: techniques of study and results of research". Indiana: University Press.

SLUCKIN W. (1966). Early experience. In B.M. Foss (Ed.) "New Horizons in Psychology". Harmondsworth: Penguin.

SOMMER R. (1967). Small group ecology. *Psychol. Bull.*, 67.

SOUTHWICK C.H. (Ed.) (1963). "Primate Social Behaviour". Princeton: Van Nostrand.

SPENCER H. (1861). "Intellectual, Moral and Physical Education". London: Williams & Norgate.

SYMPOSIUM ZOOL. SOCIETY LONDON (1962). No. 8.

SYMPOSIUM ZOOL. SOCIETY LONDON (1965). No. 14.

SZASZ T.S. (1961). "The Myth of Mental Illness". London: Secker & Warburg.

THEORDORSON G.A. & THEORDORSON A.G. (1970). "A Modern Dictionary of Sociology". London: Methuen.

TINBERGEN N. (1942). An objective study of the innate behaviour of animals. *Biblioth. Biotheor.,* I.

TINBERGEN N. (1963). On aims and methods of ethology. *Z. Tierpsych.* **20,** 410-431.

VINE I. (1970). Communication by facial-visual signals. In J.H. Crook (Ed.) "Social Behaviour in Birds and Mammals". New York: Academic Press.

WASHBURN S.L. & JAY P.C. (Eds.) (1968). "Perspectives on Human Evolution". London: Holt, Rinehart & Winston.

WATERS R.H., RETHLINGSCHAFER D.A. & CALDWELL W.E. (1960). "Principles of Comparative Psychology". London: McGraw-Hill.

WHITING J.W.M. (1968). Methods and problems in cross-cultural research. In G. Lindzey & E. Aronson (Eds.) "The Handbook of Social Psychology II". Addison-Wesley.

WHITING J.W. & CHILD I.L. (1953). "Child Training and Personality: a cross-cultural study". New Haven: Yale University Press.

Specific References

ADAMS M. (1969). The concept of Physical Education II. *Proc. Phil. Ed. Soc.*

ALDRICH-BLAKE F.P.G. (1970). Problems of social structure in forest monkeys. In J.H. Crook (Ed.) "Social Behaviour in Birds and Mammals". London: Academic Press.

ALTMAN I. & HAYTHORN W.W. (1967). The ecology of isolated groups *Beh. Sci.,* **12,** 169-182.

ALTMANN S.A. (1962). A field study of the sociobiology of Rhesus Monkeys. *Annals N.Y. Acad. Sci.,* 102.

ALTMANN S.A. (1965). Sociobiology of Rhesus Monkeys: II stochastics of social communication. *J. Theor. Biol.,* 8.

AMBROSE J.A. (1961). The development of the smiling response in early infancy. In B.M. Foss (Ed.) "Determinants of Infant Behaviour I". London: Methuen.

AMBROSE J.A. (1963). The concept of the critical period for the development of social responsiveness. In B.M. Foss (Ed.) "Determinants of Infant Behaviour II". London: Methuen.

ANDREW R.J. (1963). Evolution of facial expression. *Science,* 142.

ANDREW R.J. (1965). The origins of facial expression. *Sci. Am.,* 213.

ANTHONY D.W.J. (1966). Comparative physical education. *Phys. Ed.,* **58, 175,** 70-73.

ARDREY R. (1967). "The Territorial Imperative". London: Collins.

BASTOCK M., MORRIS D. & MOYNIHAN M. (1953). Some comments on conflict and frustration in animals. *Beh.,* **6,** 66-84.

BATESON G., JACKSON D.D., HALEY J. & WEAKLAND J. (1956). Towards a theory of schizophrenia. *Beh. Sci.,* **1,** 251-261.

BETTELHEIM B. (1959). Joey: a mechanical boy. *Sci. Am.,* 116-127.

BLURTON-JONES N.G. (1967). An ethological study of some aspects of social behaviour of children in nursery school. In D. Morris, (Ed.) "Primate Ethology". London: Weidenfeld and Nicolson.

BREWER W.D. (1951). Patterns of gesture among the Levantine Arabs. *Am. Anth.,* **53,** 232-237.

BURNS T. (1964). Non-verbal communication. *Discovery,* **25, 10,** 30-37.

BUTLER R.A. (1954). Incentive conditions which influence visual exploration. *J. Exp. Psychol.,* 48.

CAMERON N. (1959). Paranoid conditions and paranoia. In S. Ariet (Ed.) "American Handbook of Psychiatry". New York: Basic Books.

CAMPBELL D.T., KRUSKAL W.H. & WALLACE W. (1966). Seating aggregation as an index of attitude. *Sociometry,* **29,** 1-15.

CARLISLE R. (1969). The concept of physical education. *Proc. Phil. Ed. Soc.*

CASSIDY R. (1965). The cultural definition of physical education. *Quest,* **IV.**

CHANCE M.R.A. (1956). Social structure of a colony of Macaca Mulatta. *Brit. J. An. Beh.* **4.**

CHANCE M.R.A. (1971). Mother monkeys and their infants. *Sci.* **7, 1,** 35-40.

CHOMBAGI Z. (1968). Physical education and sports in Syria. *Int. Rev. Spts. Soc.* 3.

CLARK A.H., WYON S.M. & RICHARDS M.P.M. (1969). Free-play in nursery school children. *J. Child Psychol. Psychiat.,* **10,** 205-216.

COMER G. (1968). Relationships between sport and mental health. *Phys. Ed.,* **60,** 181.

CUCELOGLU D.M. (1970). Perception of facial expressions in three different cultures. *Ergonomics,* **13, 1,** 93-100.

DAMM H. (1960). Vom Wesen sogennanter Leibesübungen be: Naturvölkern *Studium Generale* **13, 1,** 3-10.

DANIELS A.S. (1966). The study of sport as an element of the culture. *Int. Rev. Spts. Soc.,* 153-165.

DAVITZ J.R. (1964). A review of research concerned with facial and vocal expressions of emotion. In J.R. Davitz (Ed.) "The Communication of Emotional Meaning". New York: McGraw-Hill.

DITTMAN A.T. (1962). The relationship between body movements and moods in interviews. *J. Consult. Psychol.,* **26,** 480.

DUNLAP H.L. (1951). Games, sports, dancing and other vigorous recreational activities and their function in Samoan culture. In J.W. Loy & G.S. Kenyon (Eds.) (1969) "Sport, Culture and Society". London: Collier-Macmillan.

FREEDMAN N. & HOFFMAN S.P. (1967). Kinetic behaviour in altered clinical states: approach to objective analysis of motor behaviour during clinical interviews. *Percept. Motor Skills,* 24.

FREDERICKSON F.S. (1969). Sports and the cultures of man. In J.W. Loy & G.S. Kenyon (Eds.) "Sport, Culture and Society". New York: Macmillan.

GLASSFORD R.G. (1970). Organisation of games and adaptive strategies of the Canadian eskimo. In G. Luschen (Ed.) "The Cross-cultural Analysis of Sport and Games". Illinois: Stipes.

GRANT E.C. (1970). Face to face. New Society 7th May.

GRAY P.H. (1958). Theory and evidence of imprinting in human infants. J. Psychol., 46, 155-166.

GUNTHER M. (1961). Infant behaviour at the breast. In B.M. Foss (Ed.) "Determinants of Infant Behaviour". London: Methuen.

HALL E.T. (1963). A system for the notation of proxemic behaviour. Am. Anth., 65, 1003-1026.

HALL K.R.L. (1962). The sexual, agonistic and derived social behaviour patterns of the wild chacma baboon, Papiu Ursinus. Proc. Zool. Soc. London., 139 (ii) 283-327.

HARLOW H. (1961). The development of affectional patterns in infant monkeys. In B.M. Foss (Ed.) "Determinants of Infant Behaviour I". London: Methuen.

HARLOW H. (1963). The maternal affectional system. In B.M. Foss (Ed.) "Determinants of Infant Behaviour II". London: Methuen.

HAYES A.S. (1964). Paralinguistics and kinesics: pedagogical perspectives. In T.A. Sebeok (Ed.) "Approaches to Semiotics". The Hague: Mouton.

HENDRY A.E. (1968). Social influences upon the early development of P.E. in England I. Phys. Ed. 60, 181.

HENDRY A.E. (1969). Social influences upon the early development of P.E. in England II. Phys. Ed., 61, 182.

HENDRY A.E. (1970). The expansion of the curriculum and physical education Brit. J.P.E. 1, 5.

HINDE R.A. (1963). The nature of imprinting. In B.M. Foss (Ed.) "Determinants of Infant Behaviour II". London: Methuen.

HINDE R.A. & ROWELL T.E. (1962). Communication by postures and facial expressions in rhesus monkey (macaca mulatta). Proc. Zoological Soc. London, 138, 1-21.

HOWELL M.I. (1969). Towards a history of sport. J. Health P.E. Recreation, March.

HUTT C. & VAIZEY M.J. (1966). Differential effects of group density in social behaviour. Nature, 209, 1371-2.

INGRAM G.I.C. (1960). Displacement activity in human behaviour. Am. Anth., 62, 6, 994-1003.

JOHANNSEN W.J. (1961). Responsiveness of chronic schizophrenics and normals to social and non-social feedback. J. Abn. Soc. Psychol., 62, 106-113.

JOHNSON W. (1967). Comparative Physical Education. The Phys. Educator, 24, 3, 117-118.

JOKL W. (1964). "Medical, Sociology and Cultural Anthropology of Sport and Physical Education". Springfield: Thomas.

JOURARD S. (1966). An exploratory study of body accessibility. Brit. J. Soc. Clin. Psychol., 5, 221-231.

JUNG C.G. (1960). "The Psychogenesis of Mental Disease". London: Routeledge & Kegan Paul.

KIRSTEIN L. (1935). "Dance". New York: Putman.

KEALIINOHOMOKU J. (1970). An anthropologist looks at ballet as a form of ethnic dance. *Impulse:* extensions of Dance, 24-33.

KENDON A. (1963). Temporal aspects of the social performance in two person encounters. Ph.D. thesis cited in M. Argyle (1969) "Social Interaction". London: Methuen.

KENYON G.S. (1968). A conceptual model for characterizing physical activity. *Res. Quart.,* **39,** **1,** 96-105.

KLOPFER P.H. (1967). Is imprinting a Cheshire Cat? *Beh. Sci.* 12.

KRAUS R. (1969). "History of the Dance in Arts and Education". New Jersey: Prentice-Hall.

KROUT M.H. (1954a). An experimental attempt to produce unconscious manual symbolic movements. *J. Genet. Psychol.,* 51.

KROUT M.H. (1954b). An experimental attempt to determine the significance of unconscious manual symbolic movements. *J. Genet. Psychol.,* 51.

KURATH G. (1950). A new method of choreographic notation. *Am. Anth.,* 52.

KURATH G. (1953). Native choreographic areas of North America. *Am. Anth.,* 55.

KURATH G. (1954). A basic vocabulary for ethnic dance descriptions. *Am. Anth.,* 56.

LABAN R. (1956). "Principles of Dance and Movement Notation". London: Macdonald & Evans.

LEEDS STUDY GROUP (1970). The concept of Physical Education. *Brit. J.P.E.,* **1,** 4.

LEFCOURT H.M. & STEFFY R.H. (1966). Sex linked censure expectancies in process and reactive schizophrenics. *J. Pers.,* **34,** 366-380.

LEFCOURT H.M. et al. (1967). Visual interaction and performance of process and reactive schizophrenics as a function of examiner's sex. *J. Pers.,* **35,** 535-546.

LERNER M.J. & FAIRWEATHER G.W. (1963). Social behaviour of chronic schizophrenics in supervised and unsupervised work groups. *J. Abn. Soc. Psychol.,* **67,** 219-225.

LIDZ T., FLECK S. & CORNELISON A. (1966). "Schizophrenia and the Family". New York: International University Press.

LOPATA L. (1968). The structure of time and the share of Physical Education in the case of industrial workers and co-operative farmers in the C.S.S.R. *Int. Rev. Spts. Soc.,* 3.

LORENZ K. (1937). Der Kumpan in der Umwelt des Vogels. English version in *The Auk,* 54.

LORENZ K. (1958). The evolution behaviour. *Sci. Am.,* reprints 412.

LOY J.W. (1968). The nature of sport: a definitional attempt. *Quest Monograph* **10,** 1-15.

MAHEU R. (1962). Sport and culture. *Int. J. Adult. Youth Ed.* UNESCO XIV, 175-6.

MARLER P. (1965). Communication in monkeys and apes. In I. De Vore (Ed.) "Primate Behaviour". New York: Holt, Rinehart & Winston.

MAY J. (1967). The relevance of historical studies in Physical Education I. *Phys. Ed.,* **59,** 178.

MAY J. (1968). The relevance of historical studies in Physical Education II. *Phys. Ed.,* **69,** 179.

MELLY G. (1965). Gesture goes classless. *New Society,* 17th June, 26-7.

MONTAGU M.F.A. (1955). Contributions of anthropology to psychosomatic medicine. *Am. J. Psychiat.,* **112,** 997-984.

MORRIS D. (1956). 'Typical intensity' and its relation to the problem of ritualization. *Beh.* **11,** 1-12.

MORRIS D. (1967). "The Naked Ape". London: Cape.

MORRIS D. (1969). "The Human Zoo". London: Cape.

MOYNIHAN M. (1967). Comparative aspects of communication in new world primates. In D. Morris (Ed.) "Primate Ethology". London: Weidenfeld & Nicolson.

NAPIER J.R. (1968). A classification of primate locomotor behaviour. In S.L. Washburn & P.C. Jay (Eds.)"Perspectives on Human Evolution". London: Holt, Rinehart & Winston.

NETTLETON B. (1967). Sport and Social values. *Phys. Ed.,* **59, 117,** 35-41.

PEASE K. (1970). Is the "Double Bind" a myth? *New Society,* 24th September.

PETRAK B. (1966). Sociology of P.E. in Czechoslovakia. *Int. Rev. Spts. Soc.,* I.

POHNDORF R.H. (1969). British rugby and American football: a fitness and cultural comparison. *Phys. Ed.,* **61,** 182.

REES D. (1966). Developments in P.E. in Wales. *Phys. Ed.,* **58,** 173.

ROBINSON J.P. (1970). Daily participation in sport across twelve countries. In G. Lüschen (Ed.) "The Cross-cultural Analysis of Sport and Games". Illinois: Stipes.

ROWELL T.E. (1967). Variability in the social organisation of primates. In. D. Morris, (Ed.) "Primate Ethology". London: Weidenfeld & Nicolson.

RUSSELL-DAVIS D. (1966). "An Introduction to Psychopathology" 2nd edition. London: Oxford University Press.

SACHS C. (1937). "World History of the Dance". New York: Norton.

SEAGOE M.V. (1962). Children's play as an indicator of cross-cultural differences. *J. Ed. Sociol.,* **36,** 278-283.

SCHOOLER C. & PARKEL D. (1966). The overt behaviour of chronic schizophrenics and its relationship to their internal state and personal history. *Psychiatry,* **29,** 67-77.

SCHUHAM A.I. (1967). The double-bind hypothesis a decade later. *T. Bull,* **68,** 409-416.

SEBEOK T.A. (1967). Discussion of communication processes. In S.A. Altmann (Ed.) "Social Communication Among Primates". Chicago: University Press.

SHIPMAN M. (1968). A sociological perspective of dance. *A.T.C.D.E. Dance section report.*

SLUCKIN W. (1964). "Imprinting and Early Learning". London: Methuen.

SOMMER R. (1959). Studies in personal space. *Sociometry,* **22,** 247-260.

SOMMER R. (1965). Further studies in small group ecology *Sociometry,* **25,** 111-116.

SOMMER R. & OSMOND H. (1962). The schizophrenic in society. *Psychiatry*, **25**, 244-255.

SULLIVAN H.S. (1955). "Conceptions of Modern Psychiatry". New York: Norton.

SUTTON-SMITH B. & ROBERTS J.M. (1970). "The Cross-cultural Analysis of Sport and Games". Illinois: Stipes.

THOMSON I. (1966). Physical Education in Scotland. *Phys. Ed.*, **58**, 173.

TIGER L. (1969). "Men in Groups". London: Nelson.

TINBERGEN N. (1960). The evolution of behaviour in gulls. *Sci. Am.*, reprints, 456.

VAN HOOFF J.A.R.A.M. (1962). Facial expressions in higher primates. *Symp. Zool. Soc. Lond.,* 8.

VAN HOOFF J.A.R.A.M. (1967). The facial displays of the catarrhine monkeys and apes. In D. Morris (Ed.) "Primate Ethology". London: Weidenfeld & Nicolson.

VEBLEN T.P. (1925). "The Theory of the Leisure Class". London: Allen & Unwin.

VEERHAULT P. (1969). A sociology of dance. *Ass. Principals of Women's P.E. Colleges,* I.M. Marsh College, Liverpool.

VERNON J. (1966). "Inside the Black Room". Harmondworth: Penguin.

VINACKE W.E. & FONG R.W. (1955). The judgement of facial expressions by three national-racial groups in Hawaii: II Oriental faces. *J. Soc. Psychol.,* **4, 41,** 185-195.

VUILLIER G. (1898). "A History of Dancing". New York: Appleton.

WALLER W. (1932). "The Sociology of Teaching". New York: Wiley.

WATSON O.M. & GRAVES T.D. (1966). Quantitative research in proxemic behaviour. *Am. Anth.,* 68.

WING J.K. (1962). Institutionalism in mental hospitals *Brit. J. Soc. Clin. Psychol.,* **1,** 38-51.

WITTKOWER E.D. (1969). Perspectives of transcultural psychiatry *Int. J. Psychiat.,* **8, 5.**

WOHL A. (1968). Fifty years of physical culture in the U.S.S.R. Reflections and conclusions. *Int. Rev. Sports Soc.,* 1.

WOLFF C. (1945). "The Psychology of Gesture". London: Methuen.

WOLFF P. (1963). Observations on the early development of smiling. In B.M. Foss (Ed.) "Determinants of Infant Behaviour II". London: Methuen.

WRIGHT R.M. (1967). Aspects of Italian P.E. *Phys. Ed.,* **58,** 174.

WYMER N. (1950). "Sport in England". London: Harrap.

ZEIGLER E.F. (1968). The idea of Physical Education in modern times. *J. Phys. Ed.*

TOPIC AREAS - DEVELOPED

PERSONALITY AND MOVEMENT BEHAVIOUR

by H.T.A. WHITING

4 I

General

Personality study is concerned with theories and factual evidence about individual differences in human behaviour. The focus in this section is on movement behaviour as a manifestation of individual differences, the source of such differences and the way in which they may be utilised in personality assessment. The field of personality study rightly belongs under the heading of Psychology and the serious student in this area would be well advised to be familiar with the following general concepts prior to developing an interest in personality and movement behaviour:-

Definitions of personality

Why we study personality—description, explanation, control, prediction

Theories of personality

Methods of personality assessment

The following books will make a useful background contribution:-

Morgan & King (1966); Hilgard & Atkinson (1967); Eysenck (1967); Cattell (1967); Hall & Lindzey (1957); Lindzey & Hall (1965)

I.1 The nature/nurture hypothesis

This topic is concerned with questions about the relative contributions of genetic endowment and environmental opportunity to personality development (Dobzhansky, 1962). Environment here includes inter-uterine conditions. In this respect, stress in pregnancy has been proposed as a possible factor in motor impairment (Stott 1959; Morris & Whiting 1971). The amount of movement behaviour in utero has been commented upon by a number of workers as a source of individual differences. The significance of movement for development in such a context has been discussed by Whiting (1967). Other workers have used the degree of spontaneity of movement behaviour in the young child as an indicator of individual differences (Bridger et al 1965; Schaffer 1966; Morris & Whiting 1971).

189

The effects of pre, para or post-natal insult (physical and chemical) on the developing brain has implications for movement impairment and hence personality development (Morris & Whiting 1971).

I.2 Dimensional Analyses

An attempt is made under this section heading to select those experimental personality theorists who have made a contribution which has particular relevance to movement behaviour—namely Witkin et al (1962); Eysenck (1967); Cattell (1967).

In terms of Eysenck's theory, the three dimension of personality—extraversion/ introversion; neuroticism & psychoticism—have a bearing as the need for the individual to seek stimulation (and hence to indulge in greater or lesser amounts of movement behaviour); to enter into social situations and hence to use movement as a means of communication and to seek particular forms of physical activity or to restrict such participation to a minimum. The physiological interpretation of some of Eysenck's findings in terms of an 'excitation/inhibition balance' and the concept of 'arousal' have particular implications for movement behaviour which are only just beginning to be made explicit. Some workers for example have pointed to the importance of kinesthetic feedback in 'arousal' terms (Bernhaut et al, 1953; Kulka 1960). The degree of conditionability of an individual reflected by his relative standing on the extraversion/introversion and neuroticism dimensions has great relevance to movement behaviour in 'fear' situations e.g. learning to swim (Whiting, 1970), rock-climbing, playing contact games etc. This concept needs further development at the present time. Ikai & Steinhaus (1961) have discussed the concept of inhibition in relation to physical performance of a quantitative nature.

Cattell & Eber's (1964) sixteen personality factor analysis of personality characteristics has received considerable usage as a means of assessing particular criterion groups engaged in particular physical activities (see Kane, 1968; Hendry, 1970 and Whiting et al, 1971) for overviews of this area. The higher-order factor derivatives of creativity, leadership and independence have significance but at this stage have not been greatly developed (Jones 1970; Jones & Whiting, 1971a, 1971b). The possibility of there existing stereotypes of particular forms of movement behaviour in criterion groups is beginning to be examined (Hendry, 1970; Jones 1970).

The contribution of Witkin et al (1962) to personality theory has been through the medium of perception. They postulate a field dependence/ independence continuum related to the ability of the individual to separate figure from ground in perceptual tasks involving in many cases whole body participation and in others movement of a more limited nature (e.g. eye movements). The concept relates to hierarchical systems of information analysers (Whiting et al, 1973) which may be related to particular cultural pressures (Wober, 1966) and to particular forms of movement experience (Whiting et al, 1973). In developmental terms their concept of differentiation/integration (Witkin et al, 1962) has implications for movement behaviour and the development of the body-concept.

1.3 Physique and temperament

The relationship between physique and temperament has a long historical connection (Kretschmer, 1925; Sheldon & Stevens, 1942; Parnell, 1958; Eysenck, 1947). Direct information on the relationships between particular movement characteristics and body-type has not been exploited to any great degree although Sheldon's description of his proposed classifications of temperament have implications which might be developed. Certainly individuals seem to associate stereotypic personality characterisitics with particular body types (Dibiase & Hjelle 1968)

1.4 Expressive movement and personality

The expression of the emotions in terms of movement, posture and gesture is not a new area of study. Darwin (1921) was one of the earlier workers in the field to consider such expressions in man and animals. The possibility of using posture and gesture in man as an indicator of individual differences has particularly been exploited by Lamb (1965) and Vernon & Allport (1933). The possibility of using 'effort' qualities of man as a relatively stable index of personality was implicit in the work of Laban (1960).. A short review and elaboration of this possibility has been presented by Jones (1970). Feldenkrais (1948) in his discussion of body and mature behaviour is a useful source of ideas. More recently, North (1972) has proposed a new approach to personality assessment through movement behaviour.

The non-experimental but influential hypotheses of Mead (1934) on the development of the self and society from participation in the social act uses 'gesture' as a key concept. Current interest in Mead's work has been provoked by the sociological and social psychological contributions of the symbolic inter-actionists—particularly Goffman (1955, 1956, 1961, 1963a, 1963b); Becker (1963); and Laing (1959, 1960, 1967).

1.5 Clinical aspects of personality study

The relationship of particular movement characteristics to psychopathology is not a new area of study (Breuer & Freud, 1968; Kanner, 1944; Nathan, 1967) although its subjective nature has perhaps limited its development. Nathan (1967) has recently presented a useful approach to diagnosis in this area utilising a flow-diagram analysis.

Bizarre movement patterns are often reported as part of certain clinical syndromes and also in relation to drug-taking and alcoholism often at a less permanent level.

1.6 Movement requirements of different personality types

This approach would not appear to have been developed to any great extent. The implication here is that particular personality 'types' would need to indulge in greater or lesser amounts of movement behaviour than the norm. This might also be generalised in a qualitative way. A link here with the concept of 'arousal' has already been suggested (see I.2). Possibilities in this area for further research and development as an extension of Eysenck's (1967) theoretical approach.

I.7 Self-concept

The contribution of movement to the development of the 'self' is confused to some extent by the generality of the concept. It is more obvious when 'daughter' concepts such as body image, body awareness etc. are utilised. Movement behaviour would seem to relate to the development of the body concept and to affect the cognitive, affective and psycho motor connotations (Whiting et al 1971).

Historically, the development of the concept of 'body concept' is well documented (Wapner & Werner, 1965; Witkin et al 1962; Schilder, 1935)

The interrelationship between body-concept, athletic participation and body-type has been investigated by Sugerman & Haronian (1964) and its relationship to physical fitness by Armstrong & Armstrong (1968).

The relative instability of the body-concept is reflected in the work of Freedman (1961), Head (1926), McKellar (1965). More recently, Francis (1968) draws attention to the effects of sensory and perceptual deprivation on the body-concept.

Witkin et al's (1962) differentiation/integration hypothesis has direct relevance to the development of the body-concept and the development of laterality as one aspect (Kephart, 1960; Jones 1970; Jones & Whiting 1971). The importance of an adequately developed body-concept is consistly reported in the literature in relation to movement impairment and compensatory education (Benyon, 1968; Kephart, 1960; Tansley, 1967; Morris & Whiting 1971).

From a skill learning point of view the relationship between information processing and body-concept has importance. Fisher (1965) for example discusses body-image as a source of selective cognitive sets. Werner & Wapner (1952); McFarland (1958); Calloway & Dembo (1958) and Hinckley & Rethlingschafer (1951) have related level of muscle tonus, degree of autonomic arousal and body-crippling to the reception and elaboration of particular stimuli. Helson (1958) also points out that bodily sensations contribute to the general adaptation level and hence have an influence on the way in which judgements are made.

Overviews of the field have been presented by Morris & Whiting (1971) and Whiting et al (1971).

Source Texts

CATTELL R.B. (1967). "The Scientific Analysis of Personality". Harmondsworth: Penguin.

DARWIN C. (1921). "The Expression of the Emotions in Man and Animals". London: Murray.

EYSENCK H.J. (1967). "The Biological Basis of Personality". Springfield: Thomas.

HALL C.S. & LINDZEY G. (1957). "Theories of Personality". New York: Wiley.

HENDRY L.B. (1970). A comparative analysis of student characteristics. Unpublished M. Ed. thesis University of Leicester.

HILGARD E.R. & ATKINSON R.C. (1967). "Introduction to Psychology". New York: Harcourt, Brace & World.

JONES M.G. (1970). Perception, personality and movement. Unpublished M. Ed. thesis University of Leicester.

KANE J.E. (1968). Personality and physical ability. Unpublished Ph. D. thesis. University of London.

KRETSCHMER E. (1925). "Physique and character". New York: Harcourt, Brace & World.

LAMB W. (1965). "Posture and Gesture: an introduction to the study of physical behaviour". London: Duckworth.

LINDZEY G. & HALL C.S. (Eds.) (1965). "Theories of Personality: primary sources and research". New York: Wiley.

MORGAN C.T. & KING R.A. (1966). "Introduction to Psychology". New York: McGraw-Hill.

PARNELL R.W. (1958). "Behaviour and Physique". London: Arnold.

SCHILDER P. (1935). "The Image and Appearance of the Human Body". London: Kegan Paul.

SHELDON W.H. & STEVENS S.S. (1942). "The Varieties of Temperament". New York: Harper.

WAPNER S. & WERNER H. (Eds.) (1965). "The Body Percept". New York: Random House.

WHITING H.T.A., HARDMAN K., HENDRY L.B. & JONES M.G. (1973). "Personality and Performance in Physical Education and Sport". London: Henry Kimpton.

WITKIN H.A., DYK R.B., FATERSON D.R. & KARP S.A. (1962). "Psychological Differentiation". New York: Wiley.

WITKIN H.A., LEWIS H.B., MACHOVER K., MEISSNER P.B. & WAPNER S. (1954). "Personality Through Perception". New York: Wiley.

VERNON P.E. & ALLPORT G. (1933). "Studies in Expressive Movement". London: MacMillan.

Specific References

ARMSTRONG H.E. & ARMSTRONG D.C. (1968). Relation of physical fitness to a dimension of body image. *Percept. Motor. Skills.*, **26**, 1173-1174.

BECKER H.S. (1963). "Outsiders". London: Collier-Macmillan.

BENYON S.D. (1968). "Intensive Programming for Slow-learners". Ohio: Merrill.

BERNHAUT J. et al (1953). Experimental contributions to the problem of consciousness. *J. Neurophysiol.,* **16,** 21-23.

BRIDGER W.H., BIRNS B.M. & BLANK M. (1965). A comparison of behavioural ratings and heart rate measurements in human neonates. *Psychosom. Med.,* **27,** 123-134.

BREUER J. & FREUD S. (1968). "Studies in Hysteria: The standard edition of the complete psychological works of Sigmund Freud: Vol II (1893-1895)". London: Hogarth.

CALLOWAY E. & DEMBO D. (1958). Narrowed attention: a psychological phenomenon that accompanies a certain physiological change. *Arch. Neurol. Psychiat.,* **79,** 74-90.

CATTELL R.B. & EBER H.W. (1964). "Handbook for 16 P.F. Questionnaire". Champaign: L.P.A.T.

DIBIASE W.J. & HJELLE L.A. (1968). Body image sterotypes and body-type preferences among male college students. *Percept. Motor. Skills.,* **27,** 1143-1146.

DOBZHANSKY F.G. (1962). "Mankind Evolving". Yale: University Press.

EYSENCK H.J. (1947). "Dimensions of Personality". London: University Press.

FELDENKRAIS (1948). "Body and Mature Behaviour". London: Kegan Paul.

FISHER S. (1965). The body image as a source of selective cognitive sets. *J. Personality,* **33,** 536-552.

FREEDMAN S.J. (1961). Sensory deprivation: facts in search of a theory. *J. Nerv. Ment. Dis.,* **132,** 17-21.

GOFFMAN E. (1955). On face-work. *Psychiatry,* **18,** 213-231.

GOFFMAN E. (1956). "The Presentation of Self in Everyday Life". Edinburgh: University Press.

GOFFMAN E. (1961). "Asylums". New York: Anchor.

GOFFMAN E. (1963a). "Behaviour in Public Places". Illinois: Free Press.

GOFFMAN E. (1963b). "Stigma". New Jersey: Prentice-Hall.

HEAD H. (1926). "Aphasia and Kindred Disorders of Speech". London: Cambridge Univ. Press.

HELSON H. (1958). The theory of adaptation level. In D.C. Beardslee & M. Wertheimer (Eds.) "Readings in Perception". Princeton: Van Nostrand.

HINCKLEY E.D. & RETHLINGSCHAFER D. (1951). Value judgments of heights of men by college students. *J. Psychol.,* **31,** 257-262.

IKAI M. & STEINHAUS A.H. (1961). Some factors modifying the expression of human strength. In "Health and Fitness in the Modern World", pub. by the Athletic Institute in association with the American College of Sports Medicine.

JONES M.G. & WHITING H.T.A. (1971a). Field dependence and laterality. Unpublished paper Physical Education Dept, University of Leeds.

JONES M.G. & WHITING H.T.A. (1971b). Field dependence and creativity. Unpublished paper Physical Education Dept., University of Leeds.

KANNER L. (1944). Early infantile autism. *J. Paediatrics,* **25,** 211-217.

KEPHART N.C. (1960). "The Slow-learner in the Classroom". Ohio: Merrill.

KULKA A., FRY C. & GOLDSTEIN F.J. (1960). Kinesthetic needs in infancy. *Amer. J. Orthopsychiat.* **30,** 306-314.

LABAN R. (1960). "The Mastery of Movement". London: Macdonald & Evans.

LAING R.D. (1959). "The Self and Others". London: Tavistock.

LAING R.D. (1960). "The Divided Self". London: Tavistock.

LAING R.D. (1967). "The Politics of Experience and the Bird of Paradise". Harmondsworth: Penguin.

McFARLAND J.H. (1958). "The Effect of Assymetrical Muscular Involvement on Visual Clarity". Paper presented at East Psych. Assoc. New York.

McKELLAR P. (1965). Thinking, remembering and imagining. In J.G. Howells (Ed.) "Modern Perspectives in Child Psychiatry". Edinburgh: Oliver & Boyd.

MEAD G.H. (1934). "Mind, Self and Society." Chicago: University Press.

MORRIS P.R. & WHITING H.T.A. (1971). "Motor Impairment and Compensatory Education". London: Bell.

NATHAN P.E. (1967). "Cues, Decisions and Diagnoses". New York: Academic Press.

NORTH M. (1972). "Personality Assessment through Movement". London: Macdonald & Evans.

SCHAFFER H.R. (1966). Activity level as a constitutional determinant of infantile reaction to deprivation. *Child. Dev.,* 37, 3, 596-602.

STOTT D.H. (1959). Evidence for pre-natal impairment of temperament in mentally retarded children. *Vita humana,* 2, 125-148.

SUGERMAN A.A. & HARONIAN F. (1964). Body-type and sophistication of body-concept. *J. Personality,* 32, 3.

TANSLEY M.E. (1967). The education of neurologically abnormal children. *Times Educ. Suppl.,* Jan 20th.

WERNER H. & WAPNER S. (1952). Toward a general theory of perception. *Psych. Rev.,* 59, 324-338.

WHITING H.T.A. (1967). Significance of movement for early development. *Rem. Gym. and Rec. Ther.,* 45, 1479.

WHITING H.T.A. (1970). "Teaching the Persistent Non-swimmer: a scientific approach". London: Bell.

WOBER M. (1966). Sensotypes. *J. Soc. Psychol.,* 70, 181-189.

TOPIC AREAS-DEVELOPED

PHYSICAL WORK METABOLISM

by J.D. BROOKE

4 J

General

Metabolism is the process by which essential nutrients are provided for the living cells and by which residual products of this activity are removed (Dorland (1968). The present review is of the whole body analysis of this metabolic chain and cellular respiration, not of the individual molecular events. For an introduction to the latter see Dagley & Nicholson (1970). For the former, general texts are Green (1968); Wright (1965); Gordon (1960); Starling & Evans (1968); and Consolazio et al (1963): Astrand and Rodahl (1970) review work physiology: a text specific to human movement is Brooke (1972); Cotes (1968) reviews lung function and Lippold (1968) is a useful programmed text on respiration: Davidson & Passmore (1969) is a compendium on human nutrition. The Handbooks of Physiology provide excellent detailed papers with particular attention being drawn to Asmussen (1965) on muscular excercise. There is a standard nomenclature for abbreviation of terms in human respiration, (Pappenheimer et al, 1950) and it is adhered to. Central control of metabolism is discussed in section K of the present text. For sports medicine see Johnson (1960); Hyman, Larson & Herrmann (1971).

Physical work is one type of environmental load. Organisms respond to environmental load seeking equilibrium (Bernard, 1885). The state of the body achieving equilibrium, Cannon (1939) called Homeostasis. With high load, compensation can occur so that the load is tolerated but with reduced overall function. Eventually with severe loads Disequilibrium occurs. This sets a finite limit to tolerance of the load. With repeated exposures to abnormal loads, human beings adapt, attempting to match their response to the demand and so regain equilibrium. These responses are subject frequently to rhythmic variation, e.g. diurnal (Adkins, 1964) and menstrual (Phillips, 1967).

J.1 Lung Function

Nunn (1969) and Cotes (1968) are sound on the general applied physiology. Brooke (1972) considers in detail respiration and exercise. Harris & Heath (1962) have an excellent text on pulmonary circulation and Caro (1966) edits a series of papers including recent developments in the field of respiration. The three volumes in section 3 of the Handbook of Physiology edited by Fenn & Rahn (1965) are excellent for detail. Brooke (1971b) is a handbook of 1000 references to respiration in exercise, divided into the work of the lung, the lung gas exchange and the tissue respiration.

Analysis of lung function at rest often fails to indicate differences in ability during exercise. Within a subject, ventilatory volume indicates lung work done and ventilatory minute volume, power. This is not a linear relationship for above approx. 45 l./min lung efficiency falls markedly, Cotes (1968), Delhez (1968). Between subjects such comparisons require statements of compliance, dV/dP. Over most of the exercise range tidal volume rises with a fall in the proportions of dead space. At severe levels of work (earlier in the less fit) the tidal volume plateaus and falls, with marked increases in ventilatory rate maintaining the increases in ventilatory volume, Brooke (1972). At this level some subjects show ventilatory muscle fatigue. Nielson (1936) suggests that at a V_E of approx. 150 l./min further increases in V_E cost more in oxygen to obtain than they provide in additional oxygen availability. Bøje (1944) and Mead (1967) propose the appropriate ventilatory combination of rate and V_T for maximum efficiency.

Krogh (1915) reported on the diffusion of gases through the lungs of man. The transfer of O_2 and CO_2 between air and blood is passively down pressure gradients. There is unequalness of transfer in different parts of the lung and West (1966) reviews involved factors such as R-L shunts, V/Q ratios and dead space air. Combined with the volume of air breathed, the gas fraction moved allows calculation of the volume of gas taken up or given off. Benedict & Cathcart (1913) and Hill (1922) made early calculations. Åstrand et al (1952; 1956; 1961) established the importance of the linear relationship between the volume of oxygen uptake ($\dot{V}o_2$) and the work load required if homeostasis is to be obtained. Non-linearity at higher work loads indicates disequilibrium and a switch from aerobic to anaerobic energy sources. Saltin & Åstrand (1967) report the highest world values for $\dot{V}o_2$ and $\dot{V}_E$. Fleisch (1959) describes the time course for stabilisation of the lung gas responses during exercise as being up to 11 minutes for full equilibrium but Cotes et al (1969) demonstrate close proximity of measures allowing 1-5 minutes for stabilisation. Prediction of physical work ability from lung responses is discussed in J.6. Kelman (1971) suggested that in the later stages of exercise the response of the organism is primarily to the removal of CO_2 and the data reported by Brooke (1971a) support this suggestion. The alveolar ventilation/-physiological dead space ratio for oxygen falls in the later stages of an exercising work load task earlier than does that for carbon dioxide. Eventually at exhaustion both are falling, and neither insufficient O_2 nor excessive CO_2 appear to be the drives for ventilation at this time (Brooke, 1971a) (see also Section C).

Åstrand & Rhyming (1954) proposed a nomogram to predict oxygen uptake from heart rate per work load. Rowell et al (1964) felt this prediction was reasonably accurate for use with endurance athletes but Sharkey (1966) studying industrial work and Davies (1968) studying an age range of humans felt that the experimental error from the prediction, 25% and 15% respectively, was too great.

Riley & Cournand (1951) made a detailed analysis of factors affecting lung gas exchange, identifying exposure time of red blood cells, gas diffusion/ml blood, shape of physiological range of dissociation curve, gas diffusion characteristics and resistance to diffusion. Bates et al (1955) measured the diffusion capacity of human lungs in ml of gas (STPD) passing per minute from alveoli to blood stream at 1mm Hg pressure gradient. Forster (1965) discusses the variability of its measurement and the effect of this on the significance of diffusion capacity calculation. Harris & Heath (1962) clearly describe the technique and how to partial the measure into alveolar-capillary membrane resistance and red cell membrane complex resistance. Anderson & Shepherd (1968) conclude that, contrary to a number of preceeding studies, when adequate experimental control is applied, diffusion capacity is not a measure that differentiates athletic from non-athletic subjects.

Sensory stimulation increases the ventilation; Henderson and Scarborough (1910) reported the ventilatory effects of pain. Flandrois et al (1967) demonstrated ventilatory increase with passive limb movement and Asmussen (1967) proposed that the arousal sensations of exhaustion distress in part account for the rise in ventilation at this time. Brooke Hamley and Stone (1970) obtained results that supported this concept. (See also Section C).

J.2. Digestion, Assimilation and Storage

Green (1968) provides a simple description and Davidson & Passmore (1969) a detailed consideration which only will be supplemented here.

The breakdown of foods is eventually to a few products suitable for absorption as nutrients, mainly, glucose or closely related simple sugars, 21 amino acids and glycerol in combination with long chain fatty acids. In addition it is necessary to make provision for vitamins, mineral salts and water. Carbohydrates provide 4 k cal/g energy using 820 ml. O_2/g, fats 9 k cal/g using 1980 ml O_2/g and proteins when used for energy and heat, 4 k cal/g after deamination. A normal daily diet might be approx. 3000 k cal split into 1,600 k cal carbohydrate, 400 k cal protein and 1000 k cal fat. There will be much variability about this average dependant on the level of physical activity. Fox & Cameron (1961) provide a simple but accurate text for the study of diet and food.

Dietary constituents can be calculated from peripheral body gas exchange if the subject is in homeostasis, Pappenheimer et al (1950). Using the Dill (1963) nomogram the metabolic non-protein respiratory quotient (RQ) can be obtained from CO_2 and O_2 exchange and from the R.Q. the k cal/l.O_2 and dietary mixture can be established. Alternatively the metabolic mixture can be obtained by also measuring the urinary nitrogen (Davidson & Passmore, 1969). In either case the energy costs of exercise can be determined, e.g. Christenson & Hansen (1939). Durnin & Passmore (1967) have made these determinations for many occupations and sports and describe techniques.

There is research concern at present about the contribution of fat and carbohydrate diets to prolonged work ability. Over hours of work, heart rates can be maintained at 180-190 beats/min and $\dot{V}o_2$ around 3-3.5 l/min in sports pursuits (Brooke & Davies, 1970b). Saltin & Hermansen (1967) report that high muscle glycogen levels after carbohydrate diets associate with extended work capacity. Preliminary reduction of body glycogen stores appeared to facilitate this, (Hermansen & Saltin, 1967). The muscle biochemistry involved is complex. By its association with cardio-vascular disease, obesity is intensely studied. It is seldom found in those who by preference or necessity live hard physical existences. Davidson & Passmore (1969) strongly recommend combining exercise and diet in weight reduction.

High levels of arousal can lead to reduced dietary intake by reduced intestinal absorption. Loss of appetite is a common symptom of anxiety neuroses and can become severe (although in some cases neuroses involve compensatory over-eating leading to obesity). Adrenal hormone release is probably involved and may link the effect of loss of appetite to a similar one occuring during hard physical exercise, when peristalsis may be inhibited or reversed.

For most sports performers the normal diet is considered adequate, although Bobb et al (1969) found a calorie deficiency in a group of sportsmen's diets. Down (1968) reviews the needs of the endurance runner. Mayer & Bullen (1960) survey nutrition and athletic performance.

J.3. Blood

Oxygen transport is primarily by association with haemoglobin in the red blood cells. Only a small proportion is physically dissolved in the blood. The dissociation curve for blood saturation with O_2 at various partial pressures is S. shaped. It is felt that this shape may be due to the four haem groups of the haemoglobin molecule picking up O_2 at different rates; the recent description of the structure of haemoglobin does not destroy this supposition; Perutz et al (1960). The curve shifts to the right during physical movement with high Pco_2, [cH+], and temperature and this facilitates the off loading of O_2 at the tissues without much altering the uptake at the lungs. Falls in the Pao_2 occur in severe exercise (Rowell et al, 1964). Most CO_2 transport is in the form of bicarbonate, with some combined with plasma proteins and haemoglobin and some in solution. $Paco_2$ also falls in severe exercise but later than the Pao_2 fall. The blood pH is normally approx. 7.40. In severe exercise this falls and may approach 7.00. Åstrand (1964) suggested that at 60-70% of the $\dot{V}o_2$ max, use of higher energy stores by anaerobic processes is indicated by the increase in blood lactates. Saiki, Margaria & Cuttica (1967) contest this and feel that significant anaerobic metabolism does not occur until $\dot{V}o_2$ max is reached. Williams, du Raan, von Rahden & Wyndham (1968) disagree. Certainly with inadequate oxygen potential lactic acid is released and alters the blood pH (Margaria et al, 1963; 1964). Wasserman et al (1964) suggest three indications of this onset of anaerobiosis in physical movement 1. increase in blood lactate concentration 2. decrease in arterial blood bicarbonate and pH and 3. increase in respiratory gas quotient. Bouhuys et al. (1966) conclude that the first provides the best determination. Benade et al (1970) note that with increasing fitness the % $\dot{V}o_2$ max at which the onset of anaerobic work occurs also rises.

Greater pH falls due to the active hydrogen ions from lactic acid and to a lesser degree carbonic acid are avoided by the buffering actions of bicarbonate ions, biphosphate ions and blood proteins including haemoglobin, which mop up the free ions. Shifts through these systems are important in the study of exhausting work. Morales & Shock (1944) describe the time course for pH low point after exercise as 4-5 minutes. This time may be reduced in athletic performers. The fall in pH is associated with the perception of pain (Borg, 1962), and tolerance to this pain is a characteristic of high work performance (Morehouse & Miller, 1967 and Brooke, Hamley & Stone 1970). (See Section C).

With increased demand for oxygen through hard training the total blood volume increases (Åstrand el al 1970), and it is reported at the time of going to press that Ekblom (1971) has demonstrated marked increases in maximum work capacity by large infusions of blood (approximating 20% of the total blood volume) to physical education students. The indirect implications require exploration.

An important blood role is heat dissipation by moving warm blood from the body core to the peripheral circulation. Also nutrients from the gut are transported in the blood from the portal vein for storage at body sites such as the liver or for use as protein in tissue maintenance and/or as glucose for muscle glycogen. Humoral control is exerted through the blood by the transport of hormones. Plasma protein resistance to disease and rejection of foreign bodies is partly formed in the white cell population. Krogh (1930) contains a delightful account of observing the amoeboid movement of a leucocyte through the capillary wall into a tissue space. In addition the electrolyte water balances of the intra-cellular and extra-cellular fluids are maintained through the plasma water pressure. The ionic balance in fluid compartments is a complex regulation (Strauss, 1957), which is important in exercise. Saltin (1964) discusses the metabolic effect of its disturbance by thermal dehydration in exercise. The amino acid balance of the body is achieved through an exchange of proteins from various parts of the body via the amino acids in the plasma protein.

J.4. Heart and Circulatory Dynamics.

Source texts are Gordon (1960) and Rosenbaum & Belknap (1959). Kelman (1971) provides a good background of general applied physiology. Heymans & Neil

(1958) provide a good review of reflexogenic control of the circulation and heart function and Blackburn (1969) is a symposium selection of papers on exercise electrocardiography. Brooke (1971c) is a duplicated handbook of 900 references on the heart and exercise, divided into (1) cardiac output (2) electrocardiography and (3) circulation and haemodynamics. Brooke (1972) considers homeostatic and adaptive mechanisms in cardiac response to physical work up to exhaustion.

The cardiac output, the volume of blood leaving the left ventricle per unit time, is the sum of the stroke volume of the heart and the rate of contraction. Both increase during exercise, Kelman (1971). Åstrand et al (1964) demonstrated that the stroke volume response preceeds the heart rate to the degree that by the time the heart rate reaches 110-115 beats/min, the stroke volume is on average approx. 90% of its maximum volume. From this level, changes in heart rate appear to indicate changes in cardiac output (Williams et al, 1962). For much of the range of response heart rate is linearly related to work load (Wahlund 1948). At high levels of response curvilinearity occurs (Brooke, Hamley & Thomason, 1968a; 1968b).

There is variability in the heart rate recorded for trained athletes at given work loads, (95% confidence level ± 3 beats) and from trial to trial (95% level ± 9 beats) (Brooke, 1968b; Brooke, Hamley & Thomason 1970).

The work done by the heart is determined by the cardiac output and the mean systolic aortic pressure. The energy expenditure is shown by the myocardial oxygen consumption. The latter is made up of three parts, due to vegetative function, to electrical activity and to pumping activity (Reeves, 1969). The last is by far the most important. Sarnoff et al (1959) has shown with the isolated supported heart that increased work by increased stroke volume little alters the myocardial $\dot{V}o_2$; increases by heart rate increase produce proportional change in this $\dot{V}o_2$; increases by pressure increase escalate the oxygen requirements of the heart. Thus with the increases due to stroke volume efficiency rises, due to heart rate increase efficiency is constant and due to pressure increase it falls. The initial use of stroke volume increase in early exercise indicates the organism's adaptation under the least energy expenditure principle, in a similar way to the lung function change of increased tidal volume versus later, less efficient, increases in ventilation rate.

Quantification of electrophysiological changes in the exercising heart of the healthy human is less well documented. The recent review of Blackburn (1969) is of great assistance. Simonson (1952) reported large R wave amplitudes in younger versus old men and Thomason, Hamley & Brooke (1969) compared the wave amplitudes for three leads of the resting ecg in female physical education and non-physical education students with only R wave amplitudes differentiating. This larger R wave in adaptation to work was maintained when a group of female swimmers also were studied. Smith (1969) reports differences between clinical and normal groups in T wave direction by mean vector analysis after a standardised exercise test. Simonson (1965) provides a visual plate differentiating normal from clinically suspect exercise ecg's. Blackburn (1969) and Taylor (1969) review developments in the exercise electrocardiogram and in exercise testing. Westura

(1969) and Lepeschkin (1969) both identify abnormalities of the S-T segment and the T wave as most commonly occurring ecg changes which may be accounted for by coronary insufficiency during exercise. Venables et al (1970) report T wave differences as correlates of anxiety. (See Section I).

Circulatory dynamics concerns the distribution of the blood in the body and the pressure load placed on the heart. As the blood moves from the left side of the heart to arteries, arterioles, capillaries and veins the pressure falls, the fall being greatest in the arterioles. This pressure may be palpated at pulse pressure points. The peripheral resistance is regulated by smooth muscle under sympathetic control in the arterioles and by blood viscosity. The central vaso-motor control of the muscle tonus is affected by general changes in Pco_2 Po_2 [cH^+], sensory stimulation, higher cortical control and arterial blood pressure feed back as described in Section C in this text. Local over-ride of this control occurs when blood flow needs to be increased as, e.g. during human movement. Ikai reports blood flow for human forearm at rest of 6.7 ml/100ml tissue/min and after exercise of 27.85 ml% per min. With training this was increased to 37 ml%/min. Increased local Pco_2, [cH^+], K^+ and temperature (in skin circulation particularly) and reduced Po_2 can cause local vasodilation and cold and angiotensin release act as strong vaso-constrictors. Some parasympathetic and sympathetic nerves cause vasodilation (Green, 1968) e.g. the nervi erigentes to the external genitalia. In exercise requiring over approx 30% of the muscle's maximum strength, occlusion of the muscle capillaries occurs during contraction (Barcroft, 1939). Systolic blood pressure rises due to hormone drive and is increased by restricted perfusion such as the occlusion described above for heavy exercise. It is suggested that this short term oxygen lack is necessary for the development of muscle strength. The venous return is assisted by a muscle pump action facilitated by adequate muscle tonus. Mental tension can effect through sympathetic innervation a restriction in peripheral circulation with increased systolic pressure. Coronary infarctions during rage indicate the pressure load that can be produced. Note that whereas in dynamic work systolic pressure rises with little diastolic change, in static work (e.g. isometric) both rise markedly. Such work places sudden high load on the cardio-vascular system as described above.

In addition to blood pressure changes, mental stimuli affect the cardiac response. Bowen (1903) first noted the anticipatory heart rate response prior to exercise. Brooke & Hamley (1970) record the range this extends to in exercise. Hickam et al (1948) report rises in cardiac output and heart rate in subjects made anxious: this disappeared some time after exercise started. Tollman (1966) reported a correlation between intensity of pain perceived by competitive swimmers in training and the level of the heart rate.

J.5. Skeletal Muscle

When human movement occurs, skeletal muscle is the source of power acting on the environment to adjust the position of the body (see also biomechanics, Section L).

Much of J.5 has been covered above. The muscle acts as an engine that converts nutrients into mechanical and heat energy. Research into these mechanisms has been done at the molecular level e.g. Huxley & Huxley (1964) and molar levels e.g. Bourne (1960); Sandow (1959). Brooke (1971b) includes 200 references to tissue respiration. Pernow & Saltin (1971) provide a modern review of muscle metabolism during exercise.

From the viewpoint of whole body metabolism primary requirements for exercising muscles are oxygen, glycogen and a salt-water balance. For growth and repair amino acids are necessary together with adequate stimulation, neural and hormonal. In exercising humans the nervous activity is usually studied by electromyography (Basmaijan, 1967). Licht (1961) reviews its use in diagnosis of muscle impairment and de Vries (1966) reports experiments using quantified emgs at tonic levels to study muscle pain from work. While common use is made of emg work as a qualitative descriptive tool, quantified emg's are little developed. Cooper (1968) reviews applications to physical education.

When the oxygen potential is adequate i.e. aerobiosis exists, the muscle chemical cycle is complete and no limit to work is apparent. With inadequate oxygen potential i.e. anaerobiosis, sufficient mechanical energy for the work is only obtained with depletion of the finite energy muscle stores (adenosine triphosphate). At this time the lack of oxidation results in the release of lactic acid from the cycle and the failure to completely resynthesise adenosine diphosphate to triphosphate. The onset of this reaction was discussed in J.3 above. This is a complex bio-chemical field, see e.g. Bell et al (1965) for an introduction. Off loading of oxygen at the muscle is facilitated on the blood side by increased $[cH^+]$, Pco_2, temperature, adrenaline and diphosphoglycerate (Benesch, 1969) and on the tissue side by the level of respiratory enzymes transporting oxygen to the mitrochondria in the muscle, (Hollosky, 1967). These are detailed research fields.

Saltin & Hermansen (1967) report the reduction of muscle glycogen with exercise, and its pre-exercise elevation by diet, see J.2. Bengstrom (1967) discusses the biochemical mechanisms underlying this and these topics are considered in detail in Pernow & Saltin (1971). It appears that biochemical changes over days are necessary for the muscle to burn more fats during heavy exercise (Karvonen, 1967).

Removal of end products of the skeletal muscle metabolism is for CO_2 at the lung, hydrogen ions the kidney, lactates the liver, water the lungs, skin and kidney, heat the skin and lungs and mechanical energy through behaviour as described below.

J.6. Behaviour Analysis.

In a factor analysis of a group of physical fitness tests Fleishman (1964) identified a number of factors of flexibility, agility and strength. The latter lies within the scope of this section. It was split into three separate factors for static, explosive and dynamic strength. Frequently in analyses of behaviour these separate variances are apparent e.g. Brooke, Clinton, Cosgrove, Dingle & Knowles (1970) found static leg dynamometry little related to kicking a stationary soccer ball or to ability to vertical jump. A common variance that could only tentatively be termed cardio-vascular endurance was identified by Fleishman. It is probable, as reported below, that the tests were not highly loaded with this type of variance.

When the work load is progressively increased the work ability of the subject increases (Lange, 1919; De Lorme, 1945; Adamson & Morgan, 1954 and Hellebrandt, 1958). If the work load decreases so does the work ability e.g. coronary inadequacy in early middle-age groups taking no active leisure pursuits.

Ekblom et al (1968) showed this type of improvement in trained subjects' cardiac outputs, oxygen uptakes and arterio-venous differences. Maxfield (1964) showed this in reduced heart rate costs of work after training. Davies & Knibbs (1970) produced small increases in aerobic power with training, but only at training work loads demanding at least 50% $\dot{V}o_2$ max. Williams et al (1968) report high levels of physiological response for up to 60 mins. in well trained subjects and Brooke & Davies (1970) extend this for work over 2 hours, see J.2. Paez et al (1967) Naughton et al (1964) and Fox et al (1964) report similar adaption for patients recovering from heart and lung disease. Note that with very high intensities of physical training myocardial lesions are reported in rats (Leon et al, 1968): the rats were also anxiety stressed, see Sections K and I: some sportsmen are also stressed by their competitive physical pursuits.

The maximum oxygen uptake ($\dot{V}o_2$ max) during increasing loads of physical work is now the reference standard for cardio-respiratory fitness (Shephard et al 1968). It is highest tested using arms and legs. By conventional testing the treadmill maximum value is approx. 3.4% higher than the step test one and 6.6% higher than the cycle ergometer $\dot{V}o_2$ max. In analysis of human movement some use is made of the oxygen transport per pulse (Dawson, 1935) and of the oxygen uptake per kilogram body weight, $\dot{V}o_2/kg$ e.g. Ishiko (1967.) For measuring work capacity Anderson (1964) has made a useful review of procedures. Brooke (1968a) has summarised by physiological parameters a set of common 'physiological' fitness tests.

The Harvard step test is frequently reported to correlate significantly with the $\dot{V}o_2/kg$ (de Vries et al, 1966; Hettinger 1961; Ishiko, 1967). Brooke, Dugmore, Humphries & King (1970) reported a correlation with the maximum $F_1 - F_Eo_2$ during an increasing work task. As the work ability range of the subjects narrows the correlation with the step test may be expected to fall. Chrastek et al (1965) studied a number of step tests and found no difference between scores from them.

Only low correlation has been found between $\dot{V}o_2$ max or $\dot{V}o_2/kg$ and the typical applied fitness battery e.g. the AAHPER tests. Even with multiple correlation the error of prediction is quite high, 17% (Drake et al 1968), 12% (Falls et al, 1966). Taylor (1944) was so pessimistic as to suggest that the only viable predictor of work performance is the work done (!) although he later Taylor (1945) reports low correlations between level of work load at exhaustion and fractional extraction of O_2.

It is clear that specificity of adaptation exists (Zuntz, 1901; Henry & Whitley 1960; Backman, 1961). The closer the laboratory test to the normal work performance, the greater the expectation of correlations.

Thus Brooke, Hamley & Thomason (1968) and Brooke (1971a) found high correlations and good prediction of laboratory and field work ability with lung gas extraction fractions $F_I - F_E O_2$ from racing cyclists working on a laboratory cycle ergometer. These gas extraction levels appear to be important in some work tasks for as well as Taylor (1945) above, Lacoste et al (1963) found them better differentiators of work ability between healthy and cardiorespiratory defect men than is $\dot{V}o_2$/work level and Fleisch (1951) reports that the gas fraction improves as the organism develops to maturity. (See Section K.) Brooke, Dugmore, Humphries & King (1970) found significant correlations between the short term work ability, gas extraction fraction and Harvard step test. Use has also been made of $\dot{V}o_2$ max/kg in successfully predicting running performances e.g. Ishiko (1967). Heart rate measures have not been as predictive. Knehr et al (1942) in well trained groups could not differentiate heart rate recovery curves. Wilmore (1968) found no differences in maximum heart rates when motivational conditions produced significant differences in work done. Brooke, Hamley & Thomason (1968a, 1968b) and Brooke & Hamley (1972) found no correlation between maximum heart rate and maximum work ability but there were significant correlations between the work ability and the degree of curvature in the heart rate-work regression in the later stage of performance of an increasing work load task.

It appears that in the application of physiological data to the description and prediction of hard or exhausting human movement it will be necessary to construct work tasks that show good resemblance to the habitual work performance and then to assess appropriate physiological variables. The next stage, now developing, is to determine accurately the comparative effects on performance capabilities of the use of different types of training or adaptive procedures.

Source Texts

ASMUSSEN E. (1965). Muscular exercise. In W.O. Fenn & H. Rahn (Eds.) "Handbook of Physiology Section 3, Vol. 2. Respiration". Washington: Amer. Physiol. Soc.

ÅSTRAND P.O. & RODAHL K. (1970). "Textbook of Work Physiology". New York: McGraw Hill.

BELL G.H., DAVIDSON J.N. & SCARBOROUGH H. (1965). "Textbook of Physiology and Biochemistry". London: Livingstone.

BROOKE J.D. (1972). "Human Performance of Exhausting Movement". London: Heinemann. (In preparation.)

CONSOLAZIO C.F., JOHNSON R.E. & PECORA L.J. (1963). "Physiological Measurements of Metabolic Functions in Man". New York: McGraw-Hill.

COTES J. (1968). "Lung Function. (2nd Ed.)" Oxford: Blackwell.

FENN W.O. & RAHN H. (Eds.) (1965). "Handbook of Physiology Section 3, Vols. 1, 2, 3, Respiration". Washington: Amer. Physiol. Soc.

GORDON B.L. (Ed.) (1960). "Clinical and Cardiopulmonary Physiology (2nd Ed.)". New York: Grune & Stratton.

GREEN J.H. (1968). "An Introduction to Human Physiology. (2nd Ed.)". London: Oxford University Press.

HYMAN A.S., LARSON L.A. & HERRMANN D.E. (1971). "The Encyclopaedia of Sport Sciences and Medicine". London: Collier-Macmillan.

JOHNSON W. (Ed.) (1960). "Science and Medicine of Exercise and Sports". New York: Harper.

LIPPOLD O. (1968). "Human Respiration: a programmed course". London: Freeman.

STARLING E.H. & EVANS L. (1968). "Principles of Human Physiology. (14th Ed.)". London: Churchill.

WRIGHT S. (1965). "Applied Physiology (11th Ed.)". London: Oxford University Press.

Specific References

ADAMSON G.T. & MORGAN R.E. (1954). Circuit training. *Phys. Ed.,* **46, 1,** 1-7.

ADKINS S. (1964). Performance, heart rate and respiration rate on the day-night continuum. *Percept. Motor. Skills.,* **18,** 409-412.

ANDERSON K.L. (1964). Measurements of work capacity. *J. Spt. Med. Phys. Fit.,* **4, 4,** 236-240.

ANDERSON T.W. & SHEPHARD R.J. (1968). The effects of hyperventilation and exercise upon the pulmonary diffusing capacity. *Respiration.,* **25,** 465-484.

ASMUSSEN E. (1967). Exercise and the regulation of ventilation. *Circ. Res. Supp.*

ÅSTRAND P.O. (1952). "Experimental Studies of Working Capacity in Relation to Sex and Age". Copenhagen: Munksgaard.

ÅSTRAND P.O. (1956). Human physical fitness with special reference to sex and age. *Physiol. Rev.,* **36,** 307.

ÅSTRAND P. O. & RHYMING I. (1954) A nomogram for calculation of aerobic capacity (physical fitness) from pulse rate during submaximal work. *J. Appl. Physiol.,* **17, 2,** 218-221.

ÅSTRAND P.O. & SALTIN B. (1961). Maximal oxygen uptake and heart rate in various types of muscular activity. *J. Appl. Physiol.,* **16,** 977-981.

ÅSTRAND P.O., CUDDY T.E., SALTIN B. & STENBERG J. (1964). Cardiac output during submaximal and maximal work. *J. Appl. Physiol.,* **19,** 168-274.

BACKMAN J.C. (1961). Specificity vs. generality in learning and performing two large muscle motor tasks. *Res. Quart.,* **32,** 1, 3.

BARCROFT H. & MILLER J.L.R. (1939). Blood flow through muscle during sustained contraction. *J. Physiol.,* **17,** 17.

BASMAIJAN J.V. (1967). "Muscles Alive. (2nd Ed.)". Baltimore: Williams & Wilkins.

BATES D.V., BOUCOT N.G. & DORMER A.E. (1955). Pulmonary diffusing capacity in man. *J. Physiol.* **129,** 237.

BENADE A.J., WYNDHAM C.H., STRYDOM N.B. & Van RENSBURG A. (1970). Physiological requirements for world class performances in endurance running. In *Proc. Int. Symp. Ed. Fisi. Desport.* Mozambique: Marques.

BENEDICT F.G. & CATHCART E.P. (1913). "Muscular Work. A Metabolic Study with Special Reference to the Efficiency of the Human Body as a Machine". Washington: Carnegie Inst.

BENESCH R. & BENESCH R.E. (1969). Intracellular organic phosphates as regulators of oxygen release by haemoglobin. *Nature,* **221,** 618.

BENGSTROM (1967). Discussion. In G. Blix (Ed.) "Nutrition and Physical Activity". Uppsala: Almquist & Wiksell.

BERNARD C. (1885). "Leçons sur les Phenoménes de la Vie Communs aux Animaux et aux Vegetaux. (2nd Ed.)". Paris: Balliére.

BLACKBURN H. (1969). The exercise electrocardiogram. Technological, procedural and conceptual development. In H. Blackburn, (Ed.) "Measurement in Exercise Electrocardiography". Illinois: Thomas.

BOBB A. et al (1969). The diet of athletes. *J. Spt. Med. Phys. Fit.,* **9,** 255.

BØJE O. (1944). Energy production, pulmonary ventilation and length of steps in well-trained runners working on a treadmill. *Acta. Physiol. Scand.,* **7,** 362-375.

BORG G. & DAHLSTROM H. (1962). The reliability and validity of a physical work test. *Acta. Physiol. Scand.,* **55,** 353-361.

BOUHUYS A., POOL J., BINKHORST R.A. & VAN LEEUWEN P. (1966). Metabolic acidosis of exercise in healthy males. *J. Appl. Physiol.,* **21, 3,** 1040-1046.

BOURNE G.H. (Ed.) (1960). "Structure and Function of Muscle". London: Oxford University Press.

BOWEN W.P. (1903). "Contributions to Medical Research". Ann Arbor: University of Michigan.

BROOKE J.D. (1968a). A review of the heart in exercise. Salford: Int. Report Physical Ed. Section, University of Salford.

BROOKE J.D. (1968b). The measurement of heart rate responses from physical exercise stress. In "Proceedings UPEA Conference". Edinburgh: The University.

BROOKE J.D. (1971a). The use of lung function measures to predict work abilities within a sports group. In J.D. Brooke (Ed.) "Lung Function and Work Capacity". Salford: The University.

BROOKE J.D. (1971b). "Reference Handbook No. 1, Respiration and Exercise". Eccles: Worthwhile Designs.

BROOKE J.D. (1971c). "Reference Handbook No. 2, Cardiac Output and Haemodynamics in Exercise". Eccles: Worthwhile Designs.

BROOKE J.D., HAMLEY E.J. & THOMASON H. (1968a). The relationship of heart-rate to physical work. *J. Physiol., 197*, 61-63.

BROOKE J.D., HAMLEY E.J. & THOMASON H. (1968b). The regression of cardiac, ventilatory and somatic measures on power output in subjects accustomed to fatigue. Paper to *"Int. Symposium on 'Stress and Fatigue', Perm. Com. and Int. Assoc. for Occu. Health"*. Paris: UNESCO.

BROOKE J.D. & DAVIES G.J. (1970). Tolerance to habitual work over hours. *Ergonomics, 13, 4,* 529.

BROOKE J.D. CLINTON N., COSGROVE I., DINGLE D. & KNOWLES J.E. (1970). The relationship between soccer kick length and static and explosive leg strength. *Br. J. Phys. Ed., 1,* 3, xvii-xviii.

BROOKE J.D., DUGMORE L., HUMPHRIES J. & KING C. (1970). The relationship between short term physical work ability, expired air oxygen fraction and the Harvard Step Test. *Br. J. Phys. Ed. 1,* 1, v-vi.

BROOKE J.D., HAMLEY E.J. & STONE P. (1970). A review of the physiological and mental state variables that predict ability at bicycle time-trial racing. In "Proceedings 18th World Congress of Sports Medicine". Oxford: (B.A.S.M. in press).

BROOKE J.D. & HAMLEY E.J. (1972). The heart-rate—physical work curve analysis for the prediction of exhausting work ability. *Med. Sci. Ex. Spt.* (in press).

BROOKE J.D., HAMLEY E.J. & THOMASON H. (1970). Variability in the measurement of exercise heart rate. *J. Sp. Med. Phys. Fit., 10,* 1, 21-26.

CANNON W.B. (1939). "The Wisdom of the Body (2nd Ed.)". New York: Norton.

CARO C.G. (Ed.) (1966). "Advances in Respiratory Physiology". London: Arnold.

CHRASTEK J.I. & STALY L. (1965). On determination of physical fitness by step up test. *J. Spt. Med. Phys. Fit. 5, 2,* 61-66.

CHRISTENSEN E.H. & HANSEN O. (1939). *Scand. Arch. Physiol., 81,* 1, 137-151.

COOPER D.F. (1968). A review of research into electromyography in relation to Physical Education. *Res. Phys. Ed., 1, 3,* 27-31.

COTES J.E. et al (1969). Human cardiopulmonary responses to exercise: comparison between progressive and steady state exercise, between arm and leg exercise and between subjects differing in body weight. *Quart. J. Exp. Physiol. 54,* 211-222.

DAGLEY S. & NICHOLSON D.E. (1970). "An Introduction to Metabolic Pathways". Oxford: Blackwell.

DAVIDSON S. & PASSMORE R. (1969). "Human Nutrition and Dietetics (4th Edn.). "Edinburgh: Livingstone.

DAVIES C.T.M. (1968). Limitations to the prediction of maximum oxygen intake from cardiac frequency measurements. *J. Appl. Physiol.,* **24, 5,** 700-706.

DAVIES C.T.M. & KNIBBS A.H. (1970). The nature of the training stimulus: Effects of intensity, frequency and duration of exercise on maximum aerobic power. In "Proceedings 18th World Congress of Sports Medicine". Oxford: B.A.S.M. (in press).

DAWSON P.M. (1935). "The Physiology of Physical Education". Baltimore: Williams & Wilkins.

DELHEZ L., DEROANNE R. & PETIT J.M. (1968). Enregistrement autonome de deux electromyogrammes respiratoire et de la fréquence cardiaque au cours de diverse activiteés physique. *Ergonomics,* **13, 4,** 529.

DILL D.B. et al (1963). Line chart for calculating RQ and true oxygen from analyses of expired air. In C.F. Consolazio et al "Physiological Measurements of Metabolic Functions in Man". New York: McGraw-Hill.

DORLAND (1968). "Medical Dictionary (24th Ed.)". London: Saunders.

DOWN M. (1968). Dietary considerations for endurance events. *J. Spt. Med.,* **4, 1,** 5-26.

DRAKE V., JONES G., BROWN J.R. & SHEPHARD R.J. (1968). Fitness performance tests and their relationship to the maximal oxygen uptake of adults. *Can. Med. Ass. J.,* **99,** 844-848.

DURNIN J.V.G.A. & PASSMORE R. (1967). "Energy, Work and Leisure". London: Heinemann.

EKBLOM B., ÅSTRAND P.O. et al (1968). Effect of training on circulatory response to exercise. *J. Appl. Physiol.,* **24, 4,** 518-528.

EKBLOM B. (1971). reported on Blood doping. *Sund. Times* 31 Oct. 30.

FALLS H.B., ISMAIL A.H. & MacLEOD D.F. (1966). Estimation of maximum oxygen uptake in adults from AAHPER Youth Fitness test items. *Res. Quart.,* **37, 2,** 192-201.

FLANDROIS R., LACOUR J.R., MAROQUIN J. & CHARLOT J. (1967). Limbs mechanoreceptors inducing the reflex hyperpnea of exercise. *Resp. Physiol.,* **2,** 335-343.

FLEISCH T. (1951). *Helvet. Med. Acta.* **18,** 23.

FLEISCH A. (1959). Les épreuves d'effort de durée moyenne chez l'homme sain. *Le Poumon et le Coeur,* **9,** 883-889.

FLEISCHMAN E.A. (1964). "The Structure and Measurement of Physical Fitness". New Jersey: Prentice Hall.

FORSTER R.E. (1965). Interpretation of measurements of pulmonary diffusing capacity. In W.O. Fenn and H. Rahn (Eds.) "Handbook of Physiology, Section 3, Vol. 2. Respiration". Washington: Amer. Physiol. Soc.

FOX B.A. & CAMERON A.G. (1961). "A Chemical Approach to Food and Nutrition". London: The University Press.

FOX M. & SKINNER J.S. (1964). Physical activity and cardiovascular health. *Am. J. Cardiol.,* **14,** 731-746.

HARRIS P. & HEATH D. (1962). "The Human Pulmonary Circulation". London: Livingstone.

HELLEBRANDT F.A. (1958). Application of the overload principle to muscle training in man. *Am. J. Physic. Med.,* **37, 5,** 278.

HENDERSON Y. & SCARBOROUGH M.M. (1910). *Am. J. Physiol.*, **25**, 385.

HENRY F.M. & WHITLEY J.D. (1960). Relationships between individual differences in strength, speed and mass in an arm movement. *Res. Quart.*, **31**, 24-33.

HERMANSEN L. & SALTIN B. (1967). *J. Appl. Physiol.*, 22.

HEYMANS C. & NEIL E. (1958). "Reflexogenic Areas of the Cardiovascular System". London: Churchill.

HICKMAN J.B., CARGILL W.H. & GOLDEN A. (1948). Cardiovascular reactions to emotional stimuli. *J. Clin. Invest.*, **27**, 290-298.

HILL A.V., LONG C.N.H. & LUPTON H. (1922/3). Muscular exercise, lactic acid and the supply and utilisation of oxygen. *Quart. J. Med.*, **16**, 135-171.

HOLLOSKY J.O. (1967). Biochemical adaptation in muscle. *J. Biol. Chem.*, **242**, 9, 2278-2282.

HUXLEY A.F. & HUXLEY H.E. (1964). A discussion on the physical and chemical basis of muscular contraction. *Proc. Roy. Soc. B.*, **160**, 434-536.

IKAI M. (1970). Basic training of aerobic work capacity. In *"Proc. Int. Symp. Ed. Fisi. Desport".* Mozambique: Marques.

ISHIKO T. (1967). Aerobic capacity and external criteria of performance. *Can. Med. Ass. J.*, **96**, 746-749.

KARVONEN (1967). Discussion. In G. Blix (Ed.) "Nutrition and Physical Activity". Uppsala: Almquist & Wiksells.

KELMAN G.R. (1971a). Cardiopulmonary function during exercise under various environmental conditions. In J.D. Brooke (Ed.) "Lung Function and Work Capacity". Salford: The University.

KELMAN G.R. (1971b). "Applied Cardiovascular Physiology". London: Butterworth.

KNEHR C.A., DILL D.B. & NEUFELD W. (1942). Physiological response to training. *Am. J. Physiol.*, **136**, 148.

KROGH M. (1915). The diffusion of gases through the lungs of man. *J. Physiol.*, **48**, 271-300.

KROGH A. (1930). "The Anatomy and Physiology of Capillaries". Yale: University Press.

LACOSTE J., ROBIN H. & BAUDOIN R. (1963). L'ergospirometrie chez les malades. *Le Poumon et le Coeur,* **19**, 6, 591.

LANGE R. (1919). "Uber Funktionelle Anpassung". Berlin: Springer Verlag.

LEON A.S. & BLOOR C.M. (1968). Effects of exercise and its cessation on the heart and its blood supply. *J. Appl. Physiol.*, **23**, 4, 485-490.

LICHT S. (Ed.) (1961). "Electro-diagnosis and Electromyography (2nd Ed.)". Baltimore: Waverley.

LORME T.L. de (1945). Restoration of muscle power by heavy resistance exercise. *J. Bone Joint Surg.*, **27**, 645-667.

LEPESCHKIN E. (1969). Physiological factors influencing the electrocardiographic responses to exercise. In H. Blackburn (Ed.) "Measurement in Exercise Electrocardiography". Illinois: Thomas.

MARGARIA R., CERETELLI P. & MARGILI F. (1964). Balance and kinetics of anaerobic energy release during strenuous exercise in man. *J. Appl. Physiol.*, **194**, 623-628.

MARGARIA R., CERETELLI P., Di PRAMPERO P.E., MASSARI C. & TORELLI G. (1963). Kinetics and mechanism of oxygen debt contraction in man. *J. Appl. Physiol.*, **18**, 371-377.

MAYER J. & BULLEN B. (1960). Nutrition and athletic performance. *Physiol. Rev.*, **40**, 369-397.

MAXFIELD M.E. (1964). Use of heart rate for evaluating cardiac strain during training in women. *J. Appl. Physiol.*, **19**, 6, 1139-1144.

MEAD J., TURNER J.M., MACKLEN P.T. & LITTLE J.B. (1967). Significance of the relationships between lung recoil and maximum expiratory flow. *J. Appl. Physiol.*, **22**, 95-108.

MORALES M.F. & SHOCK N.W. (1944). Acid anion displacement and recovery following exercise. *J. Gen. Physiol.*, **27**, 155-165.

MOREHOUSE L.E. & MILLER A.T. (1967). "Physiology of Exercise (5th Ed.)". St. Louis: C.V. MOSBY.

NAUGHTON J., BALKE B. & POORCH A. (1964). Modified work capacity studies in individuals with and without coronary artery disease. *J. Spt. Med. Phys. Fit.*, **4**, 4, 208-212.

NIELSON M. (1936). Die respirationsorbeit bei korperruke und bei muskel arbeit. *Skand. Arch. Physiol.*, **74**, 299-316.

NUNN J.F. (1969). "Applied Respiratory Physiology". London: Butterworth.

PAEZ R.N., PHILLIPSON E.A., MUSANGKOY M. & SPROULE B.J. (1967). The physiological basis of training patients with emphysema. *Am. Rev. Resp. Dis.*, **95**, 944-953.

PAPPENHEIMER J.B. et al (1950). Standardisation of definitions and symbols in respiratory physiology. *Fed. Proc.*, **9**, 602-605.

PERNOW B. & SALTIN B. (Eds.) (1971). "Muscle Metabolism During Exercise". London: Plenum.

PERUTZ M.F., ROSSMAN M.G., CULLIS A.F., MUIRHEAD H., WILL G. & NORTH A.C.T. (1960). Structure of haemoglobin. A three dimensional fourier synthesis at 5.5A resolution obtained by X-Ray analysis. *Nature*, **185**, 416-422.

PHILLIPS M. (1967). A testing procedure for studying pulse rate, weight and temperature during the menstrual cycle. *Res. Quart.*, **38**, 2, 254-262.

REEVES T.J. (1969). Panel discussion of determinants of myocardial oxygen consumption. In H. Blackburn (Ed.) "Measurement in Exercise Electrocardiograph".Illinois: Thomas.

RILEY R.L. & COURNAND A. (1951). Analysis of factors affecting partial pressures of oxygen and carbon dioxide in gas and blood of lungs. *J. Appl. Physiol.*, **4**, 77-101.

ROSENBAUM F.F. & BELKNAP E.L. (Eds.) (1959). "Work and the Heart". New York: Harper.

ROWELL L.B., TAYLOR H.L. & WANG Y. (1964). Limitations to prediction of maximal oxygen intake. *J. Appl. Physiol.*, **19**, 5, 919-927.

ROWELL L.B., TAYLOR H.L., WANG Y. & CARLSON W.S. (1964). Saturation of arterial blood with oxygen during maximal exercise. *J. Appl. Physiol.*, **19**, 284-286.

SAIKI H., MARGARIA R. & CUTTICA F. (1967). Lactic acid production in sub-maximal work. *Int. Z. Angew. Physiol. einsch. Arbeitsphysiol.*, **24**, 57-61.

SALTIN B. (1964). Aerobic work capacity and circulation at exercise in man. *Acta. Physiol. Scand.,* **62,** suppl. 230.

SALTIN B. & ÅSTRAND P.O. (1967). Maximal oxygen uptake in athletes. *J. Appl. Physiol.,* **23,** 353-358.

SALTIN B. & HERMANSEN L. (1967). Glycogen stores and prolonged severe exercise. In G. Blix (Ed.) "Nutrition and Physical Activity". Uppsala: Almquist & Wiksells.

SANDOW A. (Ed.) (1959). Second conference on muscular activity. *Ann. N.Y. Acad. Sci.* **81,** 401-509.

SARNOFF S.J., BRAUNWALD E., WELCH G.H., STAINSBY W.N., CASE R.B. & McRAY R. (1959). Hemodynamic determinants of the oxygen consumption of the heart with special reference to the tension time index. In F.F. Rosenbaum and E.L. Belknap (Eds.) "Work and the Heart". New York: Harper.

SHARKEY B.J., McDONALD J.F. & CARBRIDGE L.G. (1966). Pulse rate and pulmonary ventilation as predictors of human energy cost. *Ergonomics,* **9,** 3, 223-227.

SHEPHARD R.J., ALLEN C., BENADE A.J.S., DAVIES C.T.M., di PRAMPERO P.E., HEDMAN R., MERRIMAN J.E., MYHRE K. & SIMMONS R. (1968). The maximum oxygen uptake. *Bull W. H. O.,* **38,** 757-764.

SIMONSON E. & KEYS A. (1952). The effect of age and body weight on the electrocardiogram of healthy men. *Circulation,* **6,** 749-761.

SIMONSON E. (1965). Performance as function of age and cardiovascular disease. In A.T. Welford & J.E. Birren (Eds.) "Behaviour, Ageing and the Nervous System". Springfield: Thomas.

SMITH R.F. (1969). Quantitative exercise stress testing in the naval aviation population and in the projected Apollo spacecraft experiment MO-18. In H. Blackburn (Ed.) "Measurement in Exercise Electrocardiography". Illinois: Thomas.

STRAUSS M.B. (1957). "Body Water in Man: the acquisition and maintenance of the body fluids". London: Oxford.

TAYLOR C.L. (1944). Some properties of maximal and submaximal exercise with reference to physiological variation and the measurement of exercise tolerance. *Am. J. Physiol.,* **142,** 200-212.

TAYLOR C.L. (1945). Exercise. *Ann. Rev. Physiol.,* **7,** 599-622.

TAYLOR H.L., HASKELL W., FOX S.M. & BLACKBURN H. (1969). Exercise tests: a summary of procedures and concepts of stress testing for cardiovascular diagnosis and function evaluation. In H. Blackburn (Ed.) "Measurement in Exercise Electrocardiography". Illinois: Thomas.

THOMASON H., HAMLEY E.J. & BROOKE J.D. (1969). Changes in electrocardiograms of female physical education students. *Res. P.E.,* **1,** 4, 24-29.

TOLLMAN J. (1966). Pain, pulse and pace. *Swim. Tech.,* **3,** 1, 2-6.

VENABLES P.H. & CHRISTIE M. 1970). T-wave electrocardiogram: a potentially useful index. *Behav. Technol.,* **2,** 12.

VRIES H.A. de (1966). Quantitative emg investigation of the spasm theory of muscular pain. *Am. J. Phys. Med.,* **45,** 119-134.

VRIES H.A. de et al. (1966). Prediction of maximal oxygen intake from submaximal tests. *J. Sp. Med. Phys. Fit.* **5,** 207-214.

WAHLUND H. (1948). Determination of physical working capacity. *Acta. Med. Scand. Suppl.*, **215**, 1-78.

WASSERMAN K. & McILROY M.B. (1964). Detecting the threshold of anaerobic metabolism. *Am. J. Cardiol.*, **14**, 844-852.

WEST J.B. (1966). Oxygen transfer by the lung. In J.P. Payne & D.W. Hill (Eds.) "Oxygen Measurement in Blood and Tissues". London: Churchill.

WESTURA E.E. & RONAN J.A. (1969). Comparison of heart rate, oxygen consumption and electrocardiographic responses to submaximal step exercise and near maximal exercise on a treadmill and bicycle ergameter. In H. Blackburn (Ed.) "Measurement in Exercise Electrocardiography". Illinois: Thomas.

WILLIAMS C.G., BREDELL G.A.G., WYNDHAM C.H., STRYDOM N.B., MORRISON J.F., PETER J., FLEMING P.W. & WARD J.S. (1962). Circulatory and metabolic reaction to work in heat. *J. Appl. Physiol.*, **17**, 625-638.

WILLIAMS C.G., RAAN A.J.N. du, RAHDEN M.J. von & WYNDHAM C.H. (1968). The capacity for endurance work in healthy trained men. *Int. Z. angew. Physiol. ein Arbeitsphysiol.*, **26**, 141-149.

WILMORE J.H. (1968). Influence of motivation on physical work capacity and performance. *J. Appl. Physiol.*, **24**, 4, 459-463.

ZUNTZ N. (1901). "Physiologie des Marches". Berlin.

TOPIC AREAS-DEVELOPED

PHYSICAL GROWTH AND DEGENERATION

4 K

by J.D. BROOKE

General

This is the study of the nature of life. Growth infers movement: movement is influenced by growth. It is a complex field of work to which much research money is being directed: the knowledge available to date is not seen to inter-lock to form a comprehensible picture of the basic life processes although much useful information has been gathered at the behavioural level. Sinclair (1969) provides an over-view. For the chemistry of growth see Parport (1949). At the cellular level Ebert (1965) is a helpful background text.

K.1. Incomplete Understanding

The study of growth contains the key to account for the differentiation between living and non-living things. This section is about the changing personal physical factors that affect the movement of human beings as they progress along the continuum that ranges from the newly fertilised ovum to death. It is concerned with the maintenance of adulthood as well as the growth of the child and the degeneration of the aged. It can be expected that as laws about growth are developed they will be applicable across this span and that when interrelated will account for it. As behaviourally observable patterns of movement appear in conjunction with physical and mental growth, the study of growth is very relevant to the field of study of Human Movement.

Research has been pursued at a number of levels of enquiry. With the broadest view-point, ecological studies have been made of the relationship of growth factors to the total environment. e.g. Widdowson (1951) and the symposium proceedings reported by Kazda (1970). The constriction of the view and the clarity of the perception of the detail may increase as the level of study shifts to the behavioural [the observable changes in the whole human being e.g. Jones 1949] , to the cellular [the changes occurring to the living cell units e.g. Harris, 1964] and, to the molecular, [the chemical components of the cell unit e.g. Pauling, 1963]. In addition comparative studies in phylogeny reveal parallels with ontogeny, Tuttle and Schottelius (1969), that suggest the latter may be a more rapid repeat of the former. See also Section H.

Valid concepts from all levels are necessary if we are to understand human growth for as Weiss (1956) indicates, the term is used to denote something that is a complex, a comglomerate of intricate events. Ebert (1965) contests this, restricting growth to permanent enlargement and separating it from differentiation, reproduction or formation of tissues. The wider definition of Weiss is adopted at present. Too few valid concepts are available. Sometimes only the first level of information, data, is available. Frequently only the statement of principle is achieved, a higher level of knowledge. Understanding provided by the highest level, a valid conceptual structure, is not yet apparent. However much information is accruing [e.g., at the behavioural level, there are approx. 1200 references in Tanner, 1962 and at the micro level note the twenty nine Symposia of the Society for Developmental Biology listed in Locke, 1967]. The present description is

essentially at the second level of information. Its limited treatment of events at the microscopic study levels indicates the difficulty there is at present in distilling these to relate them to the present observable life behaviour.

K.2. Micro Level

It is impossible to uncover these important fields of molecular and cell biology further than to identify the vital study, for the range is immense. Reference must be made to the relevant academic disciplines. Weiss (1939) provides an early definitive starting point and Du Praw (1968) a modern over-view. These events at the micro level are the bases for the initiation of human movement.

In the living cell are hierarchically ordered systems of molecules and molecular groups within which lies the control of the cell, Ebert (1965) and by aggregation of the cells the direction of the behaviour of the human being (Bresnick, 1968). The constitution and organisation of these control systems is complex. Eventually if they are traced through ascending order of importance the genetic materials of deoxyribonucleic acid (DNA) and its messenger, ribonucleic acid (RNA), appear to be the reference points from which these controls derive (Davidson, 1969) and some motor impairment may derive from abnormalities at this point. (See also Section B). However these molecular events that determine behaviour (including movement), require the environment of the cell unit in order to occur. It is clear that the cell is not merely a shell to contain the components; it is dynamically involved with its components (Weiss, 1968). Similarly the micro-environment interacts with the cell, the tissue environment with the micro, organ with tissue, body with organ, external milieu with body and so on. Unfolding layers of control participating with each other down to the molecular genetic code characterise the determining factors in the growth of the human being and the development of capacity for movement. The progression of this growth results in increasing levels of complexity of function which allow wider environmental experience and the maintenance of the upward spiralling circle of development from the early embryonic life, e.g. Davies (1967), to the suggested participation with the modern technological extensions of our central nervous systems (McLuhan, 1964).

At the micro-level the manner in which the control of the cell function is achieved is not clear. (Cancer research drives for this solution). Much work has been done. It involves the egg and its fertilisation by sperm e.g. Metz (1962), the division of the resulting zygote and the subsequent regrouping of cells, e.g. DeHaan & Ebert (1964), quantification—the growth and multiplication of a type of cell and differentiation—the growth of types of cells of specific function, e.g. Barth (1964). It concerns cytoplasmic protein production by DNA-RNA and RNA-ribosome interaction, the formation and organisation of cell components such as mitochondria, ribosomes and actin-myosin groups, e.g. Loewy & Siekewitz (1963) and tissue and organ formation [for embryology Davies, 1967 and for morphogenis Torrey, 1962].

In all of these facets the interaction of the identified molecules or molecule groups with the immediate environmental conditions, chemical and physical, inanimate and animate appears critical. Much work has been done and much remains to be done. Differences in movement deriving from variations in growth should be accounted for eventually at this level. It appears to be a finite problem promising a fruitful solution.

K.3. Height, Weight, Physique

Changes in height, weight and physique affect human movement. These changes are well documented.

The construction of growth curves is a characteristic of studies of height and weight change. Measures are taken at different times and gains are recorded. The curves can be simple accumulation lines over time or can be velocity curves, lines of rates of change over time. Tanner (1962) makes use of the latter frequently. He tabulates clearly for height and weight the peak growth rate occurring in the fourth month of foetal life and the arrest in the decelerating velocity curves that occurs around the ages of 6-8 years and again more markedly during the adolescent years of 10-16. These three periods on the height-weight velocity curves provide a temporal framework to which much of the human growth cycle at the behavioural level may be referred. At the same time note that there is much individual variability and that at the pubescent stage the female growth spurt precedes the male. Asmussen (1965) draws attention to these variabilities affecting sports performance. Sinclair (1969) and Tanner (1962) well review some of the relevant studies and draw implications pertinent to the well being of 'abnormal' children at these times. The wide variability in the onset of the growth spurt points to an increase in the range of capacity for movement. From average childhood data, prediction of adult height is possible within bounds (Bayley 1956) and some prediction of weight from height is possible in the mature adult (Zuckerman 1956). It is clear that taller children start the adolescent spurt earlier as do heavier children, with the latter still being heavier in adulthood.

Rees & Eysenck (1945) carried out a factor analysis of twenty three measures of physique and found that height and weight characterised the first factor of body size. The second factor of body shape inserted some statistical structure into the study of body form and they proposed a Body Index that has been used in movement studies (Brooke, 1967; 1969 a,b; Asmussen, 1965). From the ancient period physique has been classified. Kretschmer (1938) did much preliminary modern work and Sheldon (1940) attempted to correlate with the mental condition. Kane (1969) has questioned this correlation in non-clinical groups of sportsmen. Sheldon's technique of somatotyping by three physique 7 point rating scales for ectomorphy (leanness) mesomorphy (muscle) and endomorphy (fat) has been used by Tanner, (1964) to describe Olympic athletes. The somatotype technique has been modified to a phenotype by Parnell (1958) to allow for superficial changes in the constitution due to ageing or environment.

For movement studies involving change or differences in physique there is appeal in the use of a broad but simple assessment of shape such as the Rees-Eysenck index supplemented by particular anthropometric measures as required. For growth studies this might be skeletal age from the morphological changes in bone growth centres in the wrist, Tanner & Whitehouse (1959; 1961) standards for British children. Note that in the young child changes in posture and the position of the centre of gravity are developmental indices (Medawar, 1944) and affect the development of locomotor patterns. There are also male-female differentiating growth curve characteristics (Tanner, 1962).

K.4. Growth of systems relevant to Human Movement

1. *Muscular.* Cardiac and skeletal musculature is the effector for movement of the body in the environment. The composition of muscle alters after birth with an increase in cytoplasm and reduction in water and intercellular matrix. There is probably no increase in the number of muscle fibres after the 5th foetal month although new fibres are generated in adult muscle after injury. On average, the maximum size of muscle fibres is reached in girls at age 10, together with a 10 fold increase from birth in the number of nuclei in the fibres and in boys age 14 with a 14 fold increase in nuclei. The peak of the strength velocity curve succeeds that of height by approx 14 months and of weight by 9 months, (Stoltz & Stoltz, 1951). From the early data of Whipple (1924) and Bliss (1927) and later work, e.g. Clarke et al (1962 a,b,c), it is clear that the strength gain curves follow a similar course to the height/weight curves described above. Pre-adolescence there is little difference between boys and girls in strength, but post-adolescence the boys are superior, Jones (1949). This is often attributed to changes in adrenal cortex and testes hormone release. Rich (1960) reports that fatiguability for human movement in

relation to strength is independent of sex and age, 8-17 years. This correlates with studies of $\dot{V}o_2$ max/kg and $\dot{V}o_2$ max/Heart-Pulse/kg referred to below. Note pituitary giantism, where glandular variability results in poor power-weight ratic due to the development of the skeleton beyond that of the musculature. This ratio is a determinant of the capacity for severe physical work.

Davies (1967) well describes the embryonic development of heart and blood vessels. At birth the oxygenation of blood occurs for the first time at the lungs and ceases to be through the placenta. The embryonic pulmonary by-passes are closed. Incomplete closure, particularly associated with pressure differential, leads to right-left heart shunts of de-oxygenated blood, the congenital heart condition that stunts general body growth and counter-indicates physical exercise. After birth the vessels of the heart and circulation adapt to the work load placed on them. The walls of arteries and left ventricle thicken: blood volume increases through childhood. Ikai (1970) reports progressive increase in total blood volume of the forearm (10-16 years) and no difference in effects of exercise increasing flow at the various ages. The functional efficiency of the cardio-vascular/respiratory system during physical movement assessed by oxygen uptake/kg body weight, remains relatively constant through the childhood and adolescent years (Hollman, 1965). Accordingly the heart muscle during adolescence increases in power. The resting heart rate of the 2 yr. old child is about 100 beats/min and by adulthood this will have fallen to 70 beats/min as an average value, (Iliff & Lee 1952), although in adapted endurance sportsmen it may be 40 beats/min or lower. The developmental change is usually associated with the increased assertion of vagal control. Systolic blood pressure rises through childhood with increased rise during adolescence. The timing is related to the onset of the growth spurt, thus differentiating males and females.

2. *Respiratory*. The exchange of gases involved in metabolism for movement takes place through the respiratory system. The lungs grow rapidly in size from birth and the ventilatory passageways increase in diffusive capacity. Presumably this is an adaptive response. Subsequent changes in the respiratory system most frequently follow the height/weight growth pattern. Thus splits between male and female occur at puberty for higher male haemoglobin and red blood cell levels, (Diem, 1962), lower $F_I-F_EO_2$ (Shock, 1941), and (Shock & Soley, 1939), increased vital capacity per body surface area (Åstrand, 1952) and increased larynx growth. In many of these variables a common trend has been followed by boys and girls prior to adolescence, mirroring the gradual fall in basal metabolic rate (Lewis et al, 1943) and respiratory frequency. For controlled laboratory exercise the maximum oxygen uptake per kilogram body weight is reported to be independent of sex and age up to puberty (Hollman et al, 1965; Ikai, 1970) but from puberty onwards male superiority is apparent, Åstrand (1952), the female $\dot{V}o_2$ max/kg even falling at puberty. Hollman et al (1965) did not see this fall, even temporarily, in efficiency for either sex at adolescence. Åstrand (1952) also reports increased anaerobic tolerance (pH and LA) post-puberty and a similar sex difference. The

question has to be raised of how many of these changes in functional systems supporting human movement are adaptive responses determined partly or wholly by social concepts of behaviour versus the alternative of genetic determinants.

3. *Skeletal.* This is one of the most documented growth areas. From its role in maintaining posture and organ position and by its involvement in the body weight part of the exercise power/body weight ratio, the skeletal system change is important to movement studies. The height/weight pattern indicates the common curve also for a number of skeletal components. Within this commonality there is much variation from bone to bone, ranging from skull growth, much completed by a few years after birth, to the vertebral epyphses which may not close until the mid-twenties. Radiological analyses of bone growth are used for maturity ratings, e.g. Tanner, Whitehouse & Healey (1961).

In adolescence skeletal growth spurts occur, even in the skull dimensions (Reynolds & Schoen, 1947). The order of the onset of change approximates leg length, arm length, hip width, chest breadth, shoulder breadth, trunk length, chest depth. For the limbs, acceleration of distal segment growth precedes that of proximal. From birth sex differences are apparent in skeletal maturity; girls are in advance of boys and become more so up to adultood. Sex differences in shape at adolescence indicate changes in skeletal growth as well as fat depositions. Mature male bone is also denser than comparative female tissue. Tanner (1962) provides a comprehensive overview of changes in these body measures. Sinclair (1969) devotes attention to the formation of cartilage, growth of short, long bone, skull, jaw and vertebral column and of ossification. He indicates the greater importance of injury (such as might be more likely to occur with particular types of organised movement) to the growing ends of long bones. Structural needs act to determine bone growth. Physical activity is an important determinant of bone strength (Tuttle & Schottelius, 1969). With lack of use, atrophy occurs. For embryology, see Davies (1967).

4. *Digestive.* Dental development and the eruption of deciduous and permanent teeth is the most obvious change. It affects dietary intake and modifies gastric secretions. Use is made of it for maturity ratings. Note, advanced female skeletal maturity is not seen in deciduous teeth development. Some positional changes of digestive organs occur with growth and associated behaviour changes e.g. stomach capacity and feeding; size of pelvis and position of intestines with upright posture for locomotion. At adolescence biochemical changes are observed in higher fasting blood sugar (Robinson, 1938) and the increase in free gastric acidity (HCl) from early childhood which, after a levelling off, ages 7-13, rises markedly in adolescence, with boys showing higher levels than girls (Vanzant, 1932). This male/female difference is maintained to the mid-forties. Sensitization of female sense of smell is reported to increase at puberty (Le Magnen, 1948). Tanner (1962) speculates that this is an affect of oestrogens on the olfactory apparatus. Throughout the growth phase the k cal dietary intake increases to maintain the increasing total energy output, movement of larger frames by larger musculature

involving larger input of nutrients.

5. *Genito-Urinary.* Growth in the genito-urinary system has been much studied. It is relevant in movement studies for the physical changes have cultural connotations that act to determine movement behaviour and the biochemical associates discussed below, 6, directly affect the movement involvement. Again Tanner (1962) reviews the research. For embryology see Davies (1967). The growth of the excretory system follows the pattern of the height/weight growth curves. The observable changes in the reproductive system have a basis in endocrine function. This concerns pre-natal growth such as change in foetal penis size (Spaulding, 1921), as well as post-natal change. After birth change is slight compared with the rapid growth of organs and glands at adolescence. The change in reproductive organs at that time closely relates to the general morphological growth curve (Stoltz & Stoltz, 1951). Mulcock (1954) for boys and Reynolds & Wines (1948) for girls recorded the timing of external genitalia growth. There is much variability between individuals in the onset of these changes. The early maturers have reached adult functional status before the late maturers have started to change from the child condition. The implications for sexual pre-coital and coital movements are obvious but important. However there is much less variability in the *sequence* of changes in the reproductive system. Stoltz & Stoltz (1951) show high correlations between e.g. testes, penis and pubic hair development times within individuals. The age for menarche and other maturational landmarks has reduced gradually over the past hundred years in Western Europe (Tanner 1962).

There are classification indices to enable the growth of the reproductive system to be used for maturity ratings. In males 5 point rating scales for pubic hair and genital development and in females for pubic hair and breast development are used, e.g. Delauney & Deschamps (1956), and a visual plate guide by Tanner (1962), with an additional 6th scale point for the classification of the mature pubic hair pattern that occurs in the mid-twenties or later (Dupertuis, 1945). In the mature adult the activity of the organs in sexual intercourse has recently been studied seriously in the laboratory, e.g. Masters & Johnson (1966) and in normal life behaviour, e.g. Kinsey (1949). Green (1968) in a basic undergraduate physiology text gives a scientist's account that approaches realism.

6. *Nervous.* There is considerable development of the nervous system by birth. Consequently embryonic environmental influences are important e.g. blood gas, nutrients, drugs, physical condition and the process of birth. One source of central nervous motor impairment is in embryonic and birth condition, see Section B.

By age 3 the C.N.S. has achieved more than 70% of its mature size and by age 5 more than 80%. Not all nervous tracts are fully developed at birth and in the context of human movement note that motor nerve conduction time and diameter increases up to age 3 (Thomas & Lambert, 1960). At birth the motor areas of the cortex are not excitable normally by peripheral stimulation. Pyramidal fibres do not become myelinated fully until year 2. There is a danger here of assuming function from anatomical structure but it is clear that as this neural maturity occurs and in conjunction with other maturational factors, the child develops early movement skills and walks eventually.

Complexity of the C.N.S. network increases from birth by increases in the length and number of dendrites. Adequate environmental conditions are necessary to stimulate this growth and the controversial suggestion has been made that in rich conditions further neurones are formed (Sci.Res. 1967). There is important research to be done here into the effects of a rich movement environment on C.N.S. growth. Certainly in the new born the growth of the C.N.S. cortical grey matter and its neural inter-connections is much retarded in comparison with the development of the central grey. This comparative maturity of the latter, the basal ganglia, thalamus and hypothalamus, directs the spontaneous movement activity at birth, of sucking, breathing, swallowing, crying, urinating etc., mediated through the spinal cord and brain stem (Tuttle & Schottelius, 1969). Only later does the selection of cortical input by the interaction of reticular formation, hippocampus, association areas and cortex lead to the physiological phenomenon of inhibition, facilitating the C.N.S. integration which characterises human function. There are phylogenic parallels to this ontogeny. (See Section C).

Endocrine gland growth proceeds through childhood with a growth spurt at adolescence. For the anterior pituitary, girls advance in front of boys at this stage and for the adrenals the reverse occurs. Secretion of gonadotrophin is detected at adolescence and androgen/oestrogen secretion rates split as expected for boys and girls at this point. 17 ketosteroids rise from 3 yrs and boys exceed girls at adolescence. The activity of the thyroid gland determined from serum bound iodine falls through childhood in parallel with the basal metabolic rate. These endocrine secretions act to control growth. Many are active also in controlling the physiological systems that underlie movement. There is need for longitudinal biochemical study in these areas. Note that in precocious cerebral puberty, full reproductive capacity may occur but overt mature sexual behaviour is not observed. Is this CNS under-development or 'social deprivation'?

A correlation between I.Q. and physical maturity is frequently reported e.g. height/weight correlation with IQ, (Scottish Council, 1953). Changes in the manner of processing information from empirical treatment to abstract thought are much reported, as reviewed by Ausubel (1966) and discussed in Section L. The CNS growth determining this is not known.

In this summary of the growth of systems there is a clear need for more detailed longitudinal research at the biochemical level, for assessment of the effects of various types of exercise on growth and for study of C.N.S. changes under different treatments of early movement experience.

K.5. Regulation of Growth. This section is concerned with the molar rather than the molecular level.

It has been proposed that genetic differences in individual growth patterns affect human movement. Such hypotheses raise the nature : nuture, genetic : environmental affect interaction. Research to clearly separate these is not simple. Eysenck (1971) provides a basic review of some genetic factors affecting the educational growth of different human races. With reference to human movement studies Gribe (1963) studied identical twins and siblings in terms of familial background and sports performance and concluded that there was strong inherited disposition to such achievement and that training served to act upon these constitutional characteristics. Asmussen (1965) sees similar determinants of sports ability. The main physical measures called in evidence by both researchers are body size and shape. The Birmingham School Health Service (1964) reported that the physical size of 6, 11, and 14 yr. old male and female pupils was positively related to socio-economic status of the residential district. However obesity was only related by higher incidence in poor districts in 14 yr. old girls. The environmental/genetic split was not attempted. See also Section G. The classic study of the detrimental effect of emotional stress in German orphans (Widdowson, 1951), demonstrated the effect of one environmental brake on growth, the presence of an authoritarian warden arresting the increases in child growth achieved through a boosted diet.

Dietary deficiencies retard growth in man (Howe & Schiller, 1952), and animals, (Nissen & Riesen 1949). In the normal diet the protein content provides amino acids for the growth and repair of tissues. A normal value is approx 100g/day, i.e. 400 k cal. During increased growth periods more protein is required but even in the 11-16 yr. old period this only amounts to yearly increases of an additional 44 k cal/day approximately. When deprivation stunts growth, dietary boosting often restores the growth curve back to its earlier line, the return being by decreasing increments to blend into the path of the old curve (Tanner, 1962). Some authors take this evidence of a basic path as an indication of genetic derivatives for the growth curve. It may still be however that dietary deficiencies in early development during periods of rapid growth of organs cannot be fully compensated for by later boosted diets. There is debate about the effect of a nutrient intake raised well above that of the standard diet, e.g. the increasing diet of British and American children. Yudkin (1952) suggested additions that appeared to boost growth rates in such populations. There is at present serious investigation of the possible detrimental effect of high animal fat and sugar contents in the diets of such groups, e.g. Yudkin (1964): it is a controversial topic.

Hormonal regulation of growth clearly occurs. Trophic secretions from the anterior lobe of the pituitary gland, thyroid, parathyroid, islets of Langerhans and adrenal glands are important requirements for normal growth patterns and much research is pursued in endocrinology. Green (1968) provides a simple introduction and consideration of disorders. Adolescent sex differences occur with differential secretions as described above.

K.6. Aging.

From 1890 to 1950 the average life expectancy from birth rose by 23 years; the life expectancy for 10 year olds rose by 10 years; that for 20 year olds by 9 years; for 40 year olds by 5 years and for 50 year olds by 2 years (Smithers, 1960). Senescence is clearly an important research problem today. There is female dominance in longevity, women centenarians out-numbering men by approximately 6:1. A review of research in gerontology is provided by Strehler (1964). The problem is pertinent to movement studies for with age physical skill deteriorates and the aesthetic experience of movement may be affected. See Section E.

Comfort (1970) in a simple discussion suggests there are approximately 20 current theories of ageing. The problem is the other end of the same continuum from theories about growth. Two of the more common suggestions to account for ageing are those of molecular changes in the general tissue matrix and the non-mitotic nature of the cells of the central nervous system. The regeneration of tissue is not understood. It does appear from Saunder's studies of cell death in the formation of the morphology of the chick wing (1963) that the pre-determined timing of cell death depends on the micro-environment at the cell site. It is clear that changes occur particularly in elastic tissue e.g. the skin, [Ageing can be judged by deterioration in the collagen network in exposed areas Bell et al 1965] , the eye lens (in Presbyopia, one of the most common indications of ageing from 40-45 years), the arteries, scar quality, and joints and that coupled with the deposit of calcium salts much restriction in movement function can result. In addition organ atrophy occurs, Comfort (1964). It does not appear that these changes are related directly to deterioration of gonods or ovaries (Engle, 1956). Inherited genetic characteristics for longer life span occur particularly in daughters of long-lived parents, particularly if the mother is of this type. It may be expected that the solution to these matters will lie in molecular biology.

Systemic changes occur but there is much variability from person to person (Hassler, 1965). Muscle mass and movement efficiency reduces (Tuttle & Schottelius, 1969). The decline is slight up to 70 years of age and more marked subsequently. There is a reduction in heart size and stroke volume and a lower maximum heart rate in exercise, with reduced e.c.g. amplitudes compared with younger groups (Simonson, 1952). Reduction of elasticity in large artery walls and calcium salt deposits in smaller arteries together with hypertension place highly inefficient pressure loads, (Sarnoff et al, 1959), on exercising hearts which may themselves have inadequate circulations, increased fat deposits and smaller adapted muscle masses.

Similar changes occur in the respiratory capacity. The exercise Vo_2 max/kg falls with age (Simonson, 1965), may by age 70 be 50% of the youth values (Dawson et al 1945), and is correlated with falls in psychological performance (Spieth, 1965). Breathing mechanics are limited by reduced elasticity of lung and interstitial tissue.

Diffusion capacity reduces by 50% after approx. 70 years, (Richards, 1965). Basal metabolic rate is low (Fleisch, 1951), and a predisposition to inadequacy of thermal control increases the risk of hypothermia in exposed surroundings. The red blood cell and haemoglobin levels can be low, affecting physical working capacity by lowering the anaerobic threshold. This is partly incomplete replacement of dead cells but too often inadequate dietary iron is a major factor. The level of the maximum work output can be considered one of the most important criteria of biological ageing according to Eiselt (1968).

In the ageing skeleton the rate of destruction is greater than that of replacement and a weaker brittle bone structure results. With movement and impact easier fracture occurs and healing is less good. There is also a height reduction with age, due to degenerative changes in the vertebral column.

There is a continuous loss of central nervous system neurons from early years up to death. Some of the earliest behavioural indications of ageing are the increases in multiple-choice reaction times that can be recorded from adolescence onwards, (Welford, 1965). In the aged, selective attention, set and short-term memory are less effective and Welford proposes this as one of the bases for the deterioration of function in senescence. Hassler (1965) agrees that such loss has primarily central rather than peripheral sense organ or muscle effector origin and proposes changes in the basal nucleus, central median and pallidum. The quality of perfusion and content of the cerebral blood supply is important also. Such deteriorations in CNS function reduce the quality of human movement and may be partly the result of increased physical inactivity.

Digestive changes result from reduced elasticity and hence motility of stomach and gut. With inadequate intake cannabalisation can occur from body protein and fat reserves reducing the muscle mass available for movement. Protein values as high as 1.4g/kg body weight/day may be necessary to avoid negative nitrogen balances in old people (Bell et al 1965). Obesity is not desirable. In old females Young (1963) reports fat deposits approximately 50% higher than in the young. Comfort (1971) reports experiments showing a 50% increase in mouse life span when calorie restrictions were imposed on their life diets. He tentatively extrapolated this to a possible 10-20% increase in human life span when the mechanisms are understood, perhaps over the next 15 years. An alternative to reduced calorie intake might be increased energy requirements through increased physical activity. Carbohydrate metabolism also alters with higher glucose tolerance levels. Iron levels need to be maintained. Rising cholesterols are reported for middle-age groups and debate occurs over the association with arteriosclerosis and the reduction of cholesterols by exercise (Fox & Skinner, 1964).

It appears that with adequate environmental stimuli the resulting adaptation offsets some parts of the ageing process. The relation of regular exercise to reduced risk of coronary heart disease is documented, e.g. Fox & Skinner (1964). Simonson (1952) notes differences between the e.c.g. amplitudes of young and old people similar to the differences Thomason et al (1968) found between physically trained and untrained students. Pyörälä (1968) reported significant differences in lung function between ex- cross country runners and skiers and standardised controls. Paez (1967) found increased work ability and efficiency of oxygen provision in emphysematous patients who were exercised. Eiselt (1968) noted that whereas marked falls in maximum working capacity occured at the 7th decade for normal men, athletic subjects did not show this degree of decrease until the 9th decade. Note here the physical ability of veteran sportsmen, e.g. time trial cyclists. It appears that the exercise must be of recent origin to demonstrate significant benefits for work capacity and circulation. Similar improved mental function might be expected where the environment stimulated adaptive activity requiring the focus of attention, in comparison to situations where merely vegetative function was encouraged by the absence of such stimulation. Appropriate human movement might constitute such a positive adaptative stimulation.

Much definitive work is required in this area of growth and ageing. It will need to be longitudinal and will have a large molecular biology component. For the present the behavioural data are used to indicate the need, potential, and limits in growth for human movement.

References

Source Texts

BELL G.H., DAVIDSON J.N. & SCARBOROUGH H. (1965) "Text Book of Physiology & Biochemistry". London: Livingstone.

DIEM K. (Ed.) (1962) "Documenta Geigy". Manchester: Geigy Pharmaceuticals.

DUPRAW E.J. (1968) "Cell and Molecular Biology". London: Academic Press.

EBERT J.D. (1965) "Interacting Systems in Development". London: Holt, Rinehart & Winston.

DAVIES D.V. (Ed.) (1967) "Grays Anatomy (34th Ed.)". London: Longmans.

GREEN J.H. (1968) "An Introduction to Human Physiology (2nd Ed.)". London: Oxford University Press.

PARPORT A.K. (Ed.) (1949) "Chemistry and Physiology of Growth". Princeton: University Press.

SINCLAIR D. (1969) "Human Growth After Birth". London: Oxford University Press.

TANNER J.M. (1962) "Growth at Adolescence (2nd Ed.)". Oxford: Blackwell.

WELFORD A.T. & BIRREN J.E. (Eds.) (1965) "Behaviour, Ageing and the Nervous System". Illinois: Thomas.

Specific References

ASMUSSEN E. (1965). The biological basis of sport. *Ergonomics,* **3, 2,** 137-142.

ÅSTRAND P.O. (1952) "Experimental Studies of Working Capacity in Relation to Sex and Age". Copenhagen: Munksgaard.

AUSUBEL D.P. & AUSUBEL P. (1966). Cognitive development in adolescence. *Rev. Ed. Res.,* **36, 4,** 403-413.

BARTH L.J. (1964) "Development: Selected Topics". Mass.: Addison Wells.

BAYLEY N. (1956). Growth curves of height and weight for boys and girls, scaled according to physical maturity. *J. Pediat.,* **48,** 187-94.

BIRMINGHAM HEALTH SERVICE (1964). Height, weight and obesity in children. *Education,* **123, 3190,** 510-512.

BLISS J.G. (1927). A study of progression based on sex, age and individual differences in strength and skill. *Amer. Phys. Educ. Rev.,* **32,** 85.

BRESNICK E. & SCHWARTZ A. (1968) "Functional Dynamics of the Cell". London: Academic Press.

BROOKE J.D. (1967). Extroversion, physical performance and pain perception in physical education students. *Res. Phys. Educ.,* **1, 2,** 23-30.

BROOKE J.D., HAMLEY E.J. & STONE P. (1969). Physique, grip strength, personality and achievement in physical competition in female physical education students. *Med. Sci. Spt.,* **1, 4,** 185-188.

BROOKE J.D., BRADBURY I. & MACLOCK M. (1969). A comparison of the Rees-Eysenck index of male body shape with the Rees index of female body shape in males and females. *Phys. Educ.,* **61, 184,** 64-67.

CLARKE H.H. & DEGUTES E.W. (1962). Comparison of skeletal age and various physical and motor factors with the pubescent development of 10, 13 and 16 year old boys. *Res. Quart.,* **33,** 356-68.

CLARKE H.H. & HARRISON J.C.E. (1962). Differences in physical and motor traits between boys of advanced, normal and retarded maturity. *Res. Quart.,* **33,** 13-25.

CLARKE H.H. & WICKAMS J.S. (1962). Maturity, structure, muscle strength and motor ability growth curves of boys 9-15 years of age. *Res. Quart.,* **33,** 26-39.

COMFORT A. (1964) "The Biology of Senescence (2nd Ed.)". London: Routledge & Kegan Paul.

COMFORT A. (1970). The prospects for living even longer. *Time,* 3 Aug, 42.

COMFORT A. (1971). Reported by Hutchinson, R. Science advance could add 20 years to life: Conference on control of human ageing process. *Daily Telegr.* 3 Sept., 5.

DAVIDSON E.H. (1969) "Gene Activity in Early Development". London: Academic Press.

DAWSON P.M. & HELLEBRANDT F.A. (1945). The influence of ageing in man upon his capacity for physical work and upon his cardiovascular responses to exercise. *Amer. J. Physiol.,* **143,** 420.

DeHAAN R.L. & EBERT J.D. (1964). Morphogenesis. *An. Rev. Physiol.,* **26,** 15.

DELAUNEY P. & DESCHAMPS J. (1956). Etude de la croissance staturale et ponderale des adolescents en fonction du stade pubertaire. *Biotypologie,* **17,** 217-34.

DUPERTUIS C.W., ATKINSON W.B. & ELFMAN H. (1945). Sex differences in pubic hair distribution. *Hum. Biol.,* **17,** 137-42.

EISELT E. (1968). Output of work, consumption of calories and efficiency of work in old age. *Zeit. Alterns Farschung.* **21, 3,** 233-239.

ENGLE E.T. & PINIUS G. (Eds.) (1956) "Hormones and the Ageing Process". London: Balliere, Tindall & Cox.

EYSENCK H.J. (1971) "Race, Intelligence and Education". London: Temple Smith.

FLEISCH T. (1951) *Helvet. Med. Acta.,* **18,** 23.

FOX S.M. & SKINNER J.S. (1964). Physical activity and cardio-vascular health. *Amer. J. Cardiol.,* **14,** 731-746.

GRIBE H. (1963). Genetic and constitutional bases of the development of sport skills. *Wychowanie Fizynzne i Sport,* **7,** 1.

HARRIS M. (1964) "Cell Culture and Somatic Variation". New York: Holt, Rinehart & Winston.

HASSLER R. (1965). Extrapyramidal control of the speed of behaviour and its change by primary age processes. In A.T. Welford and J.E. Birren (Eds.) "Behaviour, Ageing and the Nervous System". Illinois: Thomas.

HOLLMAN W., BOUCHARD C. & KERKENRATH G. (1965). The development of performance capacity of the cardio-pulmonary system of children and young people of 8-18 years. *Sportarst und Sportmedizin,* **16, 7,** 255-60.

HOWE P.E. & SCHILLER M. (1952). Growth responses of the school child to changes in diet and environmental factors. *J. Appl. Physiol.,* **5,** 51-61.

IKAI M. (1970). Basic training of aerobic work capacity. In "Int. Symp. Ed. Fisi Desport". Mozambique: Marques.

ILIFF A. & LEE V.A. (1952). Pulse rate, respiratory rate and body temperature of children between two months and eighteen years of age. *Child Dev.,* **23,** 237-45.

JONES H.E. (1949) "Motor Performance and Growth". Berkley: University of California Press.

KANE J. (1969). Personality and body type. *Res. Phys. Educ.,* **1, 4,** 30-38.

KAZDA S. & DENENBERG V.H. (1970) "The Post-natal Development of Phenotype". London: Butterworth.

KRETSCHMER E. (1938) "Physique and Character". Berlin: Springer.

KINSEY A.C. et al (1949). Concepts of normality and abnormality in sexual behaviour. In P.H. Hoch & J. Zubin (Eds.) "Psychosexual Development in Health and Disease". New York: Grune & Stratton.

LEWIS R.C., DUVAL A.M. & ILIFF A. (1943). Basal metabolism of normal children from 13-15 years old, inclusive. *Amer. J. Dis. Child.,* **65,** 845-57.

LOCKE M. (Ed.) (1967) "Control Mechanisms in Developmental Processes". London: Academic Press.

LOEWY A. & SIEKEWITZ P. (1963) "Cell Structure and Function". New York: Holt, Rinehart & Winston.

MAGNEN LE J. (1948). Un cas de sensibilité olfactive se presentant comme un caractère sexuel secondaire feminin. *C.R. Acad. Sci.,* **226,** 694-5.

MASTERS W.H. & JOHNSON V.E. (1966) "Human Sexual Response". Boston: Little, Brown.

McLUHAN M. (1964) "Understanding Media, the Extension of Man". London: Routledge & Kegan Paul.

MEDAWAR P.B. (1944). The shape of the human being as a function of time. *Proc. Roy. Soc. B.,* **132,** 133-141.

METZ C. (1962). Fertilization. In G.B. Moment (Ed.) "Frontiers of Modern Biology". Boston: Houghton Mifflin.

MULCOCK D. (1954). A short study of the onset of puberty in boys. *Med. Offr.,* **91,** 247-9.

NISSEN H.W. & RIESEN A.H. (1949). Retardation in onset of ossification in chimpanzee related to various environmental and physiological factors. *Anat. Rev.,* **105,** 665-75.

PAEZ P.N., PHILLIPSON E.A., MASANGKAY M. & SPROULE B.J. (1966). The physiologic basis of training patients with emphysema. *Amer. Rev. Respir. Dis.,* **95, 6,** 944-953.

PARNELL R.W. (1958) "Behaviour and Physique". London: Arnold.

PAULING L. (1963). The molecular basis of genetic defects. In M. Fishbein (Ed.) "Congenital Defects". Philadelphia: Lippincott.

PYORÅLÅ K., HEINONEN A.O. & KARVONEN M.J. (1968). Pulmonary function in former endurance athletes. *Acta Med. Scand.,* **183,** 263-273.

REES W.L. & EYSENCK H.J. (1945). A factorial study of some morphological and psychological aspects of human constitution. *J. Ment. Sci.,* **91,** 8-21.

REYNOLDS E.L. & SCHOEN G. (1947). Growth patterns of identical triplets from 8 through to 18 years. *Child Dev.,* **18,** 130-51.

REYNOLDS E.L. & WINES J.V. (1948). Individual differences in physical changes associated with adolescence in girls. *Amer. J. Dis. Child.,* **75,** 329-50.

RICH G.Q. (1960). Muscular fatigue curves of boys and girls. *Res. Quart.,* **31, 3,** 485-498.

RICHARDS D.W. (1965). Pulmonary changes due to ageing. In W.O. Fenn & H. Rahn (Eds.) "Handbooks of Physiology, Section 3, Respiration Vol. 2" Washington: Amer. Physiol. Soc.

ROBINSON S. (1938). Experimental studies of physical fitness in relation to age. *Arbeitsphysiol.* **10,** 251-323.

SARNOFF S.J., BRAUNWALD E., WELCH G.H., STAINSBY W.N., CASE R.B. & MACRUZ R. (1959). Hemodynamic determinants of the oxygen consumption of the heart with special reference to the tension time index. In F.F. Rosenbaum & E.L. Belknap (Eds.) "Work and the Heart". New York: Harper.

SAUNDERS J.W. & GOSSELING M.T. (1963). Trans-filter propogation of apical ectoderm maintenance factor in the chick embryo wing bud. *Develop. Biol.,* **7,** 64.

SCIENTIFIC RESEARCH ARTICLE (1967). Environment may induce formation of neurons after birth. *Sci. Res.,* October, 60.

SCOTTISH COUNCIL FOR RESEARCH IN EDUCATION (1953) "Social Implications of the 1947 Scottish Mental Survey". London: University Press.

SHELDON W.H. (1940) "The Varieties of Human Physique". New York: Harper.

SHOCK W.W. (1941). Age changes and sex differences in alveolar CO_2 tension. *Amer. J. Physiol.,* **133,** 610-16.

SHOCK W.W. & SOLEY M.H. (1939). Average values for basal respiratory functions in adolescents and adults. *J. Nutr.,* **18,** 143-53.

SIMONSON E. (1965). Performance as a function of age and cardiovasuclar disease. In A.T. Welford and J.E. Birren (Eds.) "Behaviour Ageing and the Nervous System". Illinois: Thomas.

SIMONSON E. & KEYS A. (1952). The effect of age and body weight on the electrocardiogram of healthy men. *Circulation,* **6,** 749-761.

SMITHERS D.W. (1960) "A Clinical Prospect of the Cancer Problem". London: Livingstone.

SPAULDING M.H. (1921). *Contr. Embryol.,* 13.

SPIETH W. (1965). Slowness of task performance and cardiovascular diseases. In A.T. Welford & I.E. Birren (Eds.) "Behaviour Ageing and the Nervous System". Illinois: Thomas.

STOLTZ H.R. & STOLTZ L.M. (1951). "Somatic Development of Adolescent Boys". London: Macmillan.

STREHLER B.L. (Ed.) (1964). "Advances in Gerontological Research". London: Academic Press.

TANNER J.M. (1964). "The Physique of the Olympic Athlete". London: Allen & Unwin.

TANNER J.M. & WHITEHOUSE R.H. (1959). Standards for skeletal maturity based on a study of 3000 British children. Inst. Child Hlth., London Univ. M.S.

TANNER J.M., WHITEHOUSE R.H. & HENLEY M.J.R. (1961). Standards for skeletal maturity based on a study of 3000 British children. II. The scoring system for all 28 bones of the hand and wrist. Inst. Child Health, Univ. London, M.S.

THOMAS J.E. & LAMBERT E.H. (1960). Ulnar nerve conduction velocity and H-reflex in infants and children. *J. Appl. Physiol.,* **15,** 1-9.

THOMASON H., HAMLEY E.J., BROOKE J.D. & BROOKE T.K. (1968). Modifications electrocardiographiques liees a la charge annuelle de travail physique chez de jeunes adultes de sexe feminin. *Travail Humain,* **31,** 3-4, 344-345.

TORREY T.W. (1962). "Morphogenesis of the Vertebrates". New York: Wiley.

TUTTLE W.W. & SCHOTTELIUS B.A. (1969). "Text Book of Physiology (16th Ed.)". Saint Louis: C.V. Mosby.

VANZANT F.R., ALVAREZ W.C., EUSTERMANN G.B., DUNN H.L. & BERKSON J. (1932). The normal range of gastric acidity from growth to old age. *Arch. Int. Med.,* **49,** 345-59.

WEISS P.A. (1939). "Principles of Development". New York: Henry Holt.

WEISS P.A. (1956). What is growth? In S.R.W. Smith, O.H. Gaebler & C.N.H. Long (Eds.) "The Hypophysiol Growth Hormone, Nature and Actions". New York: McGraw-Hill.

WEISS P.A. (1968). "Dynamics of Development". London: Academic Press.

WELFORD A.T. (1965). Performance, biological mechanisms and age: a theoretical sketch. In A.T. Welford & J.E. Birren (Eds.) "Behaviour Ageing and the Nervous System". Illinois: Thomas.

WHIPPLE O.M. (1924). "Manual of Mental and Physical Tests". Baltimore: Warwick & York.

WIDDOWSON E.M. (1951). Mental contentment and physical growth. *Lancet,* **1,** 1316-18.

YOUNG C.M., BLANDIN J., TENSUON R. & FRYER J.H. (1963). Body composition studies of "older" women, thirty to seventy years of age. *Ann. N.Y. Acad. Sci.,* **110,** 589-607.

YUDKIN J. (1952). Effect of liver supplement on growth of children. *Brit. Med. J.,* **1,** 1388-9.

YUDKIN J. (1964). *Proc. Nutr. Soc.,* **23,** 149.

ZUCKERMAN S. (1956). Growth. In W.J. Hamilton (Ed.) "Textbook of Human Anatomy". London: Macmillan.

TOPIC AREAS-DEVELOPED

BIOMECHANICAL ANALYSIS OF HUMAN MOVEMENT

by G.T. ADAMSON

General

For a better understanding of the biomechanics of human movement it is necessary to consider both 'internal' and 'external' analyses with the body at rest or in motion. This will involve the biological sciences—anatomy, physiology, anthropology and classical mechanics. Until comparatively recently 'kinesiological' texts tended to offer largely anatomical analyses which produce rather limited interpretations of human movement. Physiological texts have tended to ignore mechanical aspects. However, the recent rapid developments in electronics coupled with space research has resulted in the development of more comprehensive research equipment and literature.

The following books should provide useful background reading for the serious student wishing to study the theoretical and practical aspects of biomechanics.

Williams & Lissner (1962); Steindler (1964); Kenedi (1965); Wartenweiler et al (1971); Alexander (1969); Rasch & Burke (1967); Tricker & Tricker (1967); Basmajian (1967); Brunnstrom (1968); Bunn (1960).

L.1 Internal Analysis

This section largely examines the more mechanical/mathematical aspects of functional anatomy and physiology.

(a) *Muscle mechanics*

The necessity to analyse muscular activity in dynamic situations has resulted in increasing use of electromyography to obtain objective assessment of muscle coordination which is impossible by any other means. Integrated electromyograms are becoming more acceptable and their most promising practical application has been in the study of human locomotion with simultaneous use of cine film. Murray et al (1964) found that full extension of the knee during walking does not occur in the way that it does in standing. This as Basmajian (1967) points out throws light on the unusual finding of Sutherland (1966) that knee extension occurs after quadriceps activity has ceased. Electromyography has proved its value in the

231

assessment of dangers to the vertebral joints and ligaments by lifting with the back rather than the legs (Troup 1965; Basmajian 1967; Floyd & Silver 1955; Cyriax 1954) and in the mechanics of the hernia (Campbell & Green 1955). Morehouse & Miller (1967) suggest that the multiple electromyographic technique may possibly be employed as an objective measure of skill, performances accomplished with the least waste effort representing the highest standard of skill. Electromyography has a large part to play in the solution of the ballistic v co-contraction problem relating to striking skills such as golf swing and soccer kick (Richet 1895; Hubbard 1960; Liberson et al 1962).

(b) *Bone and Joint Kinematics*

Functional anatomy of joints needs to be analysed by methods different from those involving the use of cadavers. These include optical and electrical techniques. Grieve (1968) has presented an ingenious technique for the study of movement patterns using simultaneous limb angles at 10% stages of the movement cycle. Gollnick & Karpovich (1961) by use of electro-goniometry analysed the action of joints during running, acrobatic and swimming activities. Contini et al (1965) used interrupted light photography and accelerometry in conjunction with force plates to obtain the temporal, kinematic and kinetic characteristics of human gait. Garrett et al (1969) by use of computer graphics have attempted to demonstrate that with appropriate measurement and analysis techniques early detection of a child's movement deficiencies could possibly result in the lessening of the deficiencies through provision of appropriate learning experiences geared to the individual. Miller (1970) has developed a computer simulation model of the airborne phase of diving to ascertain the effect upon total performance of altering certain variables such as condition of take-off and the patterns of motions of the limb.

(c) *Intra-truncal pressures*

The risk and assessment of trunk overstress during resisted movement has been considered by Davis (1956, 1959); Bartelink (1957); Morris et al (1961); Eie & Wehn (1962). Davis & Troup (1964) summarise some profound physiological effects attributable to high sustained pressures within the trunk cavities. Batson (1940); Anderson (1951) and Pearce (1957) stress the hazards of sustained excess intra-abdominal pressures which have implications for athletes in training as well as in everyday living.

(d) *Haemodynamics*

Kinesiologists have conspicuously ignored this area of fluid flow in spite of its significant relevance to health. According to Fox & Hugh (1970) there is a close coincidence between the site of intra-arterial stasis, atheroma and a process known as 'boundary-layer separation', a process which can be demonstrated readily in model flows. It is suggested that this evidence of stasis supports the thrombogenitic theory of atherogenesis and that it offers an explanation of aetiology of coronary thrombosis. Fox & Hugh (1964) suggest that circulatory thrills and bruits are caused by gas cavitation against the common acceptance that turbulence per se is a cause of noise in vessels.

(e) *Body size and composition*

Study of body size and composition has relevance for physical performance. Progress on the early work of Thompson (1961) and Haldane (1933) has been made by Asmussen (1965); Smith (1970); Lietzke (1956) and Keeney (1955). Formulae relating body weight and strength supported empirically and theoretically using computer calculations have been reported by O'Carrol (1968). Adamson & Cotes (1967) have related muscle force, muscle size and composition. Superior negro performances can be attributable to differential inter-racial limb proportions (Tanner 1964). The current use of anabolic steroids for increase of body mass is related to dramatic improvement in some athletic records (Lindsay & Pickering 1970).

(f) *Work capacity and efficiency*

Energy expenditure is reported in a variety of ways in the literature and for a more complete understanding of human efficiency it is necessary for the serious student of physical education to take a more mechanistic viewpoint of physiological data. Wilkie (1960) has analysed work limits from a chemico/mechanical point of view and classified activities power output on a temporal basis. Ellis & Hubbard (1969) critically appraise physiological methods for efficiency assessment from a bioengineer's point of view.

L.2. External analysis

The analysis of the magnitude, direction and variation with time of the forces evoked from the environment by the human body during physical activity provides valuable data concerning for example efficiency of performance, prevention of injury and design of prostheses. Direct assessment of ground reaction forces through the feet by use of force plates has demonstrated the limited utility of the term power to describe the action involved in 'impulsive' activities (Adamson & Whitney, 1971). Payne et al (1968) express surprise at the significant difference between directly assessed ground reaction force/time curves and those indirectly obtained by successive differentiation technique. That the human body in coping with external forces is capable of stressing its structure beyond safe limits is evidenced by widespread occurrence of injuries (Wayne 1954; Russek 1955; Bass 1967 and Phillips (1968). Experimental approach to aetiology of injuries has been neglected in spite of the fact that a very high proportion of factory accidents occur during manual handling of loads. Concomitant assessment of ground forces at the feet by force plate and forces at the hands by strain gauge techniques suggests that in most practical lifting operations the maximum lifting force is limited by mechanical rather than muscular factors (Whitney 1958). Bioengineers have progressed by making use of estimates of environmental reaction in the determination of joint forces and assessment of phasic relationships and tension of individual muscles (Paul 1965; Rydell 1965; Hirsch 1965; Sorbie & Zalter 1965). Prediction and control of spinal injuries sustained during a variety of human body impacts by use of a mathematical model has been developed (Orne & Liu 1971). Liu (1969) has also demonstrated the non-validity of the usual description of acceleration (in 'g's) as the injury criterion in biodynamic response to shock. Reactive forces on the body during underwater activities can provide approximate means for prediction of efforts required to perform certain dynamic actions under water (Seireg et al 1971).

When forces of interaction cannot be employed, research has revealed that other methods are available to predict movement. Kane & Scher (1970) have produced quantitative results concerning the use of relative motion of parts of the body as in trampolining, diving, trapeze activities etc. for production of forces to change their orientation.

The above is an extremely limited introduction to the field of external biomechanical analysis of human movement.

Source Texts

ALEXANDER R.McN. (1968). "Animal Mechanics". London: Sidgwick & Jackson.

ANDERSON T.McL. (1951). "Human Kinetics and Analysing Body Movements". London: Heinemann.

BASMAJIAN J.V. (1967). "Muscles Alive". Baltimore: Williams & Wilkins.

BRUNNSTROM S. (1968). "Clinical Kinesiology". Philadelphia: Davis.

BUNN J.W. (1960). "Scientific Principles of Coaching". New Jersey: Prentice Hall.

GOLLNICK P.D. & KARPOVICH P.V. (1961). "Electrogoniometric Study" Massachusetts: Springfield College.

KENEDI R.M. (1965). "Biomechanics and related Bio-engineering Topics" London: Pergamon.

MOREHOUSE L.E. & MILLER A.T. (1967). "Physiology of Exercise". St. Louis: Mosby.

RASCH P.J. & BURKE R.K. (1967). "Kinesiology and Applied Anatomy". Philadelphia: Lea & Febiger.

SMITH J.M. (1970). "Mathematical Ideas in Biology". London: Cambridge University Press.

STEINDLER A. (1964). "Kinesiology of the Human Body under Normal and Pathological Conditions". Illinois: Thomas.

TANNER J.M. (1964). "The Physique of the Olympic Athlete". London: Allen & Unwin.

THOMPSON D.W. (1961). "On Growth and Form". London: Cambridge University Press.

TRICKER R.A.R. & TRICKER B.J.K. (1967). "The Science of Movement". London: Mills & Boon.

VREDENBREGT J. & WARTENWEILER J. (Eds.) (1971). "Biomechanics II". Basel: Karger.

WILLIAMS M. & LISSNER H.R. (1962). "Biomechanics of Human Motion". Philadelphia: Saunders.

Specific References

ADAMSON G.T. & COTES J.E. (1967). Static and explosive muscle force relationship to other variables. *J. Physiol.,* **189,** 76-77.

ADAMSON G.T. & WHITNEY R.J. (1971). Critical appraisal of jumping as a measure of human power. *J. Biomechanics* **3,** 208-211.

ASMUSSEN E. (1965). Biological basis of sport. *Ergonomics* **8, 2,** 137-142.

BARTELINK D.L. (1957). The role of abdominal pressure in relieving the pressure on the lumbar intervertebral discs. *J. Bone Jt. Surg* **39B,** 718-725.

BASS A. (1967). Injuries of the lower limb in football. *Proc. R. Soc. Med.,* **60,** 517-530.

BATSON O.V. (1940). The function of the vertebral veins and their roles in the spread of metastases. *Annals Surg.,* **122,** 138-149.

CAMPBELL E.J.M. & GREEN J.H. (1955). The behaviour of the abdominal muscles and the intra-abdominal pressure during quiet breathing and increased pulmonary ventilation—a study in man. *J. Physiol.* **127:** 423-426.

CONTINI R., GAGE H. & DRILLIS R. (1965). Human gait characteristics. In R.M. Kenedi (Ed.) "Bio Engineering Topics". London: Pergaman.

CYRIAX J. (1954). Advice on weight lifting. *Brit. Med. J.,* **4867,** J. Anat., **90,** 601.

DAVIS P.R. (1956). Variations of the human intra-abdominal pressure during weight-lifting in different postures. *J. Anat.,* **90, 601.**

DAVIS P.R. (1959). Posture of the trunk during the lifting of weights. *Brit. Med. J.,* **1,** 87-89.

DAVIS P.R. & TROUP J.D.C. (1964). Pressures in the trunk cavities when pulling, pushing and lifting. *Ergonomics* **7, 4,** 465-474.

EIE N. & WEHN P. (1962). Measurements of the intra-abdominal pressure in relation to weight-bearing of the lumbosacral spine *J. Oslo City Hosp.,* **2,** 205-217.

ELLIS M.J. & HUBBARD R.P. (1969). Human and mechanical factors in Ergometry Unpublished thesis, Univ. of Illinois.

FOX J.A. & HUGH A.E. (1964). An analytical investigation into the possibility of gas cavitation in the circulation. *Phys. Med. Bio.,* **9,** 359

FOX J.A. & HUGH A.E. (1970a). The precise localisation of atheroma and its association with stasis at the origin of the internal carotid artery. *Br. J. Radiol.* **43,** 377-383.

FOX J.A. & HUGH A.E. (1970b). Static zones in the internal carotid artery: correlation with boundary layer separation and stasis in model flows. *Br. J. Radiol.,* **43,** 370-376.

FLOYD W.F. & SILVER P.H.S. (1955). The function of the erectors spinal muscles in certain movements and postures in man. *J. Physiol.,* **139,** 184-203.

GARRETT G.E., GARRETT R.E., REED W.S. & WIDULE C.T. (1969). Human movement via computer graphics. Paper presented at AAHPER convention, Boston.

GRIEVE D.W. (1968). Gait patterns and the speed of walking. *Biomedical Engineering* **3,3,** 119-112.

HALDANE J.B.S. (1960). On being the right size. In J.R. Newman (Ed.) "The World of Mathematics". London: Allen & Unwin.

HIRSCH C. (1965). Forces in the hip joint. In R.M. Kenedi (Ed.) "Biomechanics and Related Bio-Engineering Topics". London: Pergamon.

HUBBARD A.W. (1960). Homokinetics in muscular function in human movement. In W.R. Johnson (Ed.) "Science and Medicine of Exercise and Sports". New York: Harper.

KANE T.R. & SCHER M.P. (1970). Human self-rotation by means of limb movements. *J. Biomechanics* **3,** 39-49.

KEENEY C.E. (1955). Relationship of body weight to strength/bodyweight ratio in championship weight-lifters. *Res. Quart.,* **26,** 54-59.

LIBERSON W.T., HOLMQUEST H.J. & HALLS A. (1962). Accelerographic study of gait. *Arch. Phys. Med.,* **43,** 547-551.

LIETZHE M.H. (1956). Relation between weight-lifting totals and body weight. *Science,* **124,** 486-487.

LINDSAY M.R. & PICKERING R.J. (1970). Bio-mechanical, physiological and psychological factors related to athletic throwing events. *Brit. J. Sports Med.* **4, 4,** 278-279.

LIU Y.K. (1969). Towards a stress criterion of injury—an example in caudo-cephalad acceleration. *J. Biomechanics,* **2,** 145-149.

MILLER D.I. (1970). A computer simulation model of the airborne phase of diving. Doctoral thesis, Pennsylvania University.

MORRIS J.M., LUCAS D.B. & BRESLER B. (1961). Role of the trunk in stability of the spine. *J. Bone Jt. Surg.* **43A,** 327-351.

MURRAY P.M., DROUGHT A.B. & KORAY R.C. (1964). Walking patterns of normal men *J. Bone Jt. Surg.* **34A** 647-650.

O'CARROL M.J. (1968). On the relation between strength and body weight. *Res. Phys. Educ.* **1, 3,** 6-11.

ORNE D. & LIU Y.K. (1971). A mathematical model of spinal response to impact. *J. Biomechanics* **4, 1,** 49-69.

PAUL J.P. (1965). Engineering analysis of forces transmitted by joints. In R.M. Kenedi (Ed.) "Biomechanics and Related Bio-Engineering Topics". London: Pergamon.

PAYNE A.H., SLATER W.J. & TELFORD T. (1968). The use of a force platform in the study of athletic activities. *Ergonomics,* **11, 2,** 123-143.

PEARCE D.J. (1957). The role of posture in lamienectomy. *Proc. Roy. Soc. Med.* **57,** 109.

PHILLIPS N. (1968). Rehabilitation at a professional football club. *Brit. J. Spts. Med.* **3,** 197-203.

RICHET M.P. (1895). Note sur la contraction du muscle quadriceps dans l'acte de donner un coup de pied, *C.R. Soc. Biol. (Paris)* **2**, 204-205.

RUSSEK A.S. (1955). Medical and economic factors relating to the compensable back injury. *Arch. Phys. Med Rehab.*, **36**, 316-323.

RYDELL N. (1965). Forces in the hip-joint. Intra vital measurements. In R.M. Kenedi (Ed.) "Biomechanics and Related Bio-Engineering Topics". London: Pergamon.

SEIREG A., BAZ A. & PATEL D. (1971). Supportive forces on the human body during underwater activities. *J. Biomechanics*, **4**, **1**, 23-30.

SORBIE C. & ZALTER R. (1965). Phasic relationship of the hip muscles in walking. In R.M. Kenedi (Ed.) "Biomechanics and Related Bio-Engineering Topics". London: Pergamon.

SUTHERLAND D.H. (1966). An electromyographic study of the plantar flexors of the ankle in normal walking on the level. *J. Bone Jt. Surg.*, **48-A**, 66-71.

THOMPSON D.W.T. (1961). "On Growth and Form". (Abridged edition by J.T. Bonner.) Cambridge: University Press.

TROUP J.D.G. (1965). Relation of lumbar spine disorder to heavy manual work and lifting. *Lancet* **1**, 857-861.

WAYNE N.D. (1954). Weight-lifting, *Brit. Med. J.*, **4870**, 1098-1099.

WHITNEY R.J. (1958). The strength of the lifting action in man. *Ergonomics* **1**, **2**, 101-127.

WILKIE D.R. (1960). Man as a source of mechanical power. *Ergonomics*, **3**, 1-8.

TOPIC AREAS-DEVELOPED

MOVEMENT ANALYSIS

4 M

by HILARY CORLETT

General

Analytical studies of human movement extend from microscopic dissection to mascroscopic classification. The form of analysis used will obviously vary according to purpose—the context or situation of the movement observed must be known. Movement is found to have structural, denotational and connotational properties (Metheny 1968) and consequently can be described at sensory (Howard & Templeton 1966); perceptual and conceptual (Mawdsley 1971) levels. For a complete analysis, synthesis of all three aspects is desirable.

In this section it is intended to consider *only* the analysis of human movement in terms of the visually identifiable data (Preston-Dunlop 1969a). This pre-supposes that the observer is perceptive to the multiplicity of movement actually taking place. For a fine analysis, several observers would be required and data evaluated comparatively. Movement is herein defined in terms of 'human body parts acting dynamically in space and time in relation to the environment' (Preston-Dunlop 1967).

Laban (1960, 1963) classified movement into four main inter-related categories—Body, Space, Effort, Relationship (Gaumer 1962). These are inter-dependent in movement as Laban's belief was that 'the whole is greater than the sum of its parts' and change in any one category would simultaneously affect other aspects. Nevertheless certain elements—such as the use of body, space and movement qualities in terms of dynamics—can be extracted, compared and contrasted with other forms of analysis (Hunt, 1964, 1968; H'Doubler 1959; Hutt & Hutt 1970a, 1970b; Seashore 1940, 1942; Argyle 1967, 1969).

Laban's classification is applicable to all forms of movement whether primarily functional, behavioural, or aesthetic in intention (Redfern 1965).

Very little research material has been published in the area of analysis of the *processes* of variation within the flow and cessation of movement, although the resultant effects of the movement in terms of performance are more easily calculable—particularly in the case of job-efforts or psychomotor skills (Gilbreth 1919; Seymour, 1966; 1968).

As movement is transient and ever-changing, it has to be observed and recorded before detailed analysis and evaluation can be undertaken. For the student, a comparison between the different notation systems will highlight the common aspects of movement. A survey of the systems is contained in Hutchinson (1966) and Layson (1967). Details of some of the currently-used systems are to be found in Hutchinson (1954); Preston-Dunlop (1969a);Knust (1958); Laban (1956); Eshkol (1968); Causley (1967); Birdwhistell (1952). Many researchers have used film in the analysis of movments (Cratty 1964; Jokl 1968). With the development of video-tape, this method of study will doubtless assume greater importance.

M.1 Elemental analysis

a. Body

The body in human movement is simultaneously the enactive tool which interacts with the environment and the reflector of personal expression (both conscious and unconscious).

The body framework for analytical purposes can be differentiated into skeletal areas, joints or surfaces. The body alphabet has been well defined by Knust (1958). Body-type has been detailed by Kretschmer (1948); Sheldon (1940) and Parnell (1958).

Behavioural studies define facial expression, hand gestures or postural attitudes as the basic units to be observed; these are reported in Argyle (1969); Carroll (1971).

At the kinesiological level movement is clarified in terms of its range of movement in the fundamental actions of bending, stretching and twisting. Joint angles and degrees of rotation are estimated in some notation systems (Eshkol-Wachmann 1968; Preston-Dunlop 1967) and the motion categories of the body are enumerated in kinetography (Knust 1958; Preston-Dunlop 1966, 1969a, 1969b). In studies of psychomotor skills, the action possibilities of the body are grouped on different continuums—e.g. Locomotor and manipulative skills (Singer 1968); *open* and *closed* skills (Poulton 1957); *fine* and *gross* skills (Seashore 1940; 1942) *simple* and *complex* (Knapp 1963). Further classifications of body movements are to be found in Munrow (1963); Arnold (1968); Broer (1960); Morison (1969); Mauldon & Layson (1965).

The degree of bodily involvement in action is significant in the posture/gesture analysis of Lamb (1965) while Laban & Lawrence (1947); Lamb (1965); North (1971) and Ullmann (1970) have studied the minute (shadow) movements. These appear as precursors or they accompany the main movements of the body and reflect inner attitude.

Efficiency of movement can be assessed from the level of co-ordination of body parts in action. It is important to note the symmetry/laterality, the leading part, parallelism in the use of limbs or opposition in different parts revealed in counter-tension. The degree of inclination of the body and the order in which the flow of movement passes through the body are factors affecting fluency of action and expressiveness of motion (Kephart 1960; Chapman 1856; Struppa 1962; Morison 1969; Laban 1960 and Preston-Dunlop 1963).

Cratty (1964) has given a comprehensive survey of studies in kinesthesis; two methods of approach are detailed, one through a psychophysical and one through a perceptual-motor. Scott (1955) reports on attempts to 'measure' kinesthesis and kinesthesis is viewed as a physiological and perceptual-illusory phenomenon by Smith (1969). Recent speculations into the relationship of 'kinesthesis', 'body awareness' and 'body concept' have been made by Dickinson (1970) and Morris & Whiting (1971).

b. Space

'Space is a hidden feature of movement and movement is a visible aspect of space' (Laban 1966). Such a tautology summarises 'space' as a dynamic sphere of changing shapes created by movement of the body, but 'space' may also be viewed as a vacuum in which objects are contained. Space actions take movement into different directions and establish 26 points of orientation in relation to the hypothetical body centre, the 27th ('personal kinesphere' (Laban 1960) 'space capsule' (Cratty 1964). Proximal-distal terms of reference are established through the extension of movement into 'general' space of the environment. Direction in kinetographical analysis is given through definition of 4 spatial keys—a) personal cross of axes, b) constant cross of axes, c) cross of the body axes, d) orientation to fixed points. These reference systems are explained in Knust (1958); Hutchinson (1966) and Preston-Dunlop (1969b).

Central, peripheral and transversal pathways are used in all forms of movement and link different points of orientation, so creating spatial patterns on the floor and in the air. The fundamental shapes (illusory) which are 'traced' in space are linear, curved, twisted or rounded; in classical ballet terminology these are known as droit, ouvert, tortillé and rond. In behavioural studies these basic patterns are important as they may be a direct reflection of an inner state of mind (Laban, 1960; Efron, 1941; Kurath, 1956). Similarly in postural studies and in drama the actual body shape in stillness is of interest—the one-dimensional linear shape with an associated 'pin-like' attitude; the two-dimensional 'wall-like' front or rounded and centrally-focused ball-shape; and the three dimensional twisted shape in which the body has many centres of interest (North, 1971a; Russell, 1969).

In his study of space harmony Laban (1966) described harmonic sequences in which he found correspondence to tetrahedric, octahedric, cubic and icosahedric forms. These trace forms are parallelled in music by major and minor scales, chords and arpeggios. Laban defined movement patterns in terms of labile and stable tendencies.

Laban's choreutic principles have been extended to the analysis of skills by

Preston-Dunlop (1967), 'breaking down the harmonic synthesis into elements of direction and angle'. She uses a 3-part distance scale and 26 directions. Such a geometric analysis is basic also to the notation system of Eshkol-Wachmann who analysed movement as circular, as a result of the build of the body, and in terms of 'degrees of these circles in one or more co-ordinates' (Hutchinson, 1967).

In the analysis of motor skill, spatial factors affecting efficiency are defined in terms of directionality, proximity, planes, location and accuracy.

Spatial preferences, as expressed in movement and relating to the amount of space used by an individual, have been assessed as indicative of personality (Cratty, 1964). Allport (1933) identified the 'areal' and 'centrifugal' factors of gestural patterns; Krout (1954) also used the latter as a term of reference in his studies of autistic children. Lamb (1965) and Bartenieff (1966) clarified spatial gesturing in terms of growing/shrinking in shape/effort analysis. Cratty (1964), Vernon (1962) and Witkin et al (1954) report a number of investigations in spatial perception in relation to personality but in general there has been little experimental verification of the relatedness of spatial qualities to other considerations in movement.

c. Rhythm.

'Metre is time divided into measurable units' and movement can be performed to fit into these units of 2 or 3 'beats' which are combined to form different basic patterns. The tempo of the whole can be changed without changing the proportional length of each unit. But rhythm is more than metre. Two basic forms of rhythmic structure—metric (derived from pulse) and non-metric, irregular, breath or free rhythm (derived from action) are clarified in H'Doubler (1946); Preston-Dunlop (1963); Laban (1960) and Humphrey (1959). It is appreciated that in human movement a sense of rhythm arises from the fusion of the weight qualities (i.e. use of accents, whether forceful or delicate) with those of time qualities (i.e. sudden or sustained).

Metric rhythms associated with stepping and jumping are fundamental to basic dance rhythms (Sachs, 1938; Dalcroze, 1921; Thackray, 1963; Leonard, 1954a, b, 1955, 1956; H'Doubler, 1946; Hayes, 1955). According to the Greeks, combinations of rhythms could arouse moods—Dorian, Phrygian and Lydian (Laban 1960) Non-metric rhythms arise in gestural movement and working actions or emotionally-charged situations in which man restrains, releases and witholds his flow of energy (Laban 1959). The stress in movement may be made at the beginning as in ballistic motion (impulse), in the middle (swing), or at the end terminating the movement with a definite accent (impact).

Laban codified four motion factors—three (weight, space and time), which have counterparts in measurable physical terms and the fourth factor (flow) which 'has to do with motivation rather than actual motion' (Preston-Dunlop, 1967). Flow is referred to as the 'control' factor (Laban & Lawrence 1947) and as 'the flux or kinetic content, the degree of lability or adaptability to change' (Ullman, 1970).

Effort is about the mover's attitude to the space-time-mass framework in which he moves (Preston-Dunlop, 1967). Two fundamental attitudes—namely attacking, contending and controlling or yielding, indulging and surrendering (Lamb, 1965)—towards each of the motion factors gives rise to eight basic elements or effort qualities—sudden/sustained, firm/light, direct/flexible, bound/free. Combinations of single elements in phrases give rise to incomplete effort basic effort actions, and effort drives. For further definition of action/sensation, see Laban (1960). In analysis the effort range of a performer can be expressed in linear effort graphs (Preston-Dunlop, 1969).

For Laban the inter-relatedness of movement and personality was portrayed in his concept of 'effort'. The 'effort' preferences of any individual can be observed and recorded in movement phrases—significance lying in the rhythmic patterning of the phrases and in the frequency of recordings of different elements (Ullman, 1970). This work is in its infancy although it has been investigated by Jones (1970) and extended by the work of Lamb (1965); Bartenieff (1966); Williams (1968); North (1971).

'Shadow' movements correlating space, weight, time and flow factors with the attentional, intentional, decisional and precisional phases of the action are also recorded as chains of movement qualities in the assessment of mental efforts (Ullmann, 1970). The relationship of Laban's concept of 'shadow movements', Birdwhistell's (1952) meta-incongruent and Goffman's (1956) theories of expression 'given off' is postulated by Layson (1969).

For any analysis to be meaningful the observer needs to note where the effort occurs in the body and also the resultant spatial patterns as in effort/shape, postural/gestural analysis of Lamb (1965). The rhythms of the major movements and accompanying (shadow) movements should be compared.

d. Relationship.

Fundamental forms of relatedness are most clearly categorised in Preston-Dunlop (1966; 1969b) and Knust (1958). Ways of relating are made through the movement of one part of the body to another body part; of one body part to the whole body; or by the motion of the individual in relation to the objects of the environment. The same forms of relationship are observed in the action and re-action of individuals in inter/intra group non-verbal communication.

Movements are related by attention, proximity, touch and weight-bearing, using the channels of communication enumerated by Argyle (1969)—namely, visual, tactile and proxemic.

In the process of relating, the following sequence occurs—paying attention, approaching, meeting, releasing and recovery. Attention is paid through looking or gesturing while approach results in staying near, surrounding, interweaving, or penetrating. Tactile relationship may be achieved when slight touch, hitting, grasping or weight-bearing occur. Release involves separation and all phrases are concluded by a recovery period, Laban (1959). This information must be related to the context of situation. Constituent parts of the above sequence are selected in the observation of children's movement behaviour in play (see section O).

The actual process of human interaction brings about different forms of movement behaviour. This is frequently noted in terms of the wider classifications of imitative response, adaptive, antagonistic, passive or submissive patterns. The rhythm of the sequence of the movement patterns can be analysed in terms of Laban's 'effort' concept. Alteration in postural attitude, focus and proximity lead to a change in spatial relationship and, as individuals draw together or move apart recognisable group shapes are formed—e.g. linear, circular, wedge, square, triangular.

M.2 Effort in work study.

In training work-operators in industry, in teaching physical skills, in athletics and games etc. the observer may look at the following effort features of the movement:- the manner in which space is utilised (linear or plastic), the way in which pressure is exerted and how time is used. The fourth factor of flow is seen as the effort made to control the 'dictated' pattern of the job (Boyd, 1969; Dewhurst, 1950; Mauldon & Redfern, 1969). The readiness to stop and adjust (bound flow) is observable in the counter-tensions in the body, while free flow permits the movement to continue without any further information.

The job sequence may be divided into parts and the load requirements of each part (in terms of space/pressure/time) assessed numerically (Preston-Dunlop, 1967). Calculations can then be made of the amounts of preferential 'effort' control compared with the amount of 'effort' control required for efficiency of the job. Lamb (1965) estimates the range of a worker's 'effort' qualities used and expresses them in ratio terms—e.g. 6 strong : 4 light. His work has shown that the postural and the gestural range may not be the same. Finer details of analysis of the preparatory, executive and recovery phases of an action have been explained in Laban & Lawrence (1947). Skill is achieved through the blending of the correct proportions of the motion factors. For a complete survey of the movement under consideration, there must be fusion of the kinetographical and the effort analysis.

M.3 Behavioural analysis.

Theories of movement expression stem from the work of Darwin (1872) and Delsarté (See Shawn 1963). The relatedness of movement/personality has been given more detailed consideration in Section I, and reference has been given in M.1. iii to a methodology of analysis developed in this area (Ullmann, 1970; North, 1971; Lamb 1965).

As basic sources of reference to the analytical studies of movement behaviour in social interaction, the student is recommended to read Argyle, (1967; 1969) and Carroll (1971). A conceptual scheme of social behaviour has been presented by

Abrahamson (1966). He elaborated on three basic mechanisms of interpersonal accommodation namely, response inhibition, vicarious learning and covert communication, in all of which movement plays a part. Movement as a non-verbal language in overt and covert communication has been closely examined in Section D & E. Cratty (1964) overviewed systems for the interpretation of movement cues and for the investigation of both innate patterns and learned cultural patterns of movement behaviour.

Analysis of dyadic and of group structure and group behaviour exists at two levels—in terms of component identifiable units and in terms of classified patterns. In the former, Argyle (1969) selected details of timing, synchronising of speech and accompanying bodily movements, while other researchers have identified units of motion—e.g. social kines (Birdwhistell, 1952; 1960). In the latter, Argyle referred to dominance, intimacy and role relationships as examples of the higher order of units in more global investigations. It can be seen that the syntactical structure of communication underlies the semantics of the language of movement. Systems for the classification and recording of movement in non-verbal behaviour have been developed by Hall (1963); Freedmann & Hoffmann (1967); Ekman & Friesen (1967); Birdwhistell (1970); Kurath (1950, 1956). Other studies of autistic children and psychiatric cases are reported in Hutt & Hutt (1970). Cratty (1964) has also outlined research methods in communication which use statistical analysis and experimental techniques. The 'gestural dance' between interactors has been analysed through film techniques by Condon & Ogston (see Argyle 1969) and Kendon (1967).

Dyadic interaction studies are numerous and well reported in Argyle (1967; 1969) and Argyle & Kendon (1967). Far less research at concrete level of analysis has been reported on the movement behaviour of small groups. Cybernetic studies by Wiener (1950) and others developing his scientific model of the man-like machines are applicable to this area. The student is recommended to use material from the discipline of sociology in the study of behaviour in small groups (Homans, 1951; Klein, 1965; Sprott, 1958). Theories of group behaviour have been postulated by Field theorists (Gardner, 1954) and by Interaction analysis (Bales, 1950; Parsons, Bales & Shils, 1953).

The total behavioural inter-action of groups is obviously affected by the nature of the group and its task and external or internal dimensions can be distinguished. In situations in which the language code is formal, gestural language is minimal but highly stylised; within 'restricted code' situations communication is partly achieved through 'natural' signs and signals. In his explanation of the relationships which arise in unstructured T-groups, Bion (1961) revealed the existence of a correlation between these patterns and the innate movement patterns of non-human primates. It is suggested that paralinguistic features, (Goffmann, 1956) are particularly relevant in this situation in influencing the interpretation of communication. The social dimensions of physical activity have been recorded in Cratty (1967). Through the use of sociometric techniques, Bales' interaction anaylsis and T-group situations, it is possible in educational spheres to study the behaviour of adolescent peer groups.

M.4   The component units which form the structure of the dance are identifiable but—as a created form—dance cannot be analysed; the essence of dance lies in its indivisible wholeness. It can be subjected to descriptive analysis but this 'needs to focus again and again on the wholeness rather than reducing it into externally related units, Sheets (1966). The phenomenological approach to dance is clarified by Sheets (1966)—in the dancers 'lived experience' of the body is incorporated a pre-reflective awareness of its temporality and spatiality. In terms of the existential analysis, qualities of force, time and space are 'virtual' or 'illusory' rather than 'actual' and may only be described as they appear in the total form (see also Langer, (1953; 1957)).

A full account of the theories of rhythmic and spatial structures of movement as expounded by leading dance exponents is not possible here. An overall survey may be found in Brown & Somner (1969); Focus (1962); Lloyd (1949) and Martin (1965). H'Doubler (1959) clarified the measurable components of movement as resistance, distance, duration, speed and force. She determined immeasurable qualities of movement from 'the way in which energy is released'. Like Graham she defined these as swing, percussive, sustained and collapse. Humphrey (1959) based her dance technique on principles of rise—fall—recovery and perceives three sources of rhythmic organisation which may be compared with those outlined (see M.1 iii). To Humphrey, dynamics was smooth or sharp, or alternating smooth and sharp, or simultaneously smooth and sharp.

Hayes (1955) and Humphrey (1959) enumerated the spatial ingredients of direction, focus, planes, density, range as fundamental to the design of a choreographed dance.

In the creation of an organic form such as the art form of dance, principles of composition (e.g. variety, unity, contrast balance, proportion) must be taken into account. These are discussed in Hayes (1955); H'Doubler (1959); Hawkins (1964); Horst (1961). Osborne (1970) elucidated the aesthetic qualities and outlined the skill of appreciation while Arnheim (1956) elaborated on the artist's perception of form and content. In the study of aesthetic qualities in man-made movement forms, Reid (1969, 1970) distinguished between the nature of aesthetic qualities and the nature of artistic elements. He suggested that there are artistic elements embedded in some sports. Identifiable criteria are used in adjudicating figure skating, for example. Anthony (1968) supported this view, although he only gave the criteria for judgement of the 'artistic' movement in some sports in general terms—such as elegance, grace, rhythm and harmony. In his concept of physical education Carlisle (1969), has taken a different view and suggests the aesthetic as the main account.

In conclusion, the significance of movement can be understood to be meaningful at different levels—as kinestruct (the visually perceivable form), or as kinecept (the perceptual form) or as kinesymbol (conceptual form) (Metheny, 1962; 1965; 1968). The student is reminded that, for a complete analysis, the sensory, perceptual and conceptual levels must be fused.

Source Texts

ARGYLE M. (1967). "The Psychology of Interpersonal Behaviour". Harmonds-worth: Pelican.
ARGYLE M. (1969). "Social Interaction". London: Methuen.
CRATTY B. (1964). "Movement Behaviour and Motor Learning". Philadelphia: Lea and Febiger.
HUTT S.J. & HUTT C. (1970). "Behaviour Studies in Psychiatry". Oxford: Pergamon.
LABAN R. (1956). "Principles of Dance and Movement Notation". London: Macdonald and Evans.
LABAN R. (1960). "Mastery of Movement". London: Macdonald and Evans.
LABAN R.(1963). "Modern Educational Dance" 2nd edn. London: Macdonald and Evans.
LABAN R. (1966). "Choreutics". London: Macdonald & Evans.
LAMB W. (1965). "Posture and Gesture". London: Duckworth.
MORRIS P.R. & WHITING H.T.A. (1971). "Motor Impairment and Compen-satory Education". London: Bell.
PRESTON-DUNLOP V. (1963). "A Handbook for Modern Educational Dance". London: Macdonald and Evans.
PRESTON-DUNLOP V. (1966). "Readers in Kinetography Laban. Books I-IV." London: Macdonald and Evans.
PRESTON-DUNLOP V. (1967). Laban's analysis of movement. *J. Phys. Ed.*, **59**, 177.
PRESTON-DUNLOP V. (1969a). A notation system for recording observable motion. *Int. J. Man-Machine Studies.* **1**, 361-386.
PRESTON-DUNLOP V. (1969b). "Practical Kinetography Laban". London: Macdonald and Evans.

Specific References

ABRAHAMSON M. (1966). "Interpersonal Accommodation". Princeton: Van Nostrand.

ALLPORT G.W. (1933). "Studies in Expressive Movement". London: Macmillan.

ANTHONY D. (1968). Sport and physical education as a means of aesthetic education. *J. Phys. Educ.*, **60**, 179.

ARGYLE M. & KENDON A. (1967). The experimental analysis of social performance. In L. Berkowitz (Ed.) "Advances in Experimental Psychology". New York: Academic Press.

ARNHEIM R. (1956). "Art and Visual Perception". London: Faber and Faber.

ARNOLD P.J. (1968). "Education, Physical Education and Personality Development". London: Heinemann.

BALES R.F. (1950). "Interaction Process and Analysis". Massachusetts: Addison - Wesley.

BARTENIEFF I. (1966). "The Effort-Shape Training Programme". New York: Dance Notation Bureau.

BOYD M. (1969). "Lacrosse: Playing and Coaching". London: Kaye and Ward.

BION W. (1961). "Experiences in Groups and Other Papers". London: Tavistock.

BIRDWHISTELL R. (1952). "Introduction to Kinesics". Louisville: University Press.

BIRDWHISTELL R. (1960). Kinesics and communication. In E. Carpenter & M. McLuhan (Eds.) "Explorations in Communication". Boston: Beacon Press.

BIRDWHISTELL R. (1970). "Kinesics and Context". Philadelphia: Pennsylvania University Press.

BROER M. (1960). "Efficiency of Human Movement". Philadelphia: Saunders.

BROOKE J. (1967). A taxonomy for human movement. *Leaflet,* **68,** 56-57.

BROWN R.C. & CRATTY B.J. (Eds.) (1969). "New Perspectives of Man in Action". New Jersey: Prentice-Hall.

BROWN M. & SOMNER B. (1969). "Movement Education: its evolution and a modern approach". Massachusetts: Addison-Wesley.

CARLISLE R. (1969). The concept of physical education. In Proceedings of the Annual Conference III. Philosophy of Education Society of Gt. Britain.

CARROLL J. (1971). A survey of some research findings from the social sciences that relate to movement behaviour. Monograph 1. Anstey College of Physical Education.

CAUSLEY M. (1967). "Benesh Notation". Bristol: Purnell.

CHAPMAN M.J. (1856). "Ling's Educational and Curative Exercises". London: Ballière.

CRATTY B.J. (1967). "Social Dimensions of Physical Activity". New Jersey: Prentice-Hall.

DALCROZE E.J. (1921). "Rhythm, Music and Education". London: Chatto & Windus.

DARWIN C.R. (1872). "The Expression of Emotions in Man and Animals". New York: Appleton.

DICKINSON J. (1970). A note on the concept of body awareness. *Brit. J. Phys. Educ.*, **I**, 34-36.

DEWHURST E. (1950). "Lawn Tennis Guaranteed". London: Pitman.

H'DOUBLER M. (1946). "Movement and its Rhythmic Structure". U.S.A. Wisconsin. Mimeograph. Kramer.

EFRON D. (1941). "Gesture and Environment". New York: King's Crown Press.

EKMAN P. & FRIESEN W. (1967). Head and body cues in the judgement of emotion: a reformulation. *Percept. Motor. Skills*, **24**, 711-724.

ESHKOL & NUL (1968). "Eshkol-Wachmann Movement Notation". Tel Aviv: Israel Music Institute.

Focus On Dance II. (1962). Washington. AAHPER.

FREEDMAN N. & HOFFMAN S.P. (1967). Kinetic behaviour in altered motor states. *Percept. Motor Skills. 24.*

GARDNER L. (Ed.) (1954). "Handbook of Social Psychology". Massachusetts: Addison-Wesley.

GAUMER D. (1962). Laban; his contribution to the world of movement. In "Focus on Dance II". Washington: A.A.H.P.E.R.

GILBRETH F.B. & GILBRETH L.M. (1919). "Applied Motion Study". New York: Macmillan.

GOFFMAN E. (1956). "The Presentation of Self in Everyday Life". Edinburgh: University Press.

HALL E.T. (1963). A system for the notation of proxemic behaviour. *Amer. Anthrop.,* **65**, 1003-1026.

HAWKINS A.M. (1964). "Creating through Dance". New Jersey: Prentice-Hall.

HAYES E. (1955). "Dance Composition and Production". New York: Ronald Press.

HOMANS G. (1951). "The Human Group". London: Routledge & Kegan Paul.

HORST L. (1961). "Modern Dance Forms". San Francisco: Impulse Pub.

HOWARD I.P. & TEMPLETON W.B. (1966). "Human Spatial Orientation". London: Wiley.

HUMPHREY D. (1959). "The Art of Making Dances". New York: Rinehart.

HUNT V. (1968). "The Biological Organisation of Man to Move". San Francisco: Impulse Pub.

HUNT V. (1964). Movement behavour: a model for action *Quest,* II.

HUTCHINSON A. (1954). "Labanotation". London: Phoenix House.

HUTCHINSON A. (1966). A survey of systems of dance notation. *L.A.M.G. Mag.* Part I Nov. 1966. Part II May 1967.

HUTT S.J. & HUTT C. (1970). "Direct Observation and Measurement of Behaviour". Springfield: Thomas.

JOKL E. (1968). "Medicine and Sport". Basel: Karger.

JONES M. (1970). Perception, personality and movement characteristics. Unpublished M. Ed. Thesis. Leicester.

KENDON A. (1967). Some functions of gaze direction in social interaction. *Acta Psych.,* **26**, 1-47.

KEPHART N.C. (1960). "The Slow Learner in the Classroom". Ohio: Merrill.

KLEIN J. (1965). "The Study of Groups". London: Routledge & Kegan Paul.

KNAPP B. (1963). "Skill in Sport". London: Routledge & Kegan Paul.

KNUST A. (1958). "Handbook of Kinetography Laban". Hamburg: Das Tanzarchiv.

KRETSCHMER E. (1948). "Physique and Character". Berlin: Springer.

KURATH G. (1950). A new method of choreographic notation. *Amer. Anthrop.,* 52.

KURATH G. (1956). Choreology and anthropology. *Amer. Anthrop.,* 58.

KROUT M.H. (1954). An experimental attempt to determine the signiticance of unconscious manual symbolic movements. *J. Genet. Psychol.,* 51.

LABAN R. (1959). The rhythm of effort and recovery. Part I. *L.A.M.G. Mag.,* **23,** 18-23.

LABAN R. (1960). The rhythm of effort and recovery, Part II. *L.A.M.G. Mag.,* **24,** 12-18.

LABAN R. & LAWRENCE F. (1947). "Effort". London: Macdonald & Evans.

LANGER S. (1953). "Feeling and Form". London: Routledge & Kegan Paul.

LANGER S. (1957). "Problems of Art". New York: Charles Scribner.

LAYSON J. (1967). A critical examination of systems of movement notation. Upublished Dip. Ed. Thesis. Manchester Univ.

LAYSON J. (1969). An introduction to some aspects of the sociology of physical education. *Proc. Ass. Wom. Coll. Phys. Educ.,* 45-53

LEONARD M. (1954a). Rhythm and dance. *L.A.M.G. Mag.* **12**

LEONARD M. (1954b). Rhythm and dance. *L.A.M.G. Mag.* **13**

LEONARD M. (1955). Rhythm and dance. *L.A.M.G. Mag.* **14**

LEONARD M. (1956). Rhythm and dance. *L.A.M.G. Mag.* **17**

LLOYD M. (1949). "The Borzoi Book of Modern Dance' New York: Alfred Knopf.

MARTIN J. (1965). "The Modern Dance". New York: Dance Horizons Inc.

MAULDON E. & REDFERN B. (1969). "Games Teaching". London: Macdonald & Evans.

MAULDON E. & LAYSON J. (1965). "Teaching Gymnastics". London: Macdonald & Evans.

MAWDSLEY H.P. (1971). A conceptual analysis of movement. *B.A.O.L.P.E. Bull.,* VIII, 39-45.

METHENY E. (1962). Symbolic forms of movement. In "Focus on Dance II". Washington. A.A.H.P.E.R.

METHENY E. (1965). "Connotations of Movement and Sport". Iowa: Brown.

METHENY E. (1968). "Movement and Meaning". New York: McGraw Hill.

MORISON R. (1969). "A Movement Approach to Educational Gymnastics". London: Dent.

MUNROW A.D. (1963). "Pure and Applied Gymnastics". Leeds: Arnold.

NORTH M. (1971a) "An Introduction to Movement Study and Teaching". London: Macdonald & Evans.

NORTH M. (1971b). "Personality and Movement". London: Macdonald & Evans.

OSBORNE H. (1970). "The Art of Appreciation". London: Oxford University Press.

PARNELL R.W. (1958). "Behaviour and Physique". London: Arnold.

PARSONS T., BALES R.F. & SHILS E.A. (1953). "Working Papers in the Therory of Action". Illinois: Free Press.

POULTON E.C. (1957). Prediction in skilled movement. *Psychol. Bull.,* **54, 6,** 447-78.

REDFERN B. (1965). "Introducing Laban Art of Movement". London: Macdonald & Evans.

REID A. (1969). "Meaning in the Arts". London: George Allen & Unwin.

REID A. (1970). Movement and meaning. *L.A.M.G. Mag.,*45, 5-31.

RUSSELL J. (1965). "Creative Dance in the Primary School". London: Macdonald & Evans.

RUSSELL J. (1969). "Creative Dance in the Secondary School". London: Macdonald & Evans.

SACHS C. (1938). "World History of the Dance". London: George Allen & Unwin.

SCOTT G. (1955). Measurement of Kinesthesis. *Res. Quart.,* **26,** 324-341.

SEASHORE R.H. (1940). Fine motor skills. *Amer. J. Psychol.,* **53,** 86-88.

SEASHORE R.H. (1942). Some relationships of fine and gross motor abilities. *Res. Quart.,* **13,** 259-274.

SEYMOUR D. (1966). "Industrial Skills". London: Pitman.

SEYMOUR D. (1968). "Skills Analysis Training". London: Pitman.

SHAWN T. (1963 repub.) "Every Little Movement". New York: Dance Horizons Inc.

SHEETS M. (1966). "The Phenomonology of Dance". Wisconsin: University Press.

SHELDON W.H. (1940). "The Varieties of Human Physique". New York: Harper.

SINGER R.N. (1968). "Motor Learning and Human Performance". New York: MacMillan.

SMITH J.L. (1969). Kinesthesis: a model for movement feedback. In R. Brown & B. Cratty "New Perspectives of Man in Action". New Jersey: Prentice Hall.

SPROTT W. (1958). "Human Groups". Harmondsworth: Pelican.

STRUPPA E. (1962). Graham; an interpretation of her technique and principles. In "Focus on Dance II". Washington: A.A.P.H.E.R.

THACKRAY R.M. (1963). "Playing for Dance". London: Novello.

ULLMANN L. (1970). Movement as an aid to understanding and development of personality. In Movement, Dance and Drama. Hull University conference report.

VERNON M.D. (1962). "The Psychology of Perception". Harmondsworth: Penquin.

WIENER M. (1950). "The Human Use of Human Beings". New York: Doubleday Anchor.

WILLIAMS S. (1968). Analysis of expression: an enquiry into the significance of human movement. Unpub. M. Ed. Thesis. University of Leicester.

WITKIN H.A. et al (1954). "Personality Through Perception". New York: Harper.

4 N TOPIC AREAS-DEVELOPED

MOVEMENT IDEAS IN PERSPECTIVE

by IDA WEBB

General

A study of movement ideas in perspective necessitates the consideration of general theories (B.1-B.9) and the application of ideas in specific areas (N.10-N.12); (See Stage 3).

It is important to consider systematically the underlying laws and principles of movement, as discovered in each chronological era, for these reflect the stages in the development of knowledge of man's understanding of his own movement. The student will need to abstract material from quoted texts and appraise it in relation to the use, the meaning and the significance of human movement. Few books have been written directly concerned with the topic under consideration but the following references will provide useful background reading; Leonard & Affleck (1914); Roper (1938); Rice & Hutchinson (1952); McIntosh (1952, 1963); Van Dalen et al (1953); Dixon et al (1957); Laban (1958); Hackensmith (1966); Smith (1968); Latchow & Egstrom (1969); Stanley (1969); Brailsford (1969); Oberteuffer & Ulrich (1970); Sweeney (1970).

The student is asked to accept the following structure as providing many starting points for further investigations in the historical study of movement. Completely comprehensive lists in each area are impracticable; selected texts have, therefore, been given to stimulate study and research into comparatively new areas of work.

N.1 Man as the centre of creation

Aristotle developed a generalised conceptual apparatus for dealing with matter and also adopted a specific line of physical doctrine (Edel 1967). His general pattern was based on the acceptance of the sense of one world; motion was continuous or infinite. Further necessary conditions were those of place, void and time. The four sublunar elements, earth, air, fire and water each had a distinctive natural movement; 'up' becoming lighter, 'down' becoming heavier; while the 'passage of heavenly bodies', was expressed in circular motion. Aristotle's central or specific concept of motion (kinesis) was based on change in quantity, quality and place. He had no unified space concept and looked at time with respect to 'before' and 'after' the movement occurred. In 'Writings on Physics', Aristotle made reference to the nature of motion and distinguished between the power of spontaneous movement and change as opposed to movement which is induced from without by force (Allan 1970). Further he divided motion into 'natural' in which there was no human interference, and 'violent', in which man intervened (Encyclopaedia Britannica Vols. 15 and 16).

Plato (1960) in 'The Laws' extended the principle of life, put forward by Aristotle; i.e. that which moves itself is primary while that which is moved is secondary. Plato concluded that in all matters the soul over-rides the body. Further he enlarged on circular motion and thoroughly examined the notion that motion and rest are not predictable of each other (Cornford, 1935; Ross, 1951) they are, in fact completely different but share in existence.

N.2 Organic evolution

The principle of natural selection put forward by Darwin (1958) and the effects of use and disuse have led not only to general changes and modifications in bodily structure but to individual variation in the manner of development. As a biped, man, has an advantage over other mammals; he is free to develop the upper trunk and limbs for prehensile capacities while the lower part of the body is used mainly for support and locomotion. Darwin initiated the genetic approach to bodily movement (Darwin 1872); he made a detailed study of expressive actions and gestures, including symbolic and sympathetic movements, of both animals and man.

N.3 Force and laws of motion

Galileo (Encyclopaedia Britannica Vol. 15 and 16) examined the phenomenon of freely falling bodies and concluded that acceleration was constant. He, therefore, disputed Aristotle's views on 'violent' motion.

Newton defined the three laws of motion, namely: inertia, acceleration, and action and reaction (Dyson 1962; Tricker & Tricker, 1967). He concluded that motion must be examined in context otherwise it is meaningless. His laws have provided the basis for the development of mechanical analysis of movement.

N.4 Movement as a means of communication

Delsarte used observation of movement as his prime method of study. He aimed to improve and perfect gestures (learned symbols) and expressive actions as they were of value in human communication. His system of 'Trinities' (A, B and C) embraced unity in time, space and motion; (A), of the vital, emotional and mental functions, (B), as expressed in eccentric, normal and concentric movement, (C). In the 'Law of

Correspondence' he related physical action to spiritual function; i.e. every movement and gesture has its own meaning or expresses some idea, feeling or emotion (Shawn 1963). His nine laws of motion made reference, to attitude, force, motion, sequence, direction, form, velocity, reaction and extension.

Lamb (1965) working again through detailed observation and analysis of body movement in terms of posture and gesture stressed the significance of patterns of movement behaviour in the assessment of personality (North 1972). The components of physical behaviour, i.e. posture/gesture, shape/effort also reveal the movement characteristics of different community cultures.

Argyle (1969) paid great attention to non-verbal cues in the patterns of social interaction. He detailed the types of cues utilised in different situations and described the areas of the body most frequently used in this form of communication.

Gesture is used either to support verbal communication or as a substitute for speech (Feldman 1959). Birdwhistell (1952; 1968) researched into the field of body motion and meaning with respect to inter-personal communication. He called the systematic study of how man communicates through body action, KINESICS. In his works he described in great detail the manneristic use of gestures and the expressive movement of all parts of the body as used in non-verbal communication. In outlining an annotation system for Kinesics he took into account the physical, physiological, psychical and cultural aspects of non-verbal communication. The three major divisions of Kinesics are: pre-Kinesics (physical basis), micro-Kinesics (isolation of kines) and social Kinesics (social interaction).

Further work on non-verbal communication has been carried out by Gehlen (n.d.), (symbolic structure), Esalen, (n.d.), (tactile sensitivity) and Hall (1959). (See also Layson (1969) and Carroll (1971) for comprehensive surveys of the literature).

N.5 Objectivity of movement

'Spectacle' was a major element in Noverre's ballets. Gesture or kinetic experience was, to him, the equivalent of vital movement. It was used as the basic abstraction in the illumination and organisation of the material content of the dance (Langer 1967). For Noverre, dance devoid of expression lacked feeling and emotion; movement through gesture revealed more than words; each action had to speak to the observer (Noverre 1930). The ballet master was charged to understand the physique of each of his dancers, to consider their natural movement and so to develop the individual's talents; only then would satisfactory results be achieved. Movement grows in size and/or intensity as the stimulation of impulses increases; it is the most prominent visual part of expression.

For Klages (1929) man was compelled to move by 'spiritual' forces extrinsic to his being; self movement was insufficient, there had to be an external pressure. Man, from observation of the form of outward movement aimed to asses inner meanings in order to 'reveal his soul' to himself. Knowledge of self was therefore acquired through knowledge of others. Both concrete and abstract concepts were constructed and emphasized in relation to objects; they were not formed in vacuo. Man has an innate capacity for expression which, to Klages, was movement; movement was used to express feelings and sensitivity in a concrete manner.

N.6 Functional activity

Buytendijk (1956) investigated movement with reference to activity (Report V. Int. C.); the individual's body was at his own disposal; he was compelled to move by a centre of spontaneous action. Therefore man having power over himself acted in a purposeful manner. Movement was meaningful from the physical or biological angle. The child plays because he is a child (Piaget 1951); this is the substance of Buytendijk's theory of 'infantile dynamics'.
Weizsacker (Heidemann 1965) viewed movement as the opposition between subject and object; it was concerned with being and moving. He theorised that the origin of movement lies within the concept of the Gestalt circle and that the latter represents the life cycle. Therefore movement is explained by movement; this theory does not allow for movement as an act of thought; as Reid (1970) says, *'meaning is embodied in the movement.'*

N.7 Scientific management

In scientific management the study of the physical capabilities and limits of a worker is combined with an economic approach to work. Taylor (1911), closely investigated the timing of actions of man and machines; he also contributed to work on standardisation of procedures in time study, the use of instruction cards and the increase of output and wages while costs were decreased. Man is driven by the fear of hunger and the need to search for profit. If material rewards are closely related to work, efforts and time units are small, response is maximum. Ultimately man functions as an appendage to the industrial machine (Etizioni, 1964; 1970).
Gilbreth & Gilbreth (1919) studied the number and type of actions in relation to conditions of bricklayers; they laid down the principles of 'motion study' and aimed to reduce unnecessary movements; each operation would therefore be more efficient and effective and less fatiguing.
The industrialist Lawrence, together with Laban, focused on the central problem, 'What is movement?' (Laban & Lawrence, 1947). They investigated the 'effort' or bodily rhythms of the individual and developed a simple method of assessment which was of value in both selection and training in relation to the job.
Scott (1947) made a detailed analysis of human movement and McClurg Anderson (1951) analysed mechanical and muscular stresses through kinetic responses. He was concerned with the purpose and suitability of movements in relation to actions. Lamb (1965) observed movement and shadow movements (Laban, 1947) as part of the interviewing procedure when appointing executives to industry. He is

concerned to place the 'right' personality in the 'right' post.

Preston-Dunlop (1969a; 1969b) has outlined a notation system for observable motion data recording of value in industry as well as in movement education.

N.8 Movement as an art form

Metheny (1965) examined the symbolic forms of movement. She expressed the meaning of the Kinecept (kinesthetic perceptual motion) through the Kinestruct (dynamic movement) in Kinesymbolic form (meaningful interpretation of movement); thus the percept was transformed into a concept.

To Langer (1957; 1967) for movement (virtual gesture) to become an art form it had to be abstracted and manifest kinesthetic feeling. The visual phenomenon enabled the observer to temporarily inhabit the performer's world; the significance of movement was revealed in both dynamic experience and audience participation. Freed from common usage, movement became symbolically expressive of human feeling. H'Doubler (1966) postulated that the fusion of inner and outer 'dance' experiences were necessary for movement to exist as an art form; that is, the organisation of movement together with an imaginative content and intellectual discipline of dance provided the essential ingredients for organic unity; man had to move and attend to his actions in order to understand and know movement.

Reid (1969; 1970) has hypothesised, movement to be an art form, must have fused into it aesthetic quality.

N.9 Elements of movement

Laban's theory and principles of movement offered man a more objective approach to the study of movement although in translation his name has become synonymous with the art of movement.

Man's movement is composed of 'effort' related to 'shape'. 'Effort' is the inner impulse that initiates and gives expression to movement; it is the visible symbol of inner attitudes and drives. 'Effort' or 'the anatomy of movement' (Ringel, 1965) is derived from the four factors of weight, space, time and flow and is itself one of four common principles of movement, namely body, space, effort, relationship. Each element has a psychological meaning and a scientific translation (Laban & Lawrence, 1947). Shape or the organisation of space gives form and structure to movement (Laban, 1966); it is the culmination of man's ability to create and see relationships (Laban, 1958). For Laban there was an inherent harmony within man's movement patterns in space (Laban, 1963); movement also provided the topic for a life time's study (Thornton, 1970). (A biography has been commissioned by the Laban Art of Movement Guild.)

N.10 Movement in an educational environment

Movement is the substance of physical training, physical education and movement education; it is the method by which contact is made with the environment. Existing literature describes the development of individual physical activities, the theory and underlying principles of each aspect and the teaching of individual sports. Few authors have drawn specific attention to the part movement itself plays in separate activities. It is something that has been taken for granted, exploited, generalised, ignored; yet it is central. General information is contained in Treves (1892); Leonard & Affleck (1914); Wood and Cassidy (1931); Roper (1938); McIntosh (1952); Rice & Hutchinson (1952); Van Dalen et al (1953); Williams (1959); Brown & Cassidy (1963); Halsey (1964); Hackensmith (1966) and Streicher (1970).

The student can trace a developing pattern which is summarised in the following quotations, 'Physical education involves exercise and movement' (Treves, 1892); 'Physical education is education through movement' (Roper, 1938); 'Through awareness in movement, man is capable of discovering his unique being' (Slusher 1967). (See also Section 1J).

Athletics

In athletics man has developed the basic skills of running, jumping and throwing and he has applied his knowledge of human movement under stringent laws of competition to produce ever improving performances in this area of physical activity.

By reference to Plato (1963); Gardiner (1930); Robinson (1955); Umminger (1963); Kitto (1963); Poole & Poole (1963); Harris (1964); Schöbel (1965); information related to the ancient Greek games and to the place and development of movement in athletic events can be abstracted. Development, change and application of knowledge of human movement can be studied in the modern era of athletics by reference to such texts as Webster (1930; 1936; 1947); Doherty (1953); Pearson (1963); Quercetani (1964); Meyer (1964). Movement in the modern olympics story can be studied by reference to Abrahams & Crump (1951): Bares (1964); Bateman (1968). Lovesey & MacNab (1969) provide a comprehensive tabulation of athletics literature.

The place and utilisation of human movement in individual events can be examined in greater detail and evaluated by reference to such texts as Bresnahan & Tuttle (1950); Dyson (1951); Cerutty (1960); Agostini (1962); Pugh & Watts (1962); Hildreth (1963); Ward (1964); Brightwell & Packer (1965); Woodeson & Watts (1966); Marlow & Watts (1970).

Dance

Man has always danced and dance as with all the arts is a reflection of culture. The Egyptians were concerned with after-life, the Greeks with the origin of the universe, the Romans with realistic mental control, exactness and precision while in the medieval era dance was suppressed; man was then obsessed with life after death. To-day youth cut's out the external features. Movement is dictated by the rhythm of sound; the loudness 'isolates' the individual, verbal communication is impossible; while the darkness serves to intensify the feelings of the dancers. Dance is the movement medium for creative experience and expression (Sachs 1937). It is symbolic and when the observer's attitude is modified through participation in the dancers' world (NAPECW Report) greater significance and communication have been achieved (Kleinman, 1964). These criteria can be applied to all forms of dance; e.g. primitive, ethnic, social, balletic, contemporary and educational. A detailed study of movement as utilised in dance and the various types of dance is given by Sachs (1937) and for America by Shawn (1959) and Kraus (1969).

Primitive man was influenced in his dance by the dances and actions of animals; the movements he copied extended from simple hopping actions to indicate movement patterns, the latter often involving the whole body while in the former the limbs were predominant. The imageless and imitative dances reflected the lack of influence or otherwise of animal dances and actions; peoples not experienced in the observation of animal movements showing little zest for life in their dances while those tribes greatly influenced danced enthusiastically. Man also copied the actions of the animals as he believed these assisted him in his hunting activities.

Emmanuel (1916); Ginner (1933) and Lawler (1964) provide a history of Hellenic dance with comments on gestures, movements, patterns and rhythms typical of early and revived Greek dance. Reference is also made to dance in Greek literature and the information that can be gained from studying the patterns and friezes on Greek vases and ancient buildings. For the ancient Greek 'Dance was necessary for complete and harmonious development' (Socrates); it was used as a training for war—Pyrrhic dance—for festivals, to mark sacred occasions, for personification of the vine and to express the universality underlying all education.

A study of movement and movements characteristic of ethnic dance (Sorell 1966); can be made from an examination of Lawson (1953); Sharp (1934); Handbooks of European National Dances (1948); Farina (1967); Thurston (1954); Kurath (1950; 1960) and numerous others that cover the dances of particular countries.

The technique of 18th century dancing can be studied by reference to Rameau (1831); while Perugini (1946) provides a survey of dance in the theatre and the evolution of ballet (Haskell 1938; Reyna 1965).

A study of movement and costume is possible by reference to Kohler (1963); Grove (1904); Scott (1892); Viullier (1898). The Eurhythmics of Dalcroze (1912; 1930) was based on the utilisation and integration of the whole body (physical, intellectual, emotional) in the development of physical co-ordinations. Bodily movement was used as the reference point for the interpretation of musical symbols and freedom in expression stimulated and encouraged the creative impulse in learning. Eurhythmics provided the framework for release from stylised movement in dance. It was followed by natural movement (Morris, 1937; Wood & Cassidy, 1931); choral dancing or movement choir work (Russell, 1958) and American Modern Dance (Lloyd, 1949; Dreier, 1933; Terry, 1969).

Horst & Russell (1961) and Terry (1963) described Duncan (see also her own book) as a romantic genius who discarded traditional clothing for light, flowing garments and danced with bare feet to regain the contact 'with life charged earth' (or restated in practical terms the 'intimacy of the body' as the instrument of dance (Duncan 1965)).

Humphrey and Weidman (Horst & Russell, 1961) utilised the principle of fall and recovery; they studied dynamics and analysed rhythm. With this background knowledge the creative exploration of movement for dance was possible (Humphrey, 1939; 1959; 1962; 1966).

Wigman, (1966) exploited freedom of expression in dance and was much influenced by the work and theories of Laban.

But Nikolais (Kraus, 1969) achieves the spectacular by dehumanising his dancers into abstractions through the use of masks, costumes and stage lighting and therefore uses movement in a very different way. Cunningham (1968) eliminates overt human motivation and imaginatively experiments in the choreography of the dance.

Laban's (1963) work has also been incorporated into the educational sphere in Great Britain and into contemporary stage dance (Winearls, 1958). His sixteen basic themes (Laban, 1963) provided the foundation for a syllabus in creative dance experience in schools (Preston-Dunlop, 1963; Russell, 1958; Bruce & Tooke, 1966) and his 'new dance language' (Ringel 1965) born out of movement itself, provided the vocabulary for 'modern' theatrical dance.

Games

In games, movement is used as a means to an end. The goal is to win, to achieve the desired end superior strength, skill and knowledge are exploited in the competitive situation; movement is developed and then used to attain particular objectives (Gomme, 1964; Strutt, 1876).

The general theories of play and the place of movement in play can be studied by reference to Huizinga (1970); Caillois (1961) and Bowen & Mitchell (1923) (See also section O). McIntosh (1952) and May (1969) provide information related to individuals whose thinking influenced and was influenced by the 'athletics and games cult' of the British Public School system.

The modern approach to the coaching of players in movement and in movement situations in games and games playing has been described in such texts as Brown (1969); Jagger (1962); Moyes (1963); Pocock (1969); M.C.C. (1952); Young (1968); Browning (1955); Hockey Association (1966); Hickey (1962); Baggalay (1966); Thomas (1961); Stratford (1963); Morgan (1953); Metzler (1967); Trotter (1965); Mauldon & Redfern (1969); Taylor (1967).

Gymnastics

Although the Swedish Gymnastics of Ling (Chapman (1856); Broman (1899); Osterberg (1897); Roberts (1905); Thulin (1936); Bukh (1925) and Lindhard (1934)) was based on scientific principles it was mechanical, technical and artificial in terms of twentieth century gymnastics (Ministry of Education (1952; 1953); Morison (1969); Mauldon & Layson (1965); Pallett (1965); London County Council (1962)). The movements were restrictive and based on routine. Ling's aesthetic, educational, medical and military gymnastics of the nineteenth century have been transformed through rhythmical educational gymnastics to modern educational gymnastics (Board of Education Syllabuses 1904; 1909; 1919; 1921; 1927; 1933). This has led to a 'freer' use of movement in gymnastic exercises and situations. Gymnastics was used as a method of discipline (Maclaren, 1885; Alexander, 1894; 1910; Bear, 1920; Clias 1823). As such in both Swedish and British Gymnastics free standing movements were to command and made in unison while in apparatus work precise patterns were laid down and related to each vault. Olympic gymnastics has been developed as a competitive sport; movements have been stylised and criteria set by which judgement of performance in style and standard can be made (Kunzle, 1956; Dunn, 1969; Hughes, 1963 and Munrow, 1963). The influence of educational theory, child-centred education and learning by doing, coincidental with the stating of universal movement principles (Laban & Lawrence, 1947) led to a re-appraisal of the function and place of gymnastics in primary and secondary schools. Guided experiment and experience replaced dictated exercises while, as an aim, the development of the individual's movement potential superceded the gaining of a healthy, obedient body; Stanley (1969); Corlett (1970).

Swimming

Swimming may be regarded as a competitive sport, as a means of saving life and as mastery of human movement in water. It is in the latter context that examination of the use of physical skill must first be made; the former area can then be tackled. Aquatics (movement play) is still very experimental and no texts devoted solely to this aspect are known. The development of individual strokes can be traced by consulting Gabrielson, Spears & Gabrielson (1960); Pollard (1963); Counsilman (1968) and Colwin (1969). Armbruster, Allen & Billingsley (1948) and Carlile (1964) give detailed accounts of movements as used in diving. Rackham (1968) has provided information related to movement in synchronised swimming and Forsberg (1963) has written on the movement aspects of modern long distance swimming.

The study of movement in educational establishments

The development of studies in human movement is in its infancy. Muybridge (1955) and Metheny (1961) put forward a new concept of physical education based on meaningful movement experiences as the unique educational contribution of this area of the school curriculum. Henry (1965) critically appraised physical education as an academic discipline. Kenyon (1968) examined the conceptualisation of sub-disciplines. Cassidy (1954) tackled the problem of physical education in curriculum development while Mawdesley (1971) has looked at a conceptual analysis of human movement.

Brown (1967) has defined a structure of knowledge of human movement and Ludwig (1968) and Brown & Cassidy (1963) have enumerated the principles of movement education. Smith (1968) has viewed movement as the first form of communication with developing movement patterns reflecting the culture of a community.

Hinks, Archbutt & Curl (1971); have described the possible criteria for future curriculum development in movement studies in higher education.

N.11 Movement in the industrial situation

Much study of human movement in terms of occupational work capacity has been made, e.g. Laban & Lawrence (1947); Broer (1966; 1968) and McClurg Anderson (1951).

A historical comparison of conditions and actions can be made by reference to Hobsbawn (1964) and Harrison (1969). Past and present methods of assessment of human work capacity and the establishment of ideal labour conditions are best examined by reference to literature on time and motion studies (Gilbreth & Gilbreth, 1919; Mundel, 1950; Etzioni, 1970; Seymour, 1966; 1968 and Singleton, 1970).

Source Texts

BRAILSFORD D. (1969). "Sport and Society". London: Routledge & Kegan Paul.

BROWN C. & CASSIDY R. (1963). "Theory in Physical Education". Philadelphia: Lea & Febiger.

DIXON J.G., MACINTOSH P., MUNROW A.D., & WILLETTS R.F. (1957). "Landmarks in the History of Physical Education". London: Routledge & Kegan Paul.

HACKENSMITH C.W. (1966). "History of Physical Education". New York:

HALSEY E. (1964). "Inquiry and Invention in Physical Education". Philadelphia: Lea & Febiger.
LABAN R. (1958). "The Mastery of Movement". London: Macdonald & Evans.
LATCHAW M. & EGSTROM G. (1969). "Human Movement". New Jersey: Prentice Hall.
LEONARD F.E. & AFFLECK G.B. (1914;1947). "History of Physical Education". Philadelphia: Lea & Febiger.
McINTOSH P.C. (1952). "Physical Education in England Since 1800". London: Bell.
McINTOSH P.C. (1963). "Sport in Society". London: Watts.
OBERTEUFFER D. & ULRICH C. (1970). "Physical Education". New York: Harper & Row.
ROPER R.E. (1938). "Movement and Thought". Glasgow: Blackie.
SMITH M. (1968). "Introduction to Human Movement". London: Addison Wesley.
STANLEY S. (1969). "Physical Education—A Movement Orientation". New York: McGraw Hill.
SWEENEY A.T. (Ed.) (1970). "Selected Readings in Movement Education". London: Addison Wesley.
TREVES F. (1892). "Physical Education". London: Churchill.
VAN DALEN D.B., MITCHELL E.D. & BENNETT B. (1953). "A World History of Physical Education". London: Prentice Hall.
WOOD T.D. & CASSIDY R.F. (1931). "The New Physical Education". New York: MacMillan.

Specific References

ABRAHAMS H. & CRUMP J. (1951). "Athletics". London: Naldrett.
AGOSTINI M. (1962). "Sprinting". London: Paul.
ALEXANDER A. (1894). "Physical Drill of all Nations". Liverpool: Philip.
ALEXANDER A. (1910). "British Physical Education for Girls". Edinburgh: McDougall.
ALLAN D.J. (1970). "The Philosophy of Aristotle". Oxford: University Press.
ARGYLE M. (1969). "Social Interaction". London: Methuen.
ARMBRUSTER D.A., ALLEN R.H. & BILLINGSLEY H.S. (1948) "Swimming and Diving". London: Kaye.
ARNOLD-BROWN A. (1962). "Unfolding Character". London: Routledge & Kegan Paul.
BAGGALAY J. (1966). "Netball for Schools". London: Pelham.
BARCS S. (1964). "The Modern Olympics Story". London: Corvina.
BATEMAN R. (1968). "The Book of the Olympic Games". London: Paul.
BEAR B.E. (1920). "The British System of Physical Education". London: Bell.
BIRDWHISTELL R.L. (1952). "Introduction to Kinesics". Louisville.
BIRDWHISTELL R.L. (1968). Kinesics. In D. Sills (Ed.) "International Encyclopaedia of Social Science". New York: Macmillan.

BOARD OF EDUCATION: (1904). "Syllabus of Physical Education for use in Public Elementary Schools". London: H.M.S.O. (1909). "The Syllabus of Physical Exercise for Public Elementary Schools". London: H.M.S.O. (1919). "Syllabus of Physical Training for Schools". London: H.M.S.O. (1921). "Syllabus of Instruction in Physical Training for Training Colleges". London: H.M.S.O. (1927). "Syllabus of Physical Training Supplement for Older Girls". "Supplement for Boys". London: H.M.S.O.
(1933). "Syllabus of Physical Training for Schools". London: H.M.S.O.

BONNET H. & MAUROIS G. (1964). "Ski—The Experts Way". London: Newnes.

BOWEN W.P. & MITCHELL E.D. (1923). "The Theory of Organised Play". New York: Baynes.

BRESNAHAN G.T. & TUTTLE W.W. (1950). "Track and Field Athletics". Philadelphia: Lea & Febiger.

BRIGHTWELL R. & PACKER A. (1965). "Sprints, Middle Distance and Relay Running". London: Kaye.

BROER M.R. (1966). "Efficiency in Human Movement". London: Saunders.

BROER M.R. (1968). "An Introduction to Kinesiology". New Jersey: Prentice Hall.

BROMAN A. (1899). "School Gymnastics on the Swedish System". London: Daniel.

BROWN C. (1967). The structure of knowledge of physical education. *Quest.*-Mono **9**,54-67.

BROWN E. (1969). "Badminton". London: Faber & Faber.

BROWN R. & CASSIDY R. (1963). "Theory in Physical Education". Philadelphia: Lea & Febiger.

BROWNING R.H.K. (1955). "A History of Golf". London: Dent.

BRUCE V. & TOOKE J. (1966). "Lord of the Dance". Oxford: Pergamon.

BUKH N. (1925). "Primary Gymnastics". London: Methuen.

BUTLER P. & K. (1968). "Judo and Self Defence for Women and Girls". London: Faber & Faber.

BUYTENDIJK (1956). "Allgemeine Theorie der Menschichen Haltung und Bewegung". Berlin/Gottingen/Heidelberg.

CAILLOIS R. (1961). "Man, Play and Games". New York: Thames & Hudson.

CARLILE F. (1964). "On Swimming". London: Pelham.

CARLQUIST M. (1955). Rhythmical Gymnastics.

CARROLL J. (1971). Monograph 1. A Survey of some research findings from the Social Sciences that relate to movement behaviour. Anstey College of Ph. Ed.

CASSIDY R. (1954). 'Curricular Development in Physical Education". New York: Harper Row.

CERUTTY P.W. (1960). "Athletics". London: Stanley Paul.

CHAPMAN M.J. (1856). "Ling's Educational and Curative Exercises". London: Ballière.

CLIAS P.H. (1823). "An Elementary Course of Gymnastic Exercise". Sherwood.

COLWIN C. (1969). "On Swimming". London: Pelham.

CORNFORD F.M. (1935). "Plato's Theory of Knowledge". London: Routledge & Kegan Paul.

CORLETT H. (1970). Modern educational gymnastics today. *Brit. J. Phys. Educ.,* **1, 2 and 3.** 36-37, 63-66.

COUNSILMAN J.E. (1968). "The Science of Swimming". London: Pelham.

CUNNINGHAM M. (1968). "Dance Perspectives No. 34". London: Murray.

DALCROZE J. (1912). "Eurhythmics". Edinburgh: Constable.

DALCROZE E. (1930). "Eurythmics, Art and Education". London: Chatto & Windus.

DARWIN C. (1872). "The Expression of the Emotions in Men and Animals". London: Murray.

DARWIN C. (1958). "The Origin of the Species". Un. of California: Mentor.

DENIS ST. R. "An Unfinished Life". London: Harrap.

DISLEY J. (1967). "Orienteering". London: Faber & Faber.

DIXON J.G., McINTOSH P., MUNROW A.D. & WILLETTS R.F. (1957). "Landmarks in the History of Physical Education". London: Routledge & Kegan Paul.

DOHERTY K. (1953). "Modern Track and Field". New Jersey: Prentice Hall.

DREIER K.S. (1933). "Shawn the Dancer". London: Dent.

DUNCAN I. (1965). "Duncan Dancer". Middleton: Wesleyan.

DUNN W.G. (1969). "Olympic Gymnastics for Schools". London: Pelham.

DYSON H.G. (1962). "The Mechanics of Athletics". London: University Press.

DYSON H.G. (1951). "Athletics for Schools". London: University Press.

EDEL A. (1967). "Aristotle. Life and Work and Selections from Writings". New York: Dell.

EMMANUEL M. (1916). "The Antique Greek Dance". London: Bodley Head.

ETZIONI A. (1964). "Modern Organisations". New Jersey: Prentice Hall.

ETZIONI A. (1970). "A Sociological Reader on Complex Organisations". London: Holt, Rinehart & Winston.

FARINA X. (1967). "Classic Dances of the Orient". New York: Crown.

FELDMAN S. (1959). "Mannerisms of Speech and Gestures in Everyday Life". New York: International University Press.

FORSBERG G. (1963). "Modern Long Distance Swimming". London: Routledge & Kegan Paul.

GABRIELSEN M.A., SPEARS B. & GABRIELSEN B.W. (1960). "Aquatics Handbook". New Jersey: Prentice Hall.

GARDINER E.N. (1930). "Athletics of the Ancient World". Oxford: Clarendon.

GILBRETH F.B. & L.M. (1919). "Applied Motion Study: a collection of papers on the efficient method to industrial preparedness". New York: Macmillan.

GINNER R. (1933). "The Revived Greek Dance". London: Methuen.

GOMME A.B. (1964). "The Traditional Games of England, Scotland and Ireland. Vol. 1 & 2". New York: Dover.

GROVE L. (1904). "Dancing". London: Longmans, Green & Co.

HACKENSMITH C.W. (1966). "History of Physical Education". New York: Harper & Row.

HALL E.T. (1959). "The Silent Language". New York: Premier Books.

HANDBOOKS OF EUROPEAN NATIONAL DANCES (1948). London: Parrish.

HARRIS H.A. (1964). "Greek Athletes and Athletics". London: Hutchinson.

HARRISON J.F.C. (1969). "Robert Owen and The Owenites in Britain and America". London: Routledge & Kegan Paul.

HASKELL A. (1938). "Ballet". Harmondsworth: Pelican.
H'DOUBLER M.N. (1966). "Dance—A Creative Art Experience". Wisconsin: University Press.
HEIDEMANN I. (1965). Philosophy of movement. In "The Adolescents of Today". Cologne: I.A.P.E.S.G.W.
HENRY F. (1965). Physical education—an academic discipline. *London: P.E.A. Leaflet* Jan-Feb.
HICKEY M. (1962). "Hockey for Women". London: Kaye.
HILDRETH P. (1963). "Athletics—How to Win". London: Heinemann.
HINKS E., ARCHBUTT S., CURL C. (1971). Movement Studies—A New Standing Committee. *Bulletin U. L. Inst. Educ.* New Series No. **23**,. 4-10.
HOBSBAWN E.J. (1964). "Labouring Men". London: Weidenfeld & Nicholson.
HOCKEY ASSOCIATION (1966). "Hockey Coaching". London: Hodder & Stoughton.
HORST & RUSSELL (1961). "Modern Dance Forms". New York: Dance Horizons.
HUGHES E. (1963). "Gymnastics for Girls". New York: Ronald.
HUIZINGA J. (1970). "Homo Ludens". London: Granada Paladin.
HUMPHREY D. (1939). "An Unfinished Life". New York: Harper.
HUMPHREY D. (1959). "The Art of Making Dances". London: Rinehart & Co.
HUMPHREY D. (1962). "Dance Observer". London: Rinehart & Co.
HUMPHREY D. (1966). "New Dance" (Autobiography). London: Rinehart & Co.
JAGGER B. (1962). "Basketball—Coaching and Playing". London: Faber & Faber.
KENYON G.S. (1968). A Conceptual model for characterizing physical activity. *Res. Quart.,* **39, 1,** 96-105.
KITTO H.D.F. (1963). "The Greeks". London: Pelican.
KLAGES L. (1929). "The Science of Character". London: Allen & Unwin.
KLEINMAN S. (1964). The significance of human movement: a phenomenological approach. *N.A.P.E.C.W. Rep.,* Washington D.C.
KNUST A. (1958). "Handbook of Kinetography Laban". Hamburg: Darstang.
KOHLER C. (1963). "A History of Costume". New York: Dover.
KRAUS R. (1969). "History of the Dance". New Jersey: Prentice Hall.
KUNZLE G.C. (1956). "Olympic Gymnastics—Freestanding". London: Barrie.
KURATH G. (1960). Panorama of dance ethnology. *Current Anthrop. Mogy.* **1**, 3.
KURATH G. (1950). A new method of choreographic notation. *Amer. Anthrop.* **52.**
LABAN R. (1956). "Principles of Dance—Movement Notation". London: Macdonald & Evans.
LABAN R. (1958). "The Mastery of Movement". London: MacDonald & Evans.
LABAN R. (1963). "Modern Educational Dance". London: Macdonald & Evans.
LABAN R. (1966). "Choreutics". London: Macdonald & Evans.
LABAN R. & LAWRENCE (1947). "Effort". London: MacDonald & Evans.
LAMB W. (1965). "Posture and Gesture". London: Duckworth.
LANGER S. (1957). "Problems of Art". New York: Scribner.
LANGER S. (1967). "Feeling and Form". London: Routledge & Kegan Paul.
LAWLER B. (1964). "The Dance in Ancient Greece". Wesleyan University Press.
LAWSON J. (1953). "European Folk Dance". London: Pitman.

LAYSON J. (1969). An introduction to some aspects of the sociology of physical education. In Association of Principals of Women's Colleges of Physical Education Conference Report.

LEONARD F. & AFFLECK G. (1947). "The History of Physical Education". London: Kimpton.

LINDHARD J. (1934). "The Theory of Gymnastics". London: Methuen.

LLOYD M. (1949). "The Borzi Book of Modern Dance". New York: Dance Horizons.

L.C.C. (1962). "Educational Gymnastics". L.C.C. London.

L.C.C. (1964). "Movement Education for Infants". L.C.C. London.

LOVESEY P. & McNAB T. (1969). "The Guide to British Track and Field Literature 1275-1968". London: Athletics Arena.

LUDWIG E.A. (1968). Toward an understanding of basic movement education in the elementary schools. *J. Health. Phys. Educ. Rec.,* **39,** 3, 28.

MACLAREN A. (1885). "A System of Physical Education Theoretical and Practical". Oxford: Clarendon.

MARLOW B. & WATTS D. (1970). "Track Athletics". London: Pelham.

M.C.C. (1952). "Cricket Coaching Book". London: Heinemann.

MAULDON E. & LAYSON J. (1965). "Teaching Gymnastics". London: Macdonald & Evans.

MAULDON E. & REDFERN H.B. (1969). "Games Teaching". London: Macdonald & Evans.

MAWDSLEY H.P. (1971). A conceptual analysis of human movement. *Bull. Phys. Educ.* **Vol. VIII,** No. 5.

MAY J. (1969). "Madame Bergman Osterberg". London: Harrap.

METHENY E. (1965). "Connotations of Movement and Sport in Dance". Iowa: Brown.

METHENY E. (1961). The unique meaning inherent in human movement. *Phys. Educ.*

METZLER P. (1967). "Advanced Tennis". Sydney: Angus & Robertson.

MEYER H.A. (1964). "Modern Athletics". London: Oxford University Press.

MINISTRY OF EDUCATION (1952). "Moving and Growing". London: H.M.S.O.

MINISTRY OF EDUCATION (1953). "Planning the Programme". London: H.M.S.O.

MORGAN J. (1953). "Squash Rackets for Women". London: Sporting.

MORISON R. (1956). "Educational Gymnastics". (Private)

MORISON R. (1960). Education Gymnastics for Secondary Schools.

MORISON R. (1969). "A Movement Approach to Educational Gymnastics". London: Dent.

MORRIS M. (1937). "Basic Physical Training". London: Heinemann.

MOYES A.G. (1963). "The Changing Face of Cricket". Sydney: Angus & Robertson.

MUNDEL M.E. (1950). "Motion and Time Study". London: Prentice Hall.

MUNROW A.D. (1963). "Pure and Applied Gymnastics". London: Arnold.

MURRAY W.H. & WRIGHT J.E.B. (1964). "The Craft of Climbing". London: Kaye.

MUYBRIDGE E. (1955). "The Human Figure in Motion". New York: Dover.

McCLURG ANDERSON T. (1951). "Human Kinetics and Analysing Body Movements". London: Heinemann.

McINTOSH P.C. (1952). "Physical Education in England Since 1800". London: Bell.

McINTOSH P.C. (1963). "Sport in Society". London: Watts.

NORTH M. (1971). "Personality Assessment Through Movement". London: Macdonald & Evans.

NOVERRE J. (1930). "Letters on Dancing and Ballets". New York: Dance Horizons.

OSTERBERG M.B. (1887, 1897, 1927). "Gymnastic Tables". London: William Andrews.

PALLETT D. (1965). "Modern Educational Gymnastics". Oxford: Pergamon.

PEARSON G.F.D. (1963). "Athletics". Edinburgh: Nelson.

PERUGINI (1946). "Mark Edward". Norwich: Jarrolds.

PIAGET J. (1951). "Play, Dreams and Imitation in Childhood". London: Routledge & Kegan Paul.

PLATO (1960). "The Laws". London: Dent.

PLATO (1963). "The Republic". London: Penguin Classic.

POCOCK P. (1969). "Bowling". London: Batsford.

POLLARD J. (1963). "Swimming". London: Kaye.

POOLE L. & G. (1963). "History of the Ancient Olympic Games". London: Vision.

PRESTON-DUNLOP V. (1963). "Handbook—Modern Educational Dance". London: Macdonald & Evans.

PRESTON-DUNLOP V. (1969). "Practical—Kinetography". London: Macdonald & Evans.

PRESTON-DUNLOP V. (1969). A notation system for recording observable motion. *Int. J. Man-Machine Studies,* **1**, 361-386.

PUGH D.L. & WATTS D.C.V. (1962). "Athletics for Women". London: Paul.

QUERCETANI R.L. (1964). "A World History of Track and Field Athletics". Oxford: University Press.

RACKHAM G. (1968). "Synchronized Swimming". London: Faber & Faber.

RAMEAU P. (1831). "The Dancing Master". Beaumont.

REID L.A. (1969). "Meaning in the Arts". London: Allen & Unwin.

REID L.A. (1970). Movement and Meaning *L.A.M.G. Mag.* **45,** 5-31.

REY J. "The World of Dance". Artici.

REYNA F. (1965). "A Concise History of Ballet". London: Thames & Hudson.

RICE E.A. & HUTCHINSON J.L. (1952). "A Brief History of Physical Education". New York: Barnes.

RINGEL F. (1965). In "Dance Scope. Effort: A Synthesis of Movement". U.S.A.: Fall.

ROBERTS E.A. (1905). "A Handbook of Free Standing Gymnastics".

ROBINSON R.S. (1955). "Sources for the History of Greek Athletics". Illinois: University Press.

ROPER R.E. (1938). "Movement and Thought". London: Blackie.

ROSS D. (1951). "Plato's Theory of Ideas". Oxford: Clarendon.

RUSSELL J. (1958). "Modern Dance in Education". London: Macdonald & Evans.
SACHS C. (1937). "World History of Dance". London: Norton.
SCHOBEL H. (1965). "The Ancient Olympic Games". London: Studio Vista.
SCOTT E. (1892). "Dancing as an Art and Pastime". London: Bell.
SCOTT M.G. (1947). "Analysis of Human Motion". New York: Crofts.
SEYMOUR W.D. (1966). "Industrial Skills". London: Pitman.
SEYMOUR W.D. (1968). "Skills Analysis Training". London: Pitman.
SHARP J.C. (Ed.) (1934). KARPELES M. "The Country Dance Books". London: Novello.
SHAWN T. (1959). "Thirty-Three Years of American Dance (1927-59) and the American Ballet". Pittsfield: Nair.
SHAWN T. (1963). "Every Little Movement". New York: Dance Horizons.
SINGLETON W.T. (1970). "Measurement of Man at Work". London: Taylor & Francis.
SKILLING B. (1963). "Basic Canoeing". London: Foulsham.
SLUSHER H.S. (1967). "Man, Sport and Existance". Philadelphia: Lea & Febiger.
SORELL W. (1966). "The Dance Has Many Faces". New York: Columbia.
STRATFORD R.B. (1963). "Netball". London: Educational.
STREICHER M. (1970). "Reshaping Physical Education". Manchester: University Press.
STRUTT J. (1876). "The Sports and Pastimes of the English People". London: Chatto & Windus.
TAYLOR E. (1967). "Woman's Hockey—Do It This Way". London: Murray.
TAYLOR FREDERICK W. (1911). "Scientific Management". Harper: New York.
TERRY W. (1963). "Isadora Duncan".
TERRY W. (1969). "Miss Ruth". New York: Dodd, Mead & Company.
THOMAS M. (1961). "Tackle Netball This Way". London: Paul.
THORNTON S. (1970). "Movement Perspective of Rudolf Laban". London: Macdonald & Evans.
THULIN J.G. "Manual of Gymnastics". Part I—Atlas.
THULIN J.G. "Manual of Gymnastics". Part II—Manual.
THULIN J.G. "Manual of Gymnastics". Part III—for little children.
THULIN J.G. (1936). "The Aims and Methods of Swedish Ling Gymnastics of the Present Day". Lasstidningens Tryckeri.
THURSTON H.A. (1954). "Scotland's Dances". London: Bell.
TRICKER R.A.R. & TRICKER B.J.K. (1967). "Science of Movement". London: Mills & Boon.
TROTTER B.J. (1965). "Volleyball for Girls and Women". New York: Ronald.
UMMINGER W. (1963). "Supermen, Heroes and Gods". London: Thames & Hudson.
VAN DALEN D.B., MITCHELL E.D. & BENNETT B. (1953). "A World History of Physical Education". New Jersey: Prentice Hall.
VAUGHAN T. (1970). "Science and Sport". London: Faber & Faber.
VUILLIER G. (1898). "A History of Dancing". London: Heinemann.
WARD A. (1964). "Modern Distance Running". London: Paul.
WEBSTER F.A.M. (1930). "Athletics of Today for Women". London: Warne.
WEBSTER F.A.M. (1936). "The Science of Athletics". London: Kaye.

WEBSTER F.A.M. (1947). "Great Moments in Athletics". London: Country Life.
WIGMAN M. (1966). "The Language of Dance". London: Macdonald & Evans.
WINEARLS J. (1958). "The Jooss–Leeder Method". London: Black.
WOODESON P.J. & WATTS D.C.V. (1966). "Schoolgirl Athletics". London: Paul.
YOUNG P.M. (1968). "A History of British Football". London: Paul.

Bibliography

AGNIEL M. (1931). "The Art of the Body". London: Batsford.
ARISTOTLE "Writings on Physics. Nature and the natural–Book II; The nature of motion–Book III; What place is–Book IV; Fifth element–movement is circular–Book I. Oxford: Clarendon Press.
ARMITAGE M. (1966). "Martha Graham". New York: Dance Horizons.
A.T.C.D.E. P.E. SECT. (1970). "Philosophy of Physical Education". B.W. Jelfs, The concept of leisure; I.M. Webb, The concept of physical education.
BADMINTON LIBRARY OF SPORTS AND PASTIMES (1904). (24 vols) London: Longmans, Green & Company.
BARRETT K. et al (1969) "Foundations for Movement". Iowa: Brown.
BJORKSTEN E. (1937, 1934). "Principles of Gymnastics for Women and Girls. Vols 1 & 11". London: Churchill.
BOYD M. (1959). "Lacrosse–Playing and Coaching". London: Kaye & Ward.
BRIGHTBILL C.K. (1960). "The Challenge of Leisure". New Jersey: Prentice Hall.
BRIGHTBILL C.K. (1961). "Man and Leisure". New Jersey: Prentice Hall.
BROWN R.C. & CRATTY G.J. (Eds.) (1969). "New Perspectives of Man in Action". New Jersey: Prentice Hall.
CASSIRER E. (1944). "An Essay on Man". Yale: University Press.
CASSIRER E. (1953). "The Philosophy of Symbolic Forms". Yale: University Press.
CORBIN H.D. (1959). "Recreation Leadership". New Jersey: Prentice Hall.
CORNFORD F.M. (1945). "The Republic of Plato". London: Oxford University Press.
COLLINGWOOD R.G. (1938). "The Principles of Art". Oxford: Clarendon.
CRATTY B.J. (1964). "Movement, Behaviour and Motor Learning". Philadelphia: Lea & Febiger.
CRATTY B.J. (1967). "Social Dimensions of Physical Activity". New Jersey: Prentice Hall.
CRATTY B.J. (1968). "Psychology and Physical Activity". New Jersey: Prentice Hall.
CROSNIER R. (1951). "Fencing with the Foil". London: Faber & Faber.
DANCE PERSPECTIVES (1966). No. 25–Doris Humphry; (1968). No. 34–Merce Cunningham; (1970). No. 41–Mary Wigman. London: Murray.
DARWIN C. (1871). "The Descent of Man". London: Murray.
DARWIN C. (1952). "The Next Million Years". London: Davis.
DISLEY J. (1962). "The Young Athletes Companion". London: Souvenir.
DUNCAN ISADORA (1928). "My Life". London: Gollancz.

EAVES G. (1969). "Diving". London: Kay & Ward.

EFRON D. (1941). "Gesture and Environment". New York: Kings Crown Press.

ELLIS H. (1923). "Dance of Life". London: Constable.

ERGONOMICS Vol. 1-Vol. 13 (1957-70). London: Taylor & Francis.

FELSHIN J. (1965). "Perspectives and Principles for Physical Education". New York: Wiley.

FOCUS ON DANCE II (1962). Report: "Theories of Movement". A.A.H.P.E.R. Washington.

GEMSCH N. & JULEN A. (1963). "Modern Ski-ing". London: Cassell.

GEORGII C.A. (1854). "Biographical Sketch of Per Henrik Ling". London: Baillière.

GOFFMAN E. (1956). "The Presentation of Self in Everyday Life". Edinburgh: University Press.

HAWKEY R.B. (1966). "Your Book of Squash". London: Faber & Faber.

HOLMSTROM A. (1949). "Swedish Gymnastics Today". Hohlmams Forlga.

HORNE D.E. (1970). "Trampolining". London: Faber and Faber.

JACKS L.P. (1931). "Education of the Whole Man". London: University Press.

JACKS L.P. (1938). "Physical Education". London: Nelson.

JAMES J.M. (1967). "Education and Physical Education". London: Bell.

JOHNSON T. (1909). "Swedish System of Physical Education". Bristol: Wright.

JONES E. (1931). "The Elements of Figure Skating". London: Allen & Unwin.

JORDAN D. (1966). "Childhood and Movement". London: Blackwell.

KNAPP B. (1963). "Skill in Sport". London: Routledge & Kegan Paul.

KURATH G. (1954). A basic vocabulary for ethnic dance descriptions. *Amer. Anthrop.,* **56.**

KUWASIMA T.S. & WELCH A.R. (1949). "Judo". London: Putnam & Company.

LA MERI (1941). "The Gesture Language of the Hindu Dance". New York: Blom.

de LASPEE H. (1856). "Calisthenics". London: Dalton.

LEATHERMAN L. (1961). "Martha Graham". London: Faber & Faber.

LLOYD-JONES H. (1962). "The Greeks". London: Watts.

LOFVING C. (1882). P.E. and its place in a system of rational education. Swann Sonnenschein.

METHENY E. (1968). "Movement and Meaning". New York: McGraw Hill.

MULAC M.E. (1961). "Leisure-time for Living and Retirement". New York: Harper.

McCAMERON W.D. & PLEASANCE P. (1963). "Education in Movement". London: Blackwell.

N.A.P.E.C.W. Report (1962-64). Aesthetic and human movement. H.P. Alkire, Making a dance — 107-111; W. Aschaffenburg, Creating the art object—music — 104-107; D.W. Ecker, Research in creative activity — 113-122; E.F. Kaelin, Being in the body — 84-103; S. Kleinman, The significance of human movement: a phenomenological approach — 123-8: H.S. Slusher, The essential function of P.E. — 129-135.

NETTLESHIP R.L. (1935). "The Theory of Education in Plato's Republic". Oxford: University Press.

NEUMEYER M.H. & NEUMEYER E.S. (1958). "Leisure and Recreation". New York: Ronald.

PATERSON A. & HALLBERG E.C. (1965). "Background Readings for Physical Education". London: Holt.

PIEPER J. (1952). "Leisure the Basis of Culture". New York: Pantheon Books.

RANDALL M.W. (1967). "Modern Ideas of Physical Education". London: Bell.

INTERNATIONAL OLYMPIC ACADEMY (1963). Report on 3rd Session. O. Broweer, The Isthmian games — 180-199; A. Brundage, Olympic philosophy — 29-39; W. Kerbs, The "gymnasium" — 164-79; P.C. McIntosh, Fitness or prowess — 89-101; P.C. McIntosh, Sport in physical education — 101-116; O. Misangyi, Sport and the student — 127-144; C. Palaeologes, Famed athletes of ancient Greece — 153-163; B. Zaulin, Laws of sport — 47-62.

RICE E.A. (1927). "A Brief History of Physical Education". New York: Barnes.

ROBERTS S. & SUTTON-SMITH P. (1962). Child training and game involvement. *Ethnology*. **1**. 166-85.

SANBORN M.A. & HARTMAN B.G. (1970). "Issues in Physical Education". Philadelphia: Lea & Febiger.

SHEETS M. (1966). "The Phenomenology of Dance". University of Wisconsin: Madison & Milwaukee.

SOLTIS J.F. (1968). "An Introduction to the Analysis of Educational Concepts". London: Addison-Wesley.

SWAIN & LEMAISTRE (1964). "Fundamentals of Physical Education". Novak: Sydney.

TIBBLE J.W. (1952). "Studies in Education: physical education and the educative process". London: Evans.

TOWNSEND H.A. & BURKE P.J. (1962). "Learning for Teachers". New York: Macmillan.

WALKER D. (1836). "Exercise for Ladies". London: Hurst.

WALKER D. (1843). "Manly Exercises". London: Hurst.

WALKER D. (1837). "Sports and Games". London: Hurst.

WALTER C.V. (1966). "The Theory and Practice of Hockey". Wellington: Reed.

WEBSTER T.B.L. (1969). "Everyday Life in Classical Athens". London: Batsford.

WILLIAMS J.F. (1964). "The Principles of Physical Education". Philadelphia: Saunders.

TOPIC AREAS - DEVELOPED

PLAY

by A.G. ROCHE

General

Such is the diversity of usage of the term 'play' that no single definition embraces the concept in all its forms. The fifty-three lengthy definitions expounded in The Oxford English Dictionary (1933) do little to crystallise and clarify the term. The present section represents an attempt to outline those aspects of the concept which have human movement as their central point.

0.1. Definitions and concepts

It is clear that many early philosophers concerned themselves with play. Rahner (1965) offers a theological expostion of the concept of eutrapelia which is based on the Thomist philosophy of man at play. Aristotle argued that the intelligent disposition of the civilised man is evidenced by his ability to pause and find amusement in life (Chase, 1911). Lodge (1947) expounds the seemingly more purposeful maturational and therapeutic values of play as expressed by Plato, and Jung & Kerenyi (1949) show how primitive notions such as the relationship between body and soul found expression in the sacred activities of play and dance.

In earlier times the common idea that play was wasteful or even wrong reflected religious convictions and has been modified only gradually, partly through the efforts of Froebel (1887) and Dewey (1959). By the turn of the century it was deemed necessary to nurture the perfectly normal tendency to play as leisure time activity was realised to be a potential educational agent.

Huizinga (1949) attempted to explore the extent to which culture embodies the character of play, largely from a philosophical rather than a psychological viewpoint. Millar (1968) does much to restore the balance by relating the seemingly paradoxical behavioural ingredients of play to those biological functions associated with ontogenesis.

The common idea that play is a relaxation from work presupposes the existence of a definite rift between the concepts of *work* and *play*. Support for this argument is provided by Dearden (1968) who sees a distinction between work and play even for the pre-school child. Caillois (1961) believes that the concepts are often indistinguishable in childhood but that a sharp division between them exists in adulthood. Play he defines as an activity which is free, uncertain, unproductive, governed by rules and, as echoed by Beneveniste (1947) and Dearden (1968), contains predominant elements of unreality. Others have commented on the inseparability of work and play (e.g., Froebel, 1887; Gulick, 1920; Jacks, 1931; Grey, 1964). For instance, Jacks argues that the dichotomy would cease to be once work was conducted in the spirit of play. This brings to mind Shaw's definition of an educational utopia where 'work was play and play was life'.

0.2. Some theories of play relevant to human movement.

One of the oldest and simplest explanations of play is the Surplus Energy Theory first postulated by Schiller in 1759, (see Schiller 1954) and elaborated by Spencer (1872). Play was regarded as an outlet for superfluous energy. Clearest support for this theory comes from the spontaneous play of animals which may assume an incidental functional role in their interaction with the environment (Kohier, 1925). Despite flaws (see Griffiths, 1935; Millar 1968), the notion of some kind of energy theory persists, and indeed has been utilised by some as a rationale for physical education.

Hall's (1915) Darwinian-inspired Recapitulation Theory, with its detailed account of the content of play stimulated interest in this area. However, few authorities today would support its thesis that ontogeny repeats phylogeny, that children are a link in the evolutionary chain from animals to men.

The Instinct-Practice Theory of Groos (1901) which stated that play is a preparation for adult life, established a teleological significance for play with respect to growth and education. Woodward's (1966) comment that the theory does not explain the random play of the young is invalid if this undifferentiated activity is regarded as functional in a developmental context.

Froebel's (1887) metaphysical notions concerning play markedly contrast with those of Hall and Groos. He saw play as a pure spiritual activity essential for the individual's harmonious development (see Lilley, 1967). This idea gained notable support from Dewey (1959), with broadly similar views being expressed by Huizinga (1949).

Play can be explained in terms of the Pleasure-Principle of Freud (1955) which states that man is motivated to seek pleasurable satisfaction and avoid painful experiences. The fact that a child tends to repeat play patterns until they are mastered can be accounted for by his Repetition-Compulsion theory. In Freudian terms, the initiation of any playful behaviour stems from the Libido, a basic instinct which forms the source of all motivation. Modernists of the Freudian school who make important contributions to the theory of play include Lowenfeld (1935), Gesell & Ilg (1946), Isaacs (1945, 1965), Gurland (1960), and Rycroft (1961).

Piaget (1951)—who regards play as an example of pure 'assimilation'—conceives of three developmental stages in play behaviour, namely; practice games, symbolic games and games with rules. A similar classification is proposed by Buhler (1935).

Learning theorists have also made a contribution to the theory of play. The influential theory of Hull (1943) was based on the two assumptions that behaviour was motivated by primary drives, and that drive reducing rewards mediated all learning. Play is persisted with because of reinforcement from a multiplicity of inocuous secondary drives which have themselves been incidentally established during the reduction of primary drives (Miller 1951). Skinner demonstrated that persistence of behaviour was a function of the rate and timing of reinforcement during learning (Skinner 1953; Ferster & Skinner 1957). On this basis, it seems that much of the play of the young child may be directly related to the presence of intermittent praise from parents and other adults.

These selected original theories are all relevant to the various facets of play manifested in terms of human movement, although they are generally abstract in nature. Few theories consider the individual's reasons (those motives of which he is aware) for his play. Such motives are especially obscure with regard to the young child who cannot communicate his thoughts, hence rendering false conclusions based on observation of his overt behaviour more likely (Millar 1968). A useful book of readings dealing with the history of play over the last fifty years is provided by Herron & Sutton-Smith (1971).

O.3 Social Play

Piaget (1951) observes that play allows the child to come to terms with the world, thereby acknowledging the link between play and socialisation. This relationship constitutes one continuous series of interactions (Helanko 1958). Its social aspects are also highlighted by Miller & Dollard (1941) who discuss it with reference to imitational behaviour.

Cowell (1960) stresses the physical element of play in social adjustment, whereas Hartley et al (1952) show how dramatic play helps in the social integration of the troubled child. McKinney (1949) notes the benefits emerging from socially-oriented play which include increased social poise, spontaneity, independence, friendships, popularity and leadership. It appears from the work of Sutton-Smith et al (1967) that with regard to competitive games, the benefits which accrue are a function of the individual's success. They found personality differences between winners and losers and between winners and those who draw.

Implicit in the present context is play's therapeutic value, which has been recognized by many workers (e.g. Klein, 1949 and Freud, 1969).

Play has been classified developmentally according to its social implications— solitary play, looking-on play, parallel play, associative play and co-operative play (Sandstrom 1968).

Whiting (1963) and Giddens (1964), while recognizing cross-cultural differences in both the informal and formal play of children, note common behavioural patterns which transcend such boundaries.

The relationship between play and more organized games and sports is also of importance. Huizinga (1949), who argues that organised sport is not pervaded with the real spirit of play, finds support from Ulrich (1968) and Koestler (1964). Man's involvement in sport is explored by Slusher (1967), whilst Berne (1964) outlines the kind of specific skills required for the successful participation in adult games. Millar (1968) notes that unorganised, spontaneous play is less in evidence in adulthood, partly because it is associated with immaturity.

A significant classification of games has been provided by Caillois (1961) who divides them into games of chance, simulation and vertigo, the last category being of the most interest to the student of human movement. Other definitions and classifications are proposed by Huizinga (1949), Chateau (1955) and Reeves (1957).

The part that play should assume in institutionalised education of the child provokes constant argument. The 'play-way' theory in education is based upon the assumption that play and games can facilitate the learning process (Bruner 1965; Meier & Duke 1966). Both Almy (1968) and Dearden (1968) support play-oriented nursery school programmes, although the latter questions the true educational value of play. Among those who have stressed play's educational merits are Isaacs (1945) and Scarfe (1962).

O.4 Motor Performance

Play is one of the main ways in which children express their growth. Because of the established importance of early movement, which can be regarded as play after about the fourth month of life, there is a need to channel this natural tendency to play towards developmentally desirable ends. That motor development is benefited by play is acknowledged by Millar (1968).

Under the directorship of Ellis, reports on the ecology of play and gross motor behaviour are becoming available from the Motor Performance and Play Research Laboratory, University of Illinois, where an attempt is being made to evolve an organismic approach to the investigation of child development (Ellis et al, 1970).

Specific references

ALMY M. (1968). Play and education. In J.L. Frost (Ed.) "Early Childhood Education Revisited". New York: Holt, Rinehart & Winston.

BENVENISTE E. (1947). Le jeu comme structure. *Deucalion,* **2**, 161.

BERNE E. (1964). "Games People Play". Harmondsworth: Penguin.

BRUNER J.S. (1965). Man: a course of study. *Educ. Serv. Inc. Quart. Rpt.,* **313**, 85-95.

BUHLER C. (1935). "From Birth to Maturity: the child and his family". London: Routledge.

CAILLOIS R. (1961). "Man, Play and Games". New York: Free Press.

CHASE D.P. (1911). "The Nicomachean Ethics of Aristotle". London: Dent.

CHATEAU J. (1955). Le jeu de l'enfant aprés trois ans, sa nature, sa discipline: introduction à la padagogie. *Etud. Psychol. Phil.* **VIII.**

COWELL C.C. (1960). The contributions of physical activity to social development. *Res. Quart.,* **31**, 286-306.

DEARDEN R.F. (1968). The concept of play. In R.S. Peters (Ed.) "The Concept of Education". London: Routledge & Kegan Paul.

DEWEY J. (1959). "Dewey on Education". New York: Teacher's College.

ELLIS M.J., WADE M.G. & BOHRER R.E. (1970). "Annual Report of the Motor Performance and Play Research Laboratory". University of Illinois.

FERSTER C.B. & SKINNER B.F. (1957). "Schedules of Reinforcement". New York: Appleton-Century-Crofts.

FREUD A. (1969). "Indications for Child Analysis: and Other Papers 1945-56". London: Hogarth.

FREUD S. (1955). Beyond the pleasure principle. "The Standard Edition of the Complete Psychological Works of Sigmund Freud. Vol. 18". London: Hogarth.

FROEBEL F.W. (1887). "The Education of Man". London: Methuen.

GESELL A. & ILG F.L. (1946). "The Child from Five to Ten". London: Hamish Hamilton.

GIDDENS A. (1964). Notes on the concepts of play and leisure. *Soc. Rev.,* **1**, 12.

GREY A. (1964). "Children at Play". New Zealand: Play Centre Association.

GRIFFITHS R. (1935). "Imagination in Early Childhood". London: Routledge & Kegan Paul.

GROOS K. (1901). "The Play of Man". London: Heinemann.

GULICK L.H. (1920). "A Philosophy of Play". New York: Scribner.

GURLAND I. (1960). Play and the growth of intelligence. *Nat. Froebel Found. Bull.,* April.

HALL S.G. (1915). "Adolescence. Vol. I.", New York: Appleton.

HARTLEY R.E., FRANK L.K. & GOLDENSON R.M. (1952). "Understanding Children's Play". New York: Columbia Univ. Press.

HELANKO R. (1958). "Theoretical Aspects of Play and Socialisation". Finland: Turku University Press.

HERRON R.E. & SUTTON-SMITH B. (1971). "Child's Play". New York: Wiley.

HUIZINGA J. (1949). "Homo Ludens". London: Routledge & Kegan Paul.

HULL C.L. (1943). "Principles of Behaviour". New York: Appleton-Century-Crofts.

ISAACS S. (1945). "Intellectual Growth in Young Children". London: Routledge.
ISAACS S. (1965). "The Nursery Years". London: Routledge & Kegan Paul.
JACKS L.P. (1931). "The Education of the Whole Man". London: University Press.
JUNG C.G. & KERÉNYI K. (1949). Essays on the science of mythology. The myth of the Divine Child and the mysteries of Elensis. "Bollingen Series 22". New York: Pantheon Books.
KLEIN M. (1949). "The Psycho-analysis of Children". London: Hogarth.
KOESTLER A. (1964). "The Act of Creation". London: Hutchinson.
KOHLER W. (1925). "The Mentality of Apes". London: Kegan Paul.
LILLEY I.M. (1967). "Friedrich Froebel: a selection of his writings". Cambridge: University Press.
LODGE R.C. (1947). "Plato's Theory of Education". London: Kegan Paul.
LOWENFELD M. (1935). "Play in Childhood". London: Gollancz.
McKINNEY F. (1949). "Psychology of Personal Adjustment". New York: Wiley.
MEIER R.L. & DUKE R.D. (1966). Game simulation for urban planning. J. Amer. Inst. Planners, 32, 3-18.
MILLAR S. (1968). "The Psychology of Play". Harmondsworth: Penguin.
MILLER N.E. (1951). Learnable drives and rewards. In S.S. Stevens (Ed.) "Handbook of Experimental Psychology". London: Wiley.
MILLER N.E. & DOLLARD J. (1941). "Social Learning and Imitation". London: Kegan Paul.
PIAGET J. (1951). "Play, Dreams and Imitation in Childhood". London: Heinemann.
RAHNER H. (1965). "Man at Play or Did You Ever Practice Eutrapelia?". London: Burns & Oates.
REEVES M. (1957). Concept of work and play. Nat. Froebel Found. Bull. No. 104.
RYCROFT C. (1961). The Kleinian viewpoint. The Observer, 9th April 1961.
SANDSTROM C.I. (1968). "The Psychology of Childhood and Adolescence". Harmondsworth: Penguin.
SCARFE N.V. (1962). Play is education. Childhood Educ., 39, 117-121.
SCHILLER F. (1954). "On the Aesthetic Education of Man". London: Routledge & Kegan Paul.
SKINNER B.F. (1953). "Science and Human Behaviour". New York: Macmillan.
SLUSHER H.S. (1967). "Man, Sport and Existence". Philadelphia: Lea & Febiger.
SPENCER H. (1872). "Principles of Psychology". London: Williams & Norgate.
SUTTON-SMITH B., ROBERTS J.M., KOZELKA R.M., CRANDALL V.T., BROVERMAN D.M., BLUM A. & KLAIBER E.L. (1967). Studies in an elementary game of strategy. Genet. Psychol. Monogr., 75, 3-42.
THE OXFORD ENGLISH DICTIONARY (1933). VIII, N.-Poy. Oxford: Clarendon.
ULRICH C. (1968). "The Social Matrix of Physical Education". New Jersey: Prentice-Hall.
WHITING B.B. (1963). "Six Cultures: studies of child rearing". New York: Wiley.
WOODWARD O.M. (1966). "The Earliest Years: the growth and development of children under five". Oxford: Pergamon.

4 P TOPIC AREAS-DEVELOPED

THE DEVELOPMENT OF MOVEMENT UNDERSTANDING IN CHILDREN

by HILARY CORLETT

General

The term 'movement behaviour' is used in this section and will be seen to encompass reflex responses, motor abilities (maturational and learned psycho-motor skills), expressive patterns of movement and conceptualised patterns of movement expression. Such patterns of movement are formed, influenced and changed through the child's physical, psychological and social experiences (Section G). Hunt (1964), refers to movement behaviour as generalised motor expression. It is the nature of the movement response which differentiates 'movement behaviour' from 'motor activity'; it implies a manner or way of behaving. The stress is laid on personal experience.

A number of methods have been used to focus attention on the development of movement, per se, and on the understanding and use of movement. Areas of research may be summarised as:

a) analysis of a single task where development is traced as maturation occurs. e.g. balance, hurdling, throwing, jumping, tumbling;
b) analysis of a battery of tests e.g. Johnson, Brace, Oseretsky, Marmolenko, Van der Lugt;
c) studies relating motor ability to mental ability.
d) behavioural studies of general and specific movement characteristics.

These studies have been applied and conducted within both longitudinal and horizontal studies to examine movement characteristics and the influence of environment on the development of motor behaviour.

Therefore in terms of actual physical accomplishment, material is plentiful but in terms of the pattern of physical behaviour, its growth and change from birth to maturity little evidence has been specifically catalogued.

The student is recommended, by way of introduction to this area, to use the following as standard texts. In the area of child development—Mussen (1960; 1970); Hurlock (1956); Garrison (1952); Stone & Church (1957); Brearley (1969);

Strang (1959). In the area of behavioural studies—Gesell et al (1933; 1946; 1948; 1955; 1956; 1960); Piaget (1951a; 1951b; 1955; 1956; 1965; 1969a.b.c.); Lovell (1962; 1968). In the area of movement behaviour—Cratty (1964; 1968); Brown & Cratty (1968); Singer (1968); Laban (1960; 1963); Preston-Dunlop (1963); Argyle (1969).

P.1. The Infant (0-7 years)

During the important first period (0-2 years) the infant moves from a neonate, functioning at a reflex level in an undifferentiated way, to a relatively coherent organisation of his actions in his immediate environment (Flavell, 1963; Foss, 1963; McCandless, 1967). Movement is the dominant form of expression at this age.

Innate patterns of movement such as crying, sucking and grasping appear in a definite sequence but may be modified through interaction with environment (Mussen, 1963; Murchison, 1933; Munn, 1962; Ambrose 1969).

The child in interacting with the animate environment first responds to smiling, tactile stimulation, noise and affection. He is nearly two years old before he differentiates accurately between human and non-human aspects of the environment. The 'instinct theory', as it is applied to the development of attachment in children is described most completely by Buhler (1930) (See also Gesell, 1948; Bridges, 1931; Murchison, 1933; Parten, 1932; 1933; Frank, 1966). The influence of different child rearing practices on the development of movement can be seen in studies by Frank (1966); Sears, Maccoby & Levin (1957) and Benedict (1953).

Emotionally the child is bound to his environment. Emotional development follows the same basic pattern as motor behaviour—i.e. from general to specific responses (Sherman, 1927; Goodenough, 1931; Bridges, 1931). Behaviour studies in psychiatry from an ethological viewpoint have been edited by Hutt & Hutt (1970), while Carroll (1971) also reports on infant behaviour studies by human ethologists.

For Lamb (1965) the physical behaviour of the baby is seen in terms of alternation of extremes of growing and shrinking, abandonment and rigidity. He views 'flow' as the main characteristic of a neonate's movement and postulates that a ratio existed between flow retention, and shape/effort development. Laban (1963) also describes the baby's first efforts as a:

> 'total stirring of the body with the fundamental leg kicking, arm hitting, and grasping actions occurring in spasmodic phrases of quick vigorous and repeated actions. The more indulging qualities arise when the infant becomes more conscious of the surrounding world, early attempts at locomotion bringing discernible differentiation in qualities of the flow factor.'

The work of Gesell et al (1933; 1946; 1948; 1955; 1960) in year by year profiles indicates that the cycle of development takes a spiral course: trends repeat themselves at ascending levels of organisation. (See also Ilg & Ames, 1955.) Basic sequences are almost universal. The development of gross motor activities is proposed to be dependent on 'learning readiness'; this in turn is related to physical maturation and the effects of physique (Sheldon, 1940; Postma, 1968).

Montessori (1914) referred to 'sensitive periods', during which time the child showed interest in exercises for the senses such as the drawing or recognition of shapes.

The motor activities of the infant have been categorised as postural, locomotive, manipulative, lateral (McCandless, 1967). Movement behaviour becomes more stable as growth decelerates (towards 10 years of age); it proceeds from reflex and generalised movement to specialisation and integration (Breckenridge, 1955; Ausubel, 1954; Bayley & Espenchade, 1941; 1944; 1950; Espenchade et al, 1967; Laban, 1963; Bruner 1969; Halverson, 1943). The whole period from the earliest explorations of the neonate through the basic specialised skills of walking throwing etc. to the sophisticated control of the 6-7 year old, is characterised by a superabundance of movement experience.

Movement and play to the infant are synonymous. His patterns of behaviour are dominated by his physical actions in which he uses his body as both a utilitarian and an expressive tool. Play is a means by which he develops and grows (Matterson, 1965). Play, to the child, is work; it is also the area for observation of his understanding and use of movement. Lowenfeld (1935) comments on the bodily activity value of play—the enjoyment of movement for its own sake. This is the essence of children's dance and body agility (Laban, 1963; Russell, 1965; Brearley et al, 1969; Stanley, 1969; Cameron & Cameron 1969).

Through movement play the human infant acquires basic drives of sex dominance and learns patterns of social interaction (Frank, 1966; Argyle, 1969). Mussen (1963) has categorised developmentally the social interactions of nursery school children as unoccupied behaviour, solitary play, onlooker behaviour, parallel or imitative play, associative and co-operative play. Millar (1968) and Isaacs (1932; 1933) have made observational studies of the social development of infants while Gesell (1948) outlined the changes in social reponses.

The cognitive development of children and the relatedness of sensory experience to same is a growing body of knowledge.

Whiting (1967) and Morris & Whiting (1971) have discussed the significance of movement in relation to social and cognitive development. Ideas of this kind form the basis of many compensatory education programmes and of the 'head-start' procedures in North America. Current knowledge cautions against the assumption that it is movement per se which is important. Interpretations such as movement leads to changes in sensory input together with increased kinesthetic feedback are more in favour.

The work of Piaget in revealing the nature of knowledge that children exhibit at different stages of development and his theory of progressive adaptation has contributed much to this area. Approaches to the understanding of Piaget's work are given in Richmond (1970); Brearley & Hitchfield (1966); Beard (1970); Flavell (1963), while Russell (1956), and Peel (1956) throw further light on children's thinking powers (see also Section O). Piaget defines three stages in this period in which thought develops as 'internalised action'—Sensori-motor (0-2 years) pre-conceptual thought (2-4 years) and intuitive thought (4-7 years). He also saw (1951b) the changes in childrens games corresponding to these stages -

practice games	-	representative
symbolic	-	make believe
constructional	-	

The findings of Gesell (1946; 1955) and Bruner (1966) in his elaboration of the enactive and iconic modes of thought can be seen to correlate with Piaget's theories.

P.2. The Junior (7-11 years)

This period of life covers both the comparatively stable years of later childhood and the unstable years of the pre-adolescent era (Strang, 1959; Mussen, 1963; Stone & Church, 1957). The former section is often incorporated in literature on childhood and the latter is included in literature on adolescence. Nevertheless it is a definable period which has been named 'the skill hungry age'. The phrase assumes a rather constrained concept of physical skill, but within its limits indicates the involvement that develops with the movement skills of the adult society—the wish to kick a soccer ball, hit with mother's tennis racquet, dance like a ballet dancer etc. At the same time, the early, more personal, movement patterns are still very much explored. The junior is active, keen, ready to learn and to master a vast range of motor and cognitive skills—he is play-centred and action-minded. Lovell (1962) gives an outline of characteristics in general terms drawn from Gesell and others.

In his intellectual development, the junior reveals an increasing ability to grasp relationships. During this stage of development 7-11 years (defined as the period of 'operational thought' by Piaget), he becomes aware of the sequence of actions in his own mind and finds he can order his own experience. Concepts are derived from first hand contact with reality (see section G). Some of the implications of Piaget's work on the teaching of physical education are discussed in Bond (1970). Evidence of creative thinking in movement patterns involved in dance, drama and art is to be found in Richmond (1949); Marshall (1968); Lowenfeld & Brittain (1947); Opie & Opie (1959; 1969); Slade (1954) and the publication by the Centre for Curriculum Renewal and Educational Development Overseas (1969).

Patterns of movement skill exhibit marked increases in proficiency in this period in a wide range of activities. It is a 'vital' stage of life (Cratty, 1964; Jersild, 1960). Children are increasingly able to differentiate in their use of the body, to combine and repeat phrases of movement and to relate action and effort quality in simple spatial orientations with other people. Principles underlying the creative approach to movement education are fully described in developmental terms through the sixteen movement themes by Laban (1963); Russell (1965); and Preston-Dunlop (1963) while these theories are substantiated in the work with children outlined in the Ministry of Education text "Moving and Growing" (1952); Brearley (1969); Bilborough & Jones (1963); Mauldon & Layson (1965) and Morison (1969).

The development and acquisition of specific skills has been detailed by Gesell et al (1933); Halverson (1966); Gutteridge (1939) and Espenchade & Eckert (1967).

Studies of play by Millar (1968); Slade (1954) Opie & Opie (1959, 1969); Huizinga (1970); Caillois (1962) provide information with reference to the development of movement imagination, of cultural influences and of socio-emotional patterns—as observed in the free and conventional forms of games. Socially it is important for the individual to be accepted by his peer group—the gang age comes into its own with the 8-9 year group. Games become more formalised due to cultural pressures and a growing understanding of 'rules', but evidence from Mauldon & Redfern (1969) reveals children's delight in practice games and in their capacity to make up their own games. Interest in the opposite sex gradually decreases through Junior stage before interest is renewed towards adolescence. Friendship is valued and indeed necessary for harmonious development to occur and there is a change from the imitative capacities of the infant to an ability to 'share' with a partner.

P.3 The Adolescent (11-16)

Adolescence has been described as the period of time when the individual seeks a 'psychological' break from the family environment. Personal and cultural influences interact and at times the multiplicity of factors that affect individuals cannot be disentangled. However, in order to guide study, a framework is of value although overlap of areas must exist.

A wealth of literature related to the adolescent period of life is available. It stems from the work of Hall (1904). Methods of approach to the study of adolescence are critically appraised in Ausubel (1954). For an overall view on sociological theory see Caser & Rosenburg (1969); Klein (1965); Fleming (1948); Sherif et al (1956); Lasswell, Burne & Aronson (1965); Lindzey & Aronson (1954); Maccoby, Newcomb & Hartley (1958) and Parsons & Bales (1955).

The work of Tanner et al (1961; 1966) has revealed from the Harpenden Growth Study and other investigations the trends towards earlier maturation in adolescence of the present generation as compared with past generations; he stressed the marked variation in tempo of growth at this age and the spiral of development. Although there are large individual differences in rates of development e.g. onset of menarche, height spurt, fertility etc. Growth patterns of the different systems of the body are detailed. Tanner's clarification of 'sunshine periods' in brain development would seem to have implications for learning in physical skill as well as for cognitive development, while Gesell's (1956) comments on visual development and the discriminative powers of adolescents may be pertinent in games playing. Other longitudinal studies have been documented in Mussen (1970). Ellis (1946; 1948) and Simmons (1944) have investigated differences with reference to age and sexual maturity and physical growth and development, respectively. (For further detail on physical growth see Section K).

As with the infant, physique has an influence on motor behaviour. Bayley & Espenchade (1941) have summarised research on motor development of children and McCandless (1970) has provided an overall view of adolescent behaviour in relation to development.

Welford (1956; 1958) has detailed factors affecting the acquisition of skill and many investigations have been made into the psychomotor abilities of pupils of secondary school age. Tests and measurements have been reported by for example Clarke (1945); McCloy & Young (1954); Mathews (1958); Campbell & Tucker (1967) and the value of physical activity has been overviewed in Research Quarterly (1964).

Developmental factors and the results from tests of motor co-ordination and specific skills have been recorded by Ismail et al (1965); Latchaw (1969); McGraw & Tolbert (1953); Espenchade (1967) and Cratty (1964).

The importance of physical ability rates more highly among boys than girls at adolescence. The 'athlete' in comparison to the popularity of scholars or 'lady's men, figured high among élite in Coleman's (1961) investigation of American adolescent status systems.

Socialization is pre-dominant in the adolescent's search for self identity; Strang (1959); important elements are role, peer group, and leisure interests (Hartup 1946; Odlum, 1957; Shears, 1933; Martens, 1970).

Mussen (1970) has edited numerous studies of peer group formation and Davitz (1964) has reported on emotional meaning within group interaction. Hartup (1946) has commented particularly on 'peer interaction and social organisation' and Newcomb, Turner & Converse (1952), have given detailed consideration to the study of individual attitudes and the processes of interpersonal interaction in group structure. The process of socialization in terms of the development of attitudes and emotional adjustments has been detailed by Hurlock (1956).

A survey of the literature concerning the influence of competition, co-operation, leadership patterns and onlookers, on the performance of children can be found in Cratty (1964). In the context of play as a socialising agent, adolescents' play becomes more structured and subject to formal rules as in games. The initiation rites and ritualistic behaviour are all part of the total picture. Mead (1934) sees games as a situation in which one learns to take the role of the 'others'. 'Natural' as a against 'chosen' leaders can be identified by their physical behaviour during the game.

The contribution of physical activity to social development has been assessed by Cowell (1960) when he collated many pieces of American research. To date, there has been no scientifically evaluated evidence published of the contribution of dance to the development of social awareness.

The choice of games and play implements have been investigated by Sutton-Smith et al (1963) as indices for differences in male/female role identification. There is very little documented evidence within sociological research of adolescent behaviour expressed in physical terms, (Loader, 1970; Ward, 1970). Information related to the actual movement behaviour patterns in social interaction must be drawn from the works of Sherif & Sherif (1964); Argyle (1969) and texts given above.

'Expressive behaviour', of adolescence (Allport, 1937), can be erratic, exaggerated and flamboyant one minute, aggressive or withdrawn and reflective the next minute. Individual differences are pronounced at this stage of development and generalisations over groups are difficult and potentially dangerous. In his search for self identity and emancipation from adult control, the adolescent is faced with many developmental tasks (Havighurst, 1953). Erikson (1950) has defined particular stages of ego-identity; he used the phrase 'feeling of being at home in one's body'. The contribution of movement to the development of self-concept is discussed in Arnold (1970) and Dennison (1967).

Kane (1966) and Whiting et al (1972) have reviewed the empiric research on personality related to sport and physical ability.

Stages in development of emotional control have been described by Gesell (1956); he views emotions as both symptoms and creative forces and in his outline of the changes in the 'total action systems' refers to the tensional outlets of the growing adolescent as 'generally bodily overflow-stamping and sniffing at 11 years, jiggling at 12 years . . .'.

Research into behavioural contagion, (Polansky et al, 1950) is limited but it would seem that the theory postulated by Lamb (1965) and Layson (1969) of effort contagion is markedly in evidence in behaviour response patterns of interaction between players and spectators in competitive situations. Transmission of gestural patterns from teacher to pupil, Layson (1969), may account for particular movement patterns within school classes in movement sessions.

Intellectually the adolescent has reached the stage of formal operations (Piaget, 1953; McCandless, 1970; Bruner, 1967 and Sigel & Hooper, 1968). He may be capable of thinking in abstract terms. Intellectual growth has been detailed by Peel (1965) and Hebb (1949) should be consulted for the study of behaviour and pyschological phenomena connected with learning and perception.

Gesell (1956) stressed the heightened intensity and receptivity of visual perception of the adolescent. 'Intense vision' has been suggested to be a stimulus to the aesthetic sense; 'it awakens creative desires'. It is noteworthy that reaction time is optimal at adolescence. Rosewarne-Jenkins (1956) relates the developmental characteristics outlined by Gesell to the movement needs and preferences of adolescent girls.

Cratty (1964) discusses 'personal equations' in movement; during this stage the adolescent adopts many roles and tries out different forms of movement behaviour. Arnold (1968) and Jones (1970) have discussed personality as it affects and is affected by movement experiences in physical education. The development of a marked 'style' in skills reflects personal movement preferences and Cox (1964) and North (1971) have used film to illustrate individual movement preferences.

The capacity to abstract from his experience may lead the adolescent to the greater appreciation of shape, form and rhythmic content of movement (aesthetic components). In creative movement the adolescent is ready for more advanced work of Themes 8-16 (Laban, 1963; Preston-Dunlop, 1963; Russell, 1969; Carroll & Lofthouse, 1969).

In the spiral development of his movement understanding, greater clarity is demanded in the formulation of the adolescent's movement sequences. Sensitivity to the qualitative components of the movement and more precision in spatial form are demonstrated. Skills also show a greater refinement in their control and fluency.

P.4 The development of movement concepts.

'Is movement the servant of cognition rather than its creator? . . .' 'cognition is an isomorph of movement' (Stone, 1969).

Concepts enable words 'to stand for whole classes or categories of objects, events or qualities' according to Lovell (1962). Definition of a concept depends on how 'the problem of knowing' is viewed (Rugg, 1963).

Piaget (1930; 1956; 1969) infers that the child's concepts of time, space and force are based upon his initial movement experiences (Beard, 1960; 1970; Hebb, 1949). Walter (1953) and Kephart (1960) also support this point of view.

Concepts have to be constructed and this process takes time (Wallace, 1965). The individual passes through various stages of cognitive and movement development and proceeds from percept to concept (Brown & Cratty, 1969; Metheny 1965; Bruner, 1956; Sigel & Hooper, 1968).

Laban defined stages of the child's development of movement understanding in terms of themes—each one representing 'a movement idea corresponding to a stage in the progressive unfolding of the feel of the movement in the young child'. He also indicated that the adolescent simultaneously is able to understand and use movement principles. Preston-Dunlop (1963) further clarified the development principle by diagrammatically illustrating the four categories of movement—body, effort, space, relationship—in a spiral pattern.

The term 'body-concept' is subject to varied interpretations and is frequently interchanged with expressions such as 'body image' and 'body awareness'. The human body is often conceived of as a physiological system which produces needs and drives. Equally it can be and is accepted as a subjective experiential system. Sheldon (1940) and Kretschmer (1948) correlated measures of body structure with personality traits but the results have been questioned.

'Body image' may be interpreted as 'an image of his own body which the individual has evolved through experience', (Whiting et al, 1972) and body awareness is a phrase currently used by physical educationists referring to the appreciation and understanding of the body as the instrument of movement and vehicle of expression in non-verbal communication. Morris & Whiting (1971) overview the work of Schilder (1950); Argyle (1969); Dickinson (1970) and others in evaluating the part played by the body-concept in the development of personality.

In the first stage the infant differentiates little between himself and the environment; everything is absolute. Piaget (1930) lists seventeen forms of relationship parallel with chronological age which exhibit the child's growing awareness of force and movement. The infant at the animistic stage endows all objects with an internal power. Gradually he passes to the stage of reciprocity when differentiation between subjective and objective aspects of the world is possible until, as adults, individuals can distinguish between personal thoughts and thinking about external objects. At first in the child's thinking, weight is the same as strength and activity. Later weight becomes relative to the medium with which it is associated and is defined by actions, performed on objects in the environment. Lovell (1962, 1968) reports on experimental research into Piaget's findings. Smedslund (1961) in particular investigated the notion of conservation of weight.

In movement terms, the child learns through his tactile and kinaesthetic senses to differentiate between firmness and strong pressure and gentle and light touch movement.

An organised knowledge of 'the body' is developed through guided experience as well as the 'trial and error' method of learning. The child learns to appreciate both objective and subjective concepts (Mauldon, 1970). The primitive postural model shows a lack of differentiation of the separate parts while western man endeavours to form a completed yet differentiated self image, which also includes objects within the surrounding space (Hunt, 1964).

Time, speed and movement are constructs; the young child however, confuses successions of events in time and the temporal intervals. These engender with their analogues in space—i.e. with the succession of points traversed in a movement and the spatial distance between.

The concept of time contains both the sense of the clock and regulated intervals and the sense of experience (Hunt, 1964). For the infant the sensation of events is personal, for the adolescent it is abstract (Flickinger & Rehage, 1949). Man becomes aware of changes, recurrences and rhythms in the environment, in his own body and in waiting for events to occur; he never directly perceives time (Stone & Church, 1957). The young child learns to appreciate time and motion, physical time, age and inner time through experience Ornstein (1969) summarises four modes of time experience as: short time, duration, temporal perspective, and simultaneity and succession. Beard (1964) carried out investigations of time among infants in relation to their use of space.

The simplest intuition of speed, as well as of movement, is based on an intuition of order (Piaget, 1956). In the child, subjective impressions do not give rise to conceptualised intuitions except when they are attached to external movements or to movement of the body, as perceived from the outside. There is not just a relationship between distance covered and time but the perception of 'overtaking'; one voluntary movement is made more effective than another; i.e. effort is involved and acceleration and deceleration are appreciated; movement is the essential ingredient of space perception. A space concept is 'a representation of the nature of space, its perceived objects and physical forces abstracted into relationships and meanings' Hunt (1964).

Although Piaget can be criticised for basing his generalisations on inadequate evidence, his many proposals merit objective investigation. Where these have been made (Bruner, 1966; Gessell, 1946; 1956) support has often been found. Piaget distinguished between perceived and conceived space (1956). To understand spatial properties and relations as opposed to noticing similarities and differences between spatially extended objects, it is necessary to involve operations which are considered as virtual or internalised actions. Lovell (1962) and likewise Laban would agree, in that they postulated that spatial perception is achieved by using one's body to create patterns. Richmond (1970) viewed space as an invisible and intangible concept which has no existence while Gesell (1956) saw space as a form of awareness, an outer experience.

Stone & Church (1957) enumerated five major stages in the development of spatial concepts—action space; body space; object space; map space; and abstract space.

Laban also perceived the development of space awareness through explorations relating the body to the environment and the establishment of near and far relationships through actions, to simple orientation and awareness of harmonious relationships in spatial forms.

The child's repertoire of movement can be seen to be formulated through his experiences in a variety of media and through his growing understanding of the principles of movement (Jordan, 1966). At the infant stage utilitarian and expressive movement patterns are undifferentiated; egocentricity prevails. As Juniors children can discriminate, represent and co-operate in movement situations. The adolescent experiences the changes in the flow of movement, learns to select and relate, to distinguish between concepts of motion outside himself as well as within his own body; he becomes aware of the illusory, qualitative and aesthetic aspects of movement as well as the concrete, measurable aspects of motion. (Sheets, 1966; Pappas, 1970; Arnheim, 1956; Read, 1958 and H'Doubler 1959.)

Movement must be meaningful whether it is described as action, expression, gesture, play, normal, natural, or a non-discursive language. Appreciation of this is most easily achieved through movement experience—i.e. 'meaning is embodied in the movement', Reid (1969). See also Postma (1968); Metheny (1962); Kleinman (1964); Rugg (1963); Langer (1953); Laban (1960).

Approaches to meaningful experience in the movement education of children are constantly being sought. Brooke (1967) outlined the need for a taxonomy and Mawdsley (1971) proposed a conceptual analysis of human movement. Webb (1972) outlined a taxonomy for the psychomotor domain; there is, however, need for further integrated approaches to the developmental aspects of human movement.

Source Texts

ARGYLE M. (1969). "Social Interaction". London: Methuen.

BREARLEY M. (Ed.) (1969). "Fundamentals in the First School". London: Blackwell.

BROWN R.C. & CRATTY B.J. (1968). "New Perspectives of Man in Action". New Jersey: Prentice Hall.

CRATTY B.J. (1964). "Movement Behaviour and Motor Learning". Philadelphia: Lea & Febiger.

CRATTY B.J. (1968). "Psychology and Physical Activity". New Jersey: Prentice-Hall.

GARRISON K.C. (1952). "Growth and Development". London: Longmans Green.

GESELL A. (1955). "The First Five Years of Life". London: Methuen.

GESELL A. & AMATRUDA C.S. (1960). "Developmental Diagnosis". New York: Hoeber.

GESELL A. & FRANCES L. (1948). "Studies in Child Development". New York: Harper.

GESELL A. & ILG F.L. (1946). "The Child From Five to Ten". London: Hamish Hamilton.

GESELL A., ILG F.L & AMES L.B. (1956). "Youth". London: Hamish Hamilton.

GESELL A. & THOMPSON H. (1933). "Infant Behaviour: its genesis and growth". New York: McGraw Hill.

HURLOCK E.B. (1956). "Child Development". New York: McGraw Hill.

LABAN R. (1960). "Mastery of Movement". London: Macdonald & Evans.

LABAN R. (1963). "Modern Educational Dance (2nd Ed.)". London: Macdonald & Evans.

LOVELL K. (1962). "Educational Psychology and Children". London: University Press.

LOVELL K. (1968). "An Introduction to Human Development". London: Macmillan.

MUSSEN P.H. (1960). "Handbook of Research Methods in Child Development". New York: Wiley & Sons.

MUSSEN P.H. (1963). "The Psychological Development of the Child". New Jersey: Prentice-Hall.

MUSSEN P.H. (1970). "Manual of Child Psychology (3rd Ed.)". New York: Wiley.

PIAGET J. (1930). "The Child's Conception of Physical Causality". London: Routledge & Kegan Paul.

PIAGET J. (1951a). "Judgement and Reasoning in the Child". London: Routledge & Kegan Paul.

PIAGET J. (1951b). "Play, Dreams and Imitation in Childhood". London: Heinemann.

PIAGET J. (1953). "The Origin of Intelligence in the Child". London: Routledge & Kegan Paul.

PIAGET J. (1955). "The Child's Construction of Reality". London: Routledge & Kegan Paul.

PIAGET J. (1956). "The Child's Conception of Space". London: Routledge & Kegan Paul.

PIAGET J. (1965). "The Moral Judgements of the Child". London: Routledge & Kegan Paul.

PIAGET J. (1969a). "The Child's Conception of Time". London: Routledge & Kegan Paul.

PIAGET J. (1969b). "The Mechanisms of Perception". London: Routledge & Kegan Paul.

PIAGET J. (1969c). "The Child's Conception of Movement and Speed". London: Routledge & Kegan Paul.

PRESTON-DUNLOP V. (1963). "A Handbook for Modern Educational Dance". London: Macdonald & Evans.

SINGER R.N. (1968). "Motor Learning and Human Performance". New York: Macmillan.

STONE L.J. & CHURCH J. (1957). "Childhood and Adolescence". New York: Random House.

STRANG R. (1959). "An Introduction to Child Study". New York: Macmillan.

Specific References

ALLPORT G.W. (1937). "Pattern and Growth in Personality". New York: Holt, Rinehart & Winston.

AMBROSE A. (Ed.) (1969). "Stimulation in Early Infancy. Proceedings of C.A.S.D.A. Study Group No. 1967". New York: Academic Press.

ARNHEIM R. (1956). "Art and Visual Perception". London: Faber.

ARNOLD P.J. (1968). "Education, Physical Education and Personality Development". London: Heinemann.

ARNOLD P.J. (1970). Physical education, creativity and the self concept. *Bull. Phys. Ed.* **VIII**, 15-19.

AUSUBEL D.P. (1954). "Theory and Problems of Adolescent Development". New York: Grune & Stratton.

BAYLEY N. & ESPENCHADE A. (1941). Motor development from birth to maturity. *Rev. Educ. Res.,* **11**.

BAYLEY N. & ESPENCHADE A. (1944). Motor development from birth to maturity. *Rev. Ed. Res.,* **14.**

BAYLEY N. & ESPENCHADE A. (1950). Motor development from birth to maturity. *Rev. Ed. Res.,* **20.**

BEARD R. (1960). An investigation of concept formation among infant school children. *Ed. Rev.* **13,** 1.

BEARD R. (1964). Further studies in cognitive developments. *ED. Rev.,* **17,** 1.

BEARD R. (1970). Educational aspects of child development. In "Conference Report on Movement Dance and Drama". Hull: The University.

BENEDICT R. (1953). "Patterns of Culture". New Amer. Lib. World Lit.

BILBOROUGH A. & JONES P. (1963). "Physical Education in the Primary School". London: The University Press.

BOND J. (1970). A review of some aspects of Piaget's theories and their possible implication for P.E. teachers in Infant and Primary schools. Unpub. Diss. Dip. P.E. University of Leeds.

BREARLEY M. & HITCHFIELD E. (1966). "A Teacher's Guide to Reading Piaget". London: Routledge & Kegan Paul.

BRECKENRIDGE M.E. & VINCENT E.L. (1955). "Child Development". Philadelphia: Saunders.

BRIDGES K.M.B. (1931). "The Social and Emotional Development of the Pre-School Child". London: Kegan Paul.

BROOKE J. (1967). A taxonomy for human movement. *Leaflet P.E.A.* **68, 7,** 56-57.

BUHLER C. (1930). The first year of life. In C. Murchison (Ed.) "The Social Behaviour of Children". New York: John Day.

BRUNER J.S. (1956). "A Study of Thinking". New York: Wiley.

BRUNER J.S. (1966). "The Process of Education". Harvard: The University Press.

BRUNER J.S. (1967). "Toward a Theory of Instruction". Harvard: The University Press.

BRUNER J.S. (1969). Eye, hand and mind. In I. Elkind (Ed.) "Studies of Cognitive Development". London: Oxford University Press.

CAILLOIS R. (1962). "Man, Play and Games". London: Thames & Hudson.

CAMERON W. & CAMERON M. (1969). "Education in Movement in the Infant School". Oxford: Blackwell.

CAMPBELL W.R. & TUCKER N.M. (1967). "An Introduction to Tests and Measurements in Physical Education". London: Bell.

CARROLL J. (1971). "A Survey of Some Research Findings from the Social Sciences that Relate to Movement Behaviour". Sutton Coldfield: Anstey College.

CARROLL J. & LOFTHOUSE P. (1969). "Creative Dance for Boys". London: Macdonald & Evans.

CASER L.A. & ROSENBERG B. (1969). "Sociological Theory (3rd Ed.)". London: Macmillan.

CENTRE FOR CURRICULUM RENEWAL AND EDUCATIONAL DEVELOP-MENT OVERSEAS (1969) "Children at School. Primary Education in Britain Today". London: Heinemann.

CLARKE H. (1945). "Application of Measurement to Health and Physical Education". New Jersey: Prentice Hall.

COLEMAN J.S. (1961). "The Adolescent Society". Glencoe (III): The Free Press.

COWELL C. (1960). The contributions of physical activity to social development. *Res. Quart.* **31**, 2.

COX B.N. (1964). Observation of movement. Unpublished M.Ed. Thesis, University of Leicester.

DAVITZ J. (Ed.) (1964). "The Communication of Emotional Meaning". New York: McGraw Hill.

DENNISON J.D. (1967). Self realisation through P.E. *A.J. Phys. Educ.* **41.**

DICKINSON J. (1970). A note of the concept of Body awareness. *J. Phys. Educ.* **1**, 2.

H'DOUBLER M. (1959). "Dance: a creative art experience". Madison: The University of Wisconsin Press.

ELLIS R.W.B. (1946). Height and weight in relation to onset of puberty in boys. *Arch. Dis. Child.* **21**, 181-188.

ELLIS R.W.B. (1948). Height and weight in relation to onset of puberty in boys. *Arch. Dis. Child.* **23**, 17-26.

ERIKSON E.M. (1950). "Childhood and Society". New York: Norton.

ESPENSCHADE A.S. & ECKERT H.M. (1967). "Motor Development". Ohio: Merrill.

FLAVELL J.H. (1963). "The Developmental Psychology of Jean Piaget". New Jersey: Van Nostrand.

FLEMING C.M. (1948). "Adolescence—Its Social Psychology". London: Routledge & Kegan Paul.

FLICKINGER A. & REHAGE K.J. (1949). Building time and space concepts. In *20th Year Book. Nat. Coun. Soc. Stud.,* 107-116.

FOSS B.M. (Ed.) (1963). "Determinants of Infants Behaviour". London: Methuen.

FRANK L.K. (1966). The cultural patterning of children. In F. Faulkner (Ed.) "Human Developement". Philadelphia: Saunders.

GOODENOUGH F.L. (1931). The experiences of emotion in infancy. *Child. Dev.,* **2**, 96-101.

GUTTERIDGE M.V. (1939). A study of motor achievements of young children. *Arch. Psychol.* **244.**

HALL G.S. (1904). "Adolescence". New York: Appleton.

HALVERSON H.M. (1943). The development of prehension in infants. In R.G. Barker, J.S. Kovnin & H.F. Wright (Eds.) "Child Development and Behaviour". New York: McGraw Hill.

HALVERSON L.E. (1966). Development of motor patterns in young children. *Quest,* VI.

HARTUP W.W. (1946). Peer interaction and social organisation. In P.H. Mussen (Ed.) "A Manual of Child Psychology. Vol. II". New York: Wiley.

HAVIGHURST R.J. (1953). "Human Development and Education". New York: Longmans Green.

HEBB D.O. (1949). "The Organisation of Behaviour". New York: Wiley.

HUIZINGA (1970). "Homo Ludens". Paladin.

HUNT V. (1964). Movement behaviour: a model for action. *Quest,* II.

HUTT S.J. & HUTT C. (1970). "Behaviour Studies in Psychiatry". Oxford: Pergamon.

ILG F.L. & AMES L.B. (1955). "Child Behaviour". New York: Harper.

ISAACS S. (1932). "The Nursery Years". London: Routledge.

ISAACS S. (1933). "Social Development in Young Children". London: Routledge.

ISMAIL A.H. & GRUBER J.H. (1965). Predictive power of co-ordination and balance items in estimating intellectual achievement. *Proc. 1st Int. Con. Psychol. Spt.* Rome.

JERSILD A. (1960). "Child Psychology". New Jersey: Prentice Hall.

JONES M. (1970). Perception, personality and movement characteristics. Unpublished M.Ed. Thesis, University of Leicester.

JORDAN D. (1966). "Childhood and Movement." Oxford: Blackwell.

KANE J. (Ed.) (1966). "Readings in Physical Education". London: Physical Education Association.

KANE J. (1969). Personality and body type. *Res. Phys. Educ.,* 1, 30-38.

KEPHART N.C. (1960). "The Slow Learner in the Classroom". Ohio: Merrill.

KLEIN J. (1965). "The Study of Groups". London: Routledge & Kegan Paul.

KLEINMAN S. (1964). The significance of human movement: a phenomenological approach. *N.A.P.E. Coll. Wom. Proc.* Michigan.

KRETSCHMER E. (1948). "Physique and Character". Berlin: Springer.

LAMB W. (1965). "Posture and Gesture". London: Duckworth.

LANGER S. (1953). "Feeling and Form". London: Routledge & Kegan Paul.

LASSWELL T.E. BURNE J.H. & ARONSON S.H. (1965). "Life in Society". Glenview: Scott, Forseman.

LATCHAW M. & EGSTROM G. (1969). "Human Movement". New Jersey: Prentice Hall.

LAYSON J. (1969). An introduction to some aspects of the sociology of physical education. *Proc. Ass. Wom. Coll. Phys. Educ.,* 45-53.

LINDZEY G. & ARONSON E. (Eds.) (1954). "Handbook of Social Psychology". Reading (Mass): Addison-Wesley.

LOADER E. (1970). Social adjustment and physical ability. In "Chelsea College of Physical Education Year Book". Eastbourne: Chelsea College.

LOWENFELD M. (1935). "Play in Childhood". London: Gollancz.

LOWENFELD V. & BRITTAIN W.L. (1947). "Creative and Mental Growth". New York: Macmillan.

MACCOBY E.E., NEWCOMB T.M. & HARTLEY E.L. (1958). "Readings in Social Psychology (3rd Ed.)". New York: Holt.

MARSHALL S. (1968). "Adventure in Creative Education". London: Pergamon.

MARTENS R. (1970). A social psychology of physical activity. *Quest* XIV.

MATHEWS D.K. (1958). "Measurement in Physical Education". Philadelphia: Saunders.

MATTERSON E.M. (1965). "Play with a Purpose for the Under-sevens." Harmondsworth: Penguin.

MAULDON E. (1970). What is physical education? *Bull. Phys. Educ.* VIII, 3, 13-20.

MAULDON E. & LAYSON J. (1965). "Teaching Gymnastics". London: Macdonald & Evans.

MAULDON E. & REDFERN H.B. (1969). "Games Teaching". London: Macdonald & Evans.

MAWDSLEY H.P. (1971). A conceptual analysis of movement. In: A world manifesto on physical education. *Bull. Phys. Educ.* VIII, 5.

MEAD G.H. (1934). "Mind, Self and Society." Chicago: University Press.

METHENY E. (1962). Symbolic forms of movement. *Focus* II.

METHENY E. (1965). "Connotation of Movement and Sport". Iowa: Brown.

MILLAR S. (1968). "The Psychology of Play". Harmondsworth: Pelican.

MINISTRY OF EDUCATION (1952). "Moving and Growing". London: H.M.S.O.

MONTESSORI M. (1914). "Dr. Montessori's Own Handbook". New York: Stokes.

MORRIS P.R. & WHITING H.T.A. (1971). "Motor Impairment and Compensatory Education". London: Bell.

MORISON R. (1969). "A Movement Approach to Educational Gymnastics". London: Dent.

MUNN N.L. (1962). "The Evolution and Growth of Human Behaviour". London: Harrap.

MURCHISON C. (Ed.) (1933). "A Handbook of Child Psychology". New York: John Day.

McCANDLESS B.R. (1967). "Children: Behaviour and Development". London: Holt, Rinehart & Winston.

McCANDLESS B.R. (1970). "Adolescents: Behaviour and Development". London: Holt, Rinehart & Winston.

McCLOY C.H. & YOUNG N.D. (1954). "Tests and Measurements in Health and in Physical Education". New York: Appleton, Century, Crofts.

McGRAW L.W. & TOLBERT J.W. (1953). Sociometric status and athletic ability of junior high school boys. *Res. Quart.* 24, 72-8.

NEWCOMB T., TURNER R.H. & CONVERSE P.E. (1952). "Social Psychology". London: Tavistock.

NORTH M. (1971). "Personality and Movement". London: Macdonald & Evans.

ODLUM D. (1957). "Journey Through Adolescence". Harmondsworth: Penguin.

OPIE I. & OPIE P. (1959). "The Lore and Language of Schoolchildren". Oxford: Clarendon.

OPIE I. & OPIE P. (1969). "Childrens Games in Street and Playground". Oxford: Clarendon.

ORNSTEIN R.E. (1969). "On the Experience of Time". Harmondsworth: Penguin.

PAPPAS G. (1970). "Concepts in Art and Education". London: Collier-Macmillan.

PARSONS T. & BALES R. F. (1955). "Family, Socialisation and Interaction Process". Glencoe (III): Free Press.

PARTEN M.B. (1932). Social participation among pre-school children. *J. Abnorm. Soc. Psychol.* 27, 243-269.

PEEL E.A. (1956). "The Psychological Basis of Education". Edinburgh: Oliver & Boyd.

PEEL E.A. (1965). Intellectual growth during adolescence. *Educ. Rev.*

POLANSKY N., LIPPITT R. & REDL F. (1950). An investigation of behavioural contagion in groups. *Hum. Relat.* **III, 4,** 319-48.

POSTMA J.W. (1968). "Introduction to the Theory of Physical Education". Cape Town: Balkema.

READ H. (1958). "Education Through Art". London: Faber.

REID A. (1969). "Meaning in the Arts". London: Allen & Unwin.

RESEARCH QUARTERLY The contributions of physical activity to human well-being. **31, 2,** II.

RICHMOND P.G. (1970). "An Introduction to Piaget". London: Routledge & Kegan Paul.

RICHMOND W.K. (1949). "Purpose in the Junior School". Bernard & Westwood.

ROSEWARNE-JENKINS M. (1956). Investigation of movement needs and preferences of secondary school girls. Unpub. Dissertation, Institute of Education, University of Cambridge.

RUGG H. (1963). "Imagination". New York: Harper & Row.

RUSSELL D. (1956). "Children's Thinking". Boston: Ginn.

RUSSELL J. (1965). "Creative Dance in the Primary School". London: Macdonald & Evans.

RUSSELL J. (1969). "Creative Dance in the Secondary School". London: Macdonald & Evans.

SCHILDER P. (1950). "The Image and Appearance of the Human Body". New York: Wiley.

SEARS R., MACCOBY E. & LEVIN H. (1957). "Patterns of Child Raising". Illinois: Row & Petersen.

SHEARS L.W. (1953). The dynamics of leadership in adolescent school groups. *B. Psychol.* **XLIV, 3,** 232-42.

SHEETS M. (1966). "The Phenemonology of Dance". Madison: Univ. of Wisconsin Press.

SHELDON W.H. (1940). "The Varieties of Human Physique". New York: Harper.

SHERIF M. & SHERIF C.W. (1956). "An Outline of Social Psychology". New York: Harper & Row.

SHERIF M. & SHERIF C.W. (1964). "Explorations into Conformity and Deviation of Adolescents". New York: Harper & Row.

SHERMAN M. (1927). The different of emotional responses in infants. *I.J. Comp. Psychol.*

SIGEL L. & HOOPER F. (1968). "Logical Thinking in Children". London: Holt, Rinehart & Winston.

SIMMONS K.I. (1944). Child growth and development II. Physical growth and development. *Monogr. Soc. Res. Child Dev.* **9,** 1.

SLADE P. (1954). "Child Drama". London: The University Press.

SMEDSLUND J. (1961). The acquisition of conversation of substance and weight in children. *Scand. J. Psychol.* **2,** 11-210.

STANLEY S. (1969). "Physical Education: a movement orientation". Toronto: McGraw Hill.

STONE R. (1969). Movement and the cognitive process. *J. Canad. Ass. Phys. Educ. Health Rec.* **35,** 3.

SUTTON-SMITH B., ROSENBERG B.G. & MORGAN E.F. (1963). Development of sex differences in play choices during pre-adolescence. *Child Dev.* **34,** 119-126.

TANNER J.M. (1961). "Education and Physical Growth". London: The University Press.

TANNER J.M., WHITEHOUSE R.H. & TAKAISHI M. (1966). Standards from birth to maturity for height weight, height velocity and weight velocity for British Children 1965. *Arch. Dis. Child.* **41,** 454-471, 613-635.

WALLACE J.G. (1965). Concept growth and the education of the child. *Nat. Found. Educ. Res. Mon.* **12.**

WALTER W.G. (1953). "The Living Brain". Harmondsworth: Pelican.

WARD E. (1970). Research in socialization and physical ability. *Br. J. Phys. Educ.* **1,** 12-14.

WEBB I. (1972). The theoretical content of Physical Education in Education and Physical Education. Conference Report: Association of Principals of Womens Colleges of Physical Education, Dartford.

WELFORD A.T. (Ed.) (1958). "Ageing and Human Skill". London: Oxford Univ. Press.

WELFORD A.T. (1966). Acquisition of skill. In J. Kane (Ed.) "Readings in Physical Education". London: P.E. Assoc.

WHITING H.T.A. (1967). Significance of movement for cognitive development. *Rem. Gym. Rec. Therapy,* **45** 14-19.

WHITING H.T.A. et al. (1973). "Personality and Performance in Physical Education and Sport". London: Henry Kimpton.

INDEX

N

O

P